Auxology: Human Growth in Health and Disorder

Proceedings of the
Serono Symposia, Volume 13

Edited by

L. Gedda
The Mendel Institute,
Piazza Galeno 5, Rome, Italy

P. Parisi
The Mendel Institute,
Piazza Galeno 5, Rome, Italy

1978

ACADEMIC PRESS   London New York San Francisco
A Subsidiary of Harcourt Brace Jovanovich, Publishers

ACADEMIC PRESS INC. (LONDON) LTD.
24-28 OVAL ROAD
LONDON NW1

U. S. Edition published by
ACADEMIC PRESS INC.
111 FIFTH AVENUE
NEW YORK, NEW YORK 10003

Copyright © 1978 by Academic Press Inc. (London) Ltd.

All rights reserved

NO PART OF THIS BOOK MAY BE REPRODUCED IN ANY FORM BY PHOTOSTAT, MICROFILM, OR BY ANY OTHER MEANS, WITHOUT WRITTEN PERMISSION FROM THE PUBLISHERS

Library of Congress Catalogue Card Number: 77-94294
ISBN: 0-12-279050-2

Printed in Great Britain by
Whitstable Litho Ltd., Whitstable, Kent

PREFACE

The field of human growth and development has attracted scientists from a wide spectrum of disciplines. Geneticists, biologists and obstetricians have probed the mysteries of intrauterine growth, anthropologists and anatomists have recorded and delineated normal growth patterns while pediatricians and endocrinologists focus on the various pathologies of abnormal growth. Nutritionists have sought to determine the influence of food upon growth under conditions of war and peace, plenty and lack. In recent years, with the great forward leap of the behavioural sciences, these have been joined by numerous psychologists, psychiatrists and sociologists as well as pediatricians who have devoted themselves to the study of behaviour patterns and their relation to age and sex. Unfortunately, meetings on one or another of the various facets of this subject have remained outside the knowledge of many of those interested since they have taken place within the framework of a highly-specific professional society. It was for this reason that Professor Luigi Gedda and Professor Paolo Parisi, both leading investigators in the field of genetics and growth from the Mendel Institute in Rome, undertook the organization of the First International Congress of Auxology. It was their aim to bring together world-known workers in these various disciplines, all with a special interest in human growth and development, thus effecting an exchange of information and fostering new, potentially-fruitful relationships. They merit our congratulations for having taken the initiative in this formidable task and for having achieved such a successful and stimulating meeting.

The subjects dealt with in the First International Congress of Auxology included growth determinants, growth patterns and standards, the chronogenetics and chronobiology of growth, the clinical aspects of growth, including diagnostic problems and therapeutic approaches, and psychosocial aspects of the growing child.

I take pleasure in presenting this volume of proceedings of the First International Congress of Auxology. It should be a particularly valuable text for all those scientists interested in growth, many of whom were unable to attend and should serve to stimulate and strengthen interdisciplinary cooperation in this fascinating challenging field.

Tel Aviv, August 1977

Z. Laron, M.D.
Professor of Pediatric Endocrinology
Sackler School of Medicine
Tel Aviv University.
Director, Institute of Pediatric and
Adolescent Endocrinology
and Israel Counselling Center for
Juvenile Diabetics
Beilinson Medical Center, Petah Tikva

CONTENTS

Preface v

1. Auxologic Determinants

The Clock of Life: Auxology and Chronogenetics
  by L. Gedda 1
Normal Growth and Neoplastic Growth: Are They Connected?
  by P.R.J. Burch 7
Intercorrelations of Growth and Maturity Characters in the Male Sex: A Factor Analysis
  by R. Fricke and R. Knussmann 23
Genetic Control of Human Growth and Ecosensitivity
  by N. Wolański 33
Growth, Nutritional Status and Food Consumption of Roman Schoolchildren
  by A. Mariani, B. Lancia, P.A. Migliaccio and D. Sorrentino .. 49

2. The Times of Auxology

Fetoscopy: Direct Visualization and Sampling of the Fetus: An Overview
  by J.M. Phillips 63
The Relationship Between Maternal Urinary Estrogens and Respiratory Instability in the Human Neonate
  by A. Steinschneider, R.W. Abdul-Karim, S.N. Beydoun and M. Pavy 71
Early Postnatal Growth Evaluation in Full-Term, Preterm, and Small-for-Dates Infants
  by F. Falkner 79
Hormonal Control of Pubertal Development
  by D. Gupta 87

Puberty in Patients with Growth Hormone Deficiency
  by A. Pertzelan, R. Kauli, Z. Zadik, I. Blum and Z. Laron .. 101

3. Human Growth Standards and Auxologic Variance

Human Growth Standards: Construction and Use
  by J.M. Tanner 109
The Fitting of Longitudinal Growth Data of Man
  by E. Marubini 123
United States Growth Charts
  by A.F. Roche and P.V.V. Hamill 133
Variability Between Populations in Growth During Childhood: Comparative Auxology of Childhood
  by P.B. Eveleth 139
Some Characteristics of the Postwar Secular Growth in the Netherlands
  by J.C. van Wieringen 153
Racial Variations in Body Height and Weight
  by A.M. Budy and M.P. Mi 157
Growth of Children in Arctic Conditions and in Northern European Countries: Genetic, Nutritional, and Socioeconomic Aspects
  by H.K. Åkerblom 167
Biological Response to Social Change: Acceleration of Growth in Czechoslovakia
  by M. Prokopec and V. Lipková 175
Changes in Body Measurements and Proportion of Children, Based on Körmend Growth Study
  by O.G. Eiben 187
Nutrition and Growth Performance in an Italian Rural Environment
  by A. Ferro-Luzzi, A. D'Amicis, A.M. Ferrini and G. Maiale .. 199

4. Clinical Auxology

Growth Hormone and Testosterone Interactions
  by M. Zachmann 207
Hormonal and Metabolic Responses to Fast: Influence of Growth in Normal Children and in Pathological Conditions
  by J.L. Chaussain, P. Georges, G. Olive and J.C. Job 213
Evidence for the Social Inheritance of Obesity in Childhood and Adolescence
  by S. Garn, S. Bailey, P. Cole, and I.T.T. Higgins .. 217
Primary Growth Deficiencies: Diagnostic Problems
  by G. Segni and P. Mastroiacovo 225
Growth Disturbances in Hypothalamic Pituitary Diseases
  by Z. Laron, Z. Zadik, A. Pertzelan and A. Roitman 239
Human Growth Hormone, Immunoreactive Insulin, and Somatomedin Activity in Achondroplasia
  by G. Giordano, E. Foppiani, F. Minuto, R. Cocco, M. Di Cicco, A. Barreca, F. Caiazza and F. Morabito 253

Human Growth Hormone, Immunoreactive Insulin, and
Somatomedin Activity in Turner's Syndrome
    by F. Morabito, F. Caiazza, A. Barreca, M. Di Cicco, F. Minuto,
    E. Foppiani and G. Giordano   261

Gonadotropins, Prolactin and Thyrotropin (Baseline Values and
Pituitary Responses) in Turner's Syndrome
    by G. Valenti, G.P. Ceda, S. Bernasconi, E. Tarditi, P. Chiodera,
    P.P. Vescovi, A. Banchini and U. Butturini   269

Results of the Treatment with Growth Hormone in a Case of
Goldenhar-Like Syndrome with Deletion of the Short Arm of
Chromosome 18
    by G. Aicardi, L. Buffoni, A. Naselli and A. Tarateta   279

5. Auxology, Psychology, and Psychosocial Problems

Family and School: Subsystems in Interaction
    by C. Pontalti and L. Ancona   287

Correlation Between the Age of Onset of a Chronic Illness and the
Child's Psychological Adaptation to It
    by J. Appelboom-Fondu   295

Psychological and Rehabilitation Aspects of Short Stature and
Delayed Puberty
    by A. Galatzer, E. Rosenblith and Z. Laron   303

Subject Index   313

1. AUXOLOGIC DETERMINANTS

Chronogenetic, Genetic, and Environmental Factors in Human Growth

THE CLOCK OF LIFE: AUXOLOGY AND CHRONOGENETICS

L. Gedda

*The Gregor Mendel Institute for Medical Genetics and Twin Studies
Rome, Italy*

SUMMARY

Human growth and development — and, more generally, human life — are basically controlled by the chronologic system that characterizes the genetic blueprint of any individual. Such individual endogenous time, the study of which is the object of *chronogenetics*, underlies any biomedical time, hence, the processes of growth and development and their variability, the study of which is the object of *auxology*.

Auxology represents a stage of the medical thought that has been separately attained in different areas of scientific knowledge. In genetics, to start with, where the hereditary units we now refer to as 'genes' were called by Gregor Mendel 'factors' — a term that implies an operative concept, i.e., a functional time. In order to express this concept better, the term 'phenogenesis' was introduced in genetics, meaning the operative translation of the genetic program, i.e., the realization of the hereditary blueprint in time.

General biology, considering the life cycle, has described the succession of forms that characterize the gradual attainment of the adult standard in every species, and has called it 'ontogenesis'.

Medicine, in turn — especially in areas such as physiology, obstetrics and gynecology, perinatology, pediatrics, endocrinology, and experimental psychology — makes use of the words 'growth' and 'development' to refer to the stage progression of the organism through to its final values.

Considering that the genetic perspective does not sufficiently account for the study of the environmental factors in phenogenesis, that the ontogenetic perspective is not open to the study of the interacting diseases, and, finally, that the terms, growth, or development, are much too general to account for normality and pathology, some clarification appeared to be needed. Such a clarification consists in the fact that the anthropological sciences in general, and those I have previous mentioned in particular, should consider time as a fundamental parameter to understand better the normal and pathologic phenomena of man. What I mean is that a dynamic interpretation of the phenomena of life, and especially of their complex, corresponding to somatopsychic individuality, should be considered together with the static phenomena, limited in their time variability, that have thus far dominated medicine. As Etienne-Jules Marey used to say, we should move from photography on to chronophotography, i.e., to the dynamic projection of life.

It was with the aim to deal better with this area of study, as is now high time, and to define it, together with its methods, as well as to stimulate the training of specialists and an exchange of information, that we decided that the term, Auxology, should be emphasized, and that an international meeting should be promoted. Our initiative was met by enthusiastic international support and resulted in the present Congress, which was planned by the Medico-Scientific Committee of the Italian Auxologic Center, that I chair since its establishment, by an apposite Organizing Committee, and by its Secretary General, Professor Parisi, to whom I particularly wish to express my appreciation.

I should like to bring my contribution to auxology by developing the concept that the fundamental module of auxology, that on which all others are based and meet, is time.

It is a truism that the auxology of man involves a part of his life's calendar. Because the calendar marks astral time, auxology is in a continuous relation with the chronology marked by the sun, the moon, and generally with the time of the physical universe.

The study of these relations between life and the exogenous time is the object of a branch of study called 'chronobiology'. I wish to mention it for its importance and because of its specific value in auxology, not only for the rhythms on which it is based but also for its physico-chemical and biological contents that the physical time succession offers to the human organism, as happens, for instance, with the seasonal rhythm of the pollens.

Auxology thus has to consider the time marked by the calendar in years, months, days, and day and night hours. It is in this time that life develops, as well as that environmental 'continuum' that conditions life, but, as far as we know, does not create life. The exogenous time is, therefore, the life's common denominator of a variety of numerators that represent the life of the species and of the individuals.

What is important to stress is that this exogenous time interacts with another time, endogenous or genetic in nature, which is related to the primary cause of life and on which auxology is based. This genetic time characterizes the individual biologic life of every human being, since everyone inherits it randomly with his genes and randomly transmits it to his offspring.

To give an example, the fact that, in the recent Seveso tragedy, the chloracne struck 400 schoolchildren over a total of 7,157 subjects examined, clearly shows that given subjects and a given age range are more susceptible than others to given environmental poisons. Therefore, toxicology and pharmacology should also consider the genetic time when evaluating the effects of active substances. And the same applies to all other branches of medicine.

The individual endogenous time is the preeminent biomedical time and therefore underlies the individual auxologic variance. This is the time to which I have referred in the title of my opening lecture, "The Clock of Life". This clock should be synchronized on the local time, and it appears that the synchronizing center of the two times might be the mysterious pineal gland. But these are secondary adaptations. What should be stressed is the fact that the biological life of a human being is not primarily controlled by his watch, but by the chronologic system that is contained in his genes.

It is from monozygotic (MZ) twins that we have learnt of the existence of a temporal dimension of the gene. MZ twins are not simply identical: they are identical at any given time; in other words, they exhibit a synchronous timing and thus point to the existence of genotypes the action of which is parallel, i.e., follows the same time pattern. This we can see in our daily practice and research at the Mendel Institute on the about 15,000 twin pairs of our register. But the layman too may perhaps be struck by somewhat similar considerations: may be on such occasions as that of the 19-year-old MZ twins who, as happened last March, won the Sun Valley slalom with a difference of 39 hundredths of a second.

The auxologic times are the most clearcut in twins, because of the number of stages involved. The MZ twins' synchronism is even more characteristic when it is referred to pathologic events, for which the hereditary nature of endogenous time may no longer be related to the genome in general, but to a specific gene mutation.

A number of cases recently brought to our attention may, better than any theoretical consideration, illustrate the existence of a genetics of time, i.e., of chronogenetics. The Dental Clinic of the University of Rome has recently observed the case of two 38-year-old MZ twins simultaneously affected by a bilateral mandibular cyst originated by the noneruption of lower canine teeth, i.e., primarily caused by an isochronic operative block at a given auxologic stage. In a hospital in Central Italy (Marche), two MZ twin women are assisted because of a juvenile progressive muscular dystrophy started at the age of 15 years, 15 days apart one from the other; they are now 35 years old and show overlapping deformities. At Bondeno di Ferrara live two MZ twin women, both affected by syringomyelia, with an overlapping evolution and mirror imaging.

Another case is from Rome and refers to two 9-year-old MZ twin girls, Alessandra and Roberta. On 16 February this year, while playing tennis, Alessandra, who is left-handed, fell and broke her left radius with a greenstick fracture. On 12 March, while playing with her brother, Roberta, who is right-handed, fell and also broke her left radius with a greenstick fracture. It would appear that the fragility of the left radius may have been simultaneously determined in the phenotype of the two twins, thus producing a chronogenetic phenomenon that might appear unreal, were it not for the X-ray finding and for the existence of similar cases, such as that of the two MZ twin girls from Iona, in the Swiss canton of Saint Gallen, Dieter and Wolfy, who broke their right leg while skiing last year, 24 hours apart one from the other.

An interesting chronogenetic perspective stems from the study of the inheritance of quantitative traits in relation to maternal age, In a collaborative study,

Lints, from the Louvain Catholic University, and Parisi, from the Mendel Institute, have shown the inheritance of the Total Finger Ridge Count to increase with maternal age. The results, based on the fingerprint study of 279 twin pairs, are now being confirmed on a much larger sample. Moreover, similar results have been obtained with respect to the sternopleural cheta number in *Drosophila melanogaster* and to the number of caudal fin rays in a fish, *Lebistes reticulatus*. These results clearly point to the existence of a temporal variability of maternal inheritance, which might perhaps be related to the nonnuclear heredity of the female gamete.

MZ twins thus provide us with sufficient evidence to conclude that the gene must possess a temporal dimension, i.e., an informative potential that determines its life span and exhaustion.

Supplementary evidence stems from other twin studies, such as the one carried out by Barbati-Crouzet at the Mendel Institute, concerning a sample of twin pairs, aged 2 to 14 years and still presenting epicanthus — a mark of immaturity in Caucasion populations, more frequent in twins that in singletons. The reduction of the epicanthus after birth and until disappearance is progressive and overlapping in all the MZ (10/10) and in 5/13 DZ twin pairs, thus showing a complete genetic control on the timing of an auxologic subclinical phenomenon.

Pedigree studies also point to the existence of inherited biological times, showing the duration of both normal and especially abnormal information to be inherited according to the mechanisms of Mendelian genetics. Therefore, in auxology, genetics does not start with population differences, but much more deeply, with the hereditary blueprint. Let us see why.

Numerous phenomena explored in the most diverse fields of human, animal, and plant biology, allow us to state that genetic variability does not merely concern the quality and quantity of the information, but the *stability* of the gene as well, this being variable and determining a variable informative time. We have called *ergon* the stability of the gene and *chronon* its corresponding life span, i.e., the potential time of functioning of the gene's primary product. Whereas the ergon is unique, i.e., genotypic, the chronon exists under two forms, a genotypic, hence potential form, and a phenotypic, hence actual form.

The stability of a gene is first of all a function of its molecular structure and, therefore, also of the type of synonyms used in the coding system. This factor of stability has been called 'synonymy' by Gedda and Brenci. In fact, the different ratio of Adenine-Thymine to Guanine-Cytosine (AT/GC), determined by synonymous codons, is reflected, through the number of hydrogen bonds joining the two hemimolecules, in a different chemico-physical stability of the entire molecule. In other words, because of synonymy, DNA segments that code for one and the same information may differ in AT/GC contents and consequently possess a different chemico-physical stability. Moreover, the recent finding of a position effect of a given base confirms that a different information stability corresponds to a given structure of the molecule.

A second stability factor is given by the number of sequences that code for the same information. This phenomenon, known in molecular genetics under the name of 'redundance', is reflected on the stability of the gene in such a way that, under the same mutagenic action, the more numerous the sequences coding for information, the more stable the gene and the more protracted its informative potential.

A third stability factor consists in the interaction between the structure of the information and the 'repair' systems developed to repair those gene damages that the internal or external environment may produce.

Thus, there are three chronogenetic stability factors we have so far hypothesized: synonymy, redundance, and repair. This is the causal tripod underlying the functioning of the system that controls the fourth dimension of the gene and to which we have given the name, *ergon/chronon system*. Thus far, we could only probe the immense area of normal and abnormal heredity in the perspective of the ergon/chronon system, but we already have enough evidence to say that the time parameter characterizes the individual gene both in the case of simple and multifactorial inheritance. The family study at the sibship level, moreover, has allowed us to confirm the decay of gene stability: in fact, when pathologic traits are repeated in a sibship, their time of onset appears to be inversely proportional to birth rank, i.e., the onset is earlier the later the conception. This reminds us of the age of the gamete, formed before birth with the germinal line, and of the fact that the aging of the gamete and of its genes, because of a loss of stability, is a temporal and hereditary, i.e., chronogenetic phenomenon.

If, as it happens, life is influenced by two times, one exogenous and the other endogenous, auxology, which is concerned with the temporal 'escalation' of the human biotype, should consider them both and jointly.

Genetic time, precisely because it is genetic, i.e., hereditary, is deeply rooted in the family space and may be verified at the genealogical level. We should therefore introduce in auxology the study of the time of onset of auxologic determinants and pathology in the parents and ancestors of our subjects, and possibly in their collaterals as well.

The first step should be the identification of primary genetic traits with respect to secondary ones, involving different genes and interactions between heredity and environment. Once the endogenous causal picture is defined, chronogenetics will allow us not only to formulate clinically, i.e., individually, the diagnosis of auxologic normality or pathology, but also to proceed further, towards the development of a preventive individual auxology that is the practical goal of general auxology. An 'auxologic risk' might thus be calculated, which would turn into a prognosis in those cases in which a pathology is present or likely.

Prognostic medicine in general and preventive medicine in particular are still practised starting from clinical and laboratory data and then referring to a control population. This procedure is largely influenced by the intuition of the doctor, that medicine has for centuries covered with a magic expression, the clinical eye. Far from underestimating the great value of this intuition based on medical experience, it seems to me, however, that the time has come for a more objective, individual prevention in the various fields of medicine, including auxology. But this may only be possible when medical genetics has developed the goal and methods of chronogenetics.

If the gene possesses an informative potentiality, i.e., ergon, and this potentiality — in given conditions of genotypic, somatic, and exogenous environment — produces a chronon, i.e., a progressive exhaustion of the information, then there are three ways that research may follow. The first way consists in using the information on the gene extinction in the ancestors to infer the gene's informative potential in the candidate. The second way consists in the study of the gene's exhaustion curve through the dosage of its primary product at given times and subsequent extrapolation for the future behavior. Finally, the third way, which I think is the one that chronogeneticists will preferably follow, consists in having recourse to that fruitful champion of biologic individuality, the lymphocyte, and determine *in vitro* the specific reaction of the given gene in order to study its characteristics in terms of stability and decay.

Prevention is a dominant idea in modern medicine and it should be of special importance to the auxologist, who may not simply prescribe, e.g., an appropriate diet, or indicate a general need for physical exercise or the danger of habits such as smoking, etc.. These and other indications should instead be integrated with the subject's dynamic constitution, i.e., with the exploration of his chronogenetic future.

In his novel, "Le vent dans les arbres", Jean-Claude Andro uses this beautiful image: "Un arbre n'occupe pas seulement l'espace. Il est plein d'une sevè venue des profondeurs." I would also use it, and say that man, especially in the auxologic period, does not simply occupy a space, but is filled with a lymph, i.e., a potential, that genetics endeavours to define and to dose, and thus put in the hands of the auxologist the keys of the future.

NORMAL GROWTH AND NEOPLASTIC GROWTH: ARE THEY CONNECTED?

P.R.J. Burch

*The University of Leeds, Department of Medical Physics
Leeds General Infirmary, Leeds LS1 3EX, England*

SUMMARY

The basis of the author's unified theory of growth and age-dependent disease — neoplastic and nonneoplastic — is described and illustrated with reference to retinoblastoma. It is shown that the chronogenetics of two types of retinoblastoma, in large series from England and Wales, the United States, and France, conform to the statistical criteria of an autoaggressive disease. It is concluded that Type 1, involving autosomal dominant predisposition, is initiated by a single somatic mutation in a central growth-control stem cell. About 50% of Type 1 cases are bilateral. Type 2, involving polygenic predisposition, is initiated by three somatic mutations in a central growth-control stem cell, and is usually unilateral.

I. INTRODUCTION

Growth can be defined as a net increase in size or mass. In the mammal, growth during prenatal and postnatal phases depends mainly on an increase in cell number. Hypertrophy — the increase in the size of cells, such as neurones — makes only a small contribution to overall growth. From the formation of the zygote, some 20 years elapse before the mature human acquires a full complement of around 10^{14} cells.

In the fully-grown organism, many tissues are in a state of dynamic equilibrium, although the rate of cell turnover differs widely from tissue to tissue.

When effete cells are being continually destroyed or shed — as, for example, in the erythrocytic, granulocytic, various epithelial and epidermal systems — at least one stage of *asymmetrical* mitosis maintains the steady state. Thus, a stem cell has to divide to give, on the average, one cell that persists as a stem cell and another that divides to give descendants that will make good the loss of effete cells.

During growth, the absolute number of cells in, for example, the basal layer of the epidermis increases, and to effect this, *symmetrical* mitoses occur in which a single basal cell divides to yield two basal daughter cells. When cytodifferentiation has been completed, this form of *symmetrical* mitosis is essential to growth. During embryogenesis, a stem cell may divide to give two stem cells of dissimilar differentiation.

Whereas normal growth is a programmed sequence that serves the anatomical and physiological requirements of the whole organism, neoplastic growth usually conflicts with normal biological needs. Nevertheless, many tumours in experimental animals — spontaneous, induced and transplanted — grow to a limiting size. The growth curve can be described by a simple Gompertz function: an initially exponential increase in size gives way to a progressively diminishing growth-rate, and ultimately, to a constant size and therefore no net growth (Laird, 1964, 1965; McCredie *et al.*, 1965; Sullivan and Salmon, 1972). From such observations and the finding that the growth of the whole organism is described by the same type of Gompertz function, Laird (1965) argued that a tumour does not grow as a population of individual cells but as though it were a single organism. Many other observations and experiments — reviewed elsewhere (Burch, 1976) — indicate that neoplasia is often not so much *un*controlled growth but, from the point of view of the organism, *mis*controlled growth of aberrant cells. Brand (1976) stresses that an understanding of carcinogenesis is likely to remain out of reach as long as the mechanisms of the normal regulation of growth remain obscure.

My collaborators and I have developed the thesis that, after a certain stage of embryogenesis, normal growth is regulated homoeostatically by a central mesenchymal system and that spontaneous autoaggressive disease, including naturally occurring neoplastic growth, is initiated by gene mutations that occur randomly in the comparator stem cells of this *central* control system. A mutant comparator stem cell propagates a clone of descendant cells — a *forbidden clone* in Burnet's (1959) terminology — and these mutant clonal cells, or their secreted humoral products, attack target cells carrying complementary recognition proteins (see Fig.1). In neoplastic disease, one or more transformed target cells are stimulated to divide and to propagate a uni- or multifocal tumour. The growth of natural neoplasms can be regarded as the consequence of an "incorrect", nonphysiological, setting of the homoeostat, brought about by mutation of the comparator. To some extent, we can separate this "error" in growth control from the dysfunction and invasiveness that characterizes malignant cells.

In this chapter I describe our unified theory of growth and age-dependent disease and show its application to an early-onset neoplasm, retinoblastoma.

II. THE CHRONOGENETICS OF DISEASE

From studies of the chronogenetics of numerous diseases, neoplastic, preneoplastic and nonneoplastic, and from the systematics of the anatomical distri-

AUTOAGGRESSIVE DISEASE : GENERAL SCHEME

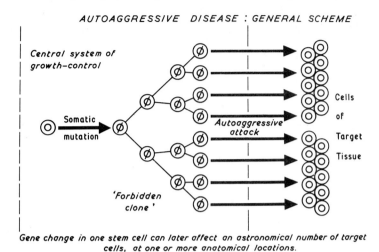

Gene change in one stem cell can later affect an astronomical number of target cells, at one or more anatomical locations.

Fig.1 Basic scheme of autoaggressive disease. When a stem cell of the central system of growth control acquires an appropriate complement of somatic gene mutations, it propagates a *forbidden clone* of similarly-mutant descendant cells. These cells, or their secreted humoral mutant products, attack those cells of the target tissue that carry complementary recognition macromolecules. Target cells may be located at one, or multiple, anatomical sites.

bution of lesions, I have proposed the following generalizations (Burch, 1968, 1976):

(a) A narrowly-defined disease that exhibits a reproducible age-pattern conforming to equation (1) below, should be regarded as an *autoaggressive* disorder, confined to persons with a specific genetic predisposition.

(b) An autoaggressive disease results from the attack of one or more (n) centrally-derived *forbidden clones* of cells, or their humoral products, on the same number (n) of distinctive mosaic elements of a target tissue. The anatomical location of mosaic elements is genetically determined.

(c) A forbidden clone is initiated by the spontaneous occurrence of one or more (r) specific gene mutations in a stem cell of the central system of growth control.

(d) The average rate of a specific initiating somatic mutation is constant from within about ± 0.1 yr of birth, throughout postnatal growth and to the end of the life span.

(e) Between the occurrence of the last random initiating mutation and the first onset of the associated autoaggressive disease an interval or *latent period* elapses.

Provided certain conditions hold — and they frequently do — the age-specific initiation-rate, dP/dt, of a specific autoaggressive disease at initiation age t, is given by the following general stochastic equation (Burch, 1968, 1976):

$$dP/dt = \left\{ nrkSt^{r-1} \exp(-kt^r) \right\} \left\{ 1-\exp(-kt^r) \right\}^{n-1} \qquad (1)$$

where S is the proportion of the population studied that is genetically predisposed to the disease at birth; and k is a kinetic constant related to the average rate, m, of each of the r initiating events by $k = Lm^r$, where L is the (constant) number of growth-control stem cells at somatic mutational risk.

In applying equation (1) to the chronogenetics of retinoblastoma, we need to consider two special cases:

When $n = 1$ we have the Weibull equation:

$$dP/dt = rkSt^{r-1} \exp(-kt^r) \ . \tag{2}$$

When $n = 1$ and $r = 1$ we have the simple negative exponential equation:

$$dP/dt = kS \exp(-kt) \ . \tag{3}$$

Importance of Early-onset Autoaggressive Disease to Theory

Autoaggressive diseases that have appreciable incidence during infancy, childhood and adolescence show that k is effectively constant during the postnatal phase of growth. This invariance was a most unexpected finding because the size of most cell populations increases during postnatal growth. From the constancy of k values, for different values of r, it follows that L, the number of cells at somatic mutational risk, also stays constant during postnatal growth.

This implication became intelligible when Burwell (1963) proposed that the intrinsic or physiological function of one part of the lymphoid system is the central regulation of growth. In any negative feedback (homoeostatic) system of growth control, the central controlling organ needs a *comparator*, or "fixed yardstick", against which the size of the target tissue can be "measured". We concluded, for reasons described in Section IV, that the fixed number of stem cells at somatic mutational risk in autoaggressive disease serve normally as the comparator in the central system of growth control (Burch and Burwell, 1965; Burch, 1968, 1976).

A contrast between superficially similar phenomena should be emphasized: (i) the reproducibility of the age-distribution of onset of autoaggressive diseases; and (ii) the regularities of age-dependent developmental phenomena. Phenomenon (i) follows from the invariance of the parameters n and r and the constancy of k. These parameters determine the reproducible shape of the age-pattern. However, because the initiation of an autoaggressive disease is a stochastic process, the disease will not manifest at the same age, even in a monozygotic twin pair, except by chance. (But, if nr is large, the age-distribution of onset will be sharply peaked.) By contrast, developmental phenomena follow a sequential, deterministic programme that is the antithesis of a stochastic process. Given similar environments, and once any discrepancy in birth sizes has been overcome, the two members of a monozygotic twin pair attain well-defined developmental stages, such as the menarche, at an almost identical age, which is genetically determined.

III. RETINOBLASTOMA AS AN EARLY-ONSET MALIGNANT AUTOAGGRESSIVE DISEASE

Ideas on the genetics, aetiology and pathogenesis of retinoblastoma have flourished over the past decade (Schappert-Kimmijser *et al.*, 1964; Nielsen and Goldschmidt, 1968; Knudson, 1971; Czeizel and Gárdonyi, 1974; Matsunaga,

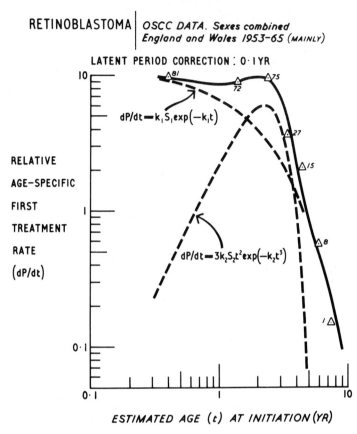

Fig.2 Relative age-specific first treatment rates, sexes combined, for retinoblastoma in England and Wales, mainly for the period 1953-65, in relation to estimated age at initiation (Stewart, 1969). For clarity, data are represented by points at the middle of the age range, rather than by a histogram. Log-log scales. Uncertainties over ascertainment, the time period for collection of cases, and the size of the population at risk, preclude the calculation of absolute rates and hence of S_1 and S_2. An average latent period of 0.1 yr is assumed to elapse between the completion of initiation and diagnosis; the zero of the age scale is assumed to be birth. See also Table I.

1976; Matsunaga and Ogyu, 1976; Bonaïti-Pellie *et al.*, 1976). Genetically, two forms of the disease are recognized, autosomal dominant *familial*, and so-called *sporadic*. Anatomically, unilateral and bilateral tumours are distinguished. The number of neoplastic foci in an affected eye is often difficult or impossible to estimate.

Of special interest is Knudson's (1971) proposal " ... that retinoblastoma is a cancer caused by two mutational events". In the autosomal dominant inherited form of the disease, the first mutation is said to occur in the germ cell line and the second in a somatic cell. In the "sporadic" and supposedly noninherited form of the disease, Knudson (1971) postulates that both mutations occur in somatic cells. This "two-mutation" hypothesis resembles an earlier view of mine concerning the equivalence, under some circumstances, of germinal and somatic (especially embryonal) mutations (Burch, 1962).

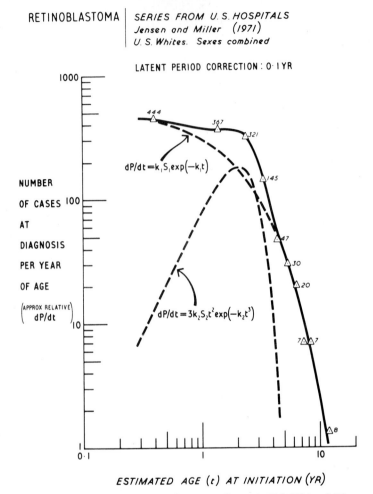

Fig.3 Number of cases of retinoblastomas, by age at diagnosis, U.S. White children, sexes combined, 1960-67, in relation to estimated age at initiation (Jensen and Miller, 1971). The number of cases per year of age will be approximately proportional to the age-specific diagnostic rate. Otherwise, as for Fig.2.

The age-distribution of onset of cases in a series from England and Wales (Stewart, 1969), two U.S. series (Jensen and Miller, 1971) and one from France (Bonaïti-Pellie *et al.*, 1976) show immediately that a key aspect of Knudson's hypothesis cannot be sustained (Figs 2 to 5). For each of the four series, the overall age-pattern can be accurately described by the sum of two stochastic functions, which are particular versions of the general equation (1):

$$dP/dt = k_1 S_1 \exp(-k_1 t) \tag{4}$$

in which $n = 1$, $r = 1$, the disease being initiated by a single somatic mutation, and

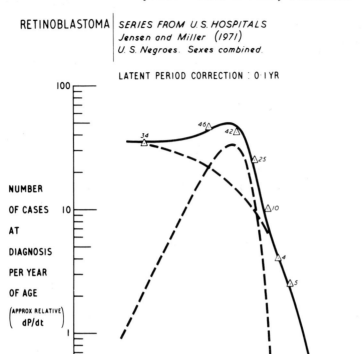

Fig. 4 As for Fig. 3. Data for U.S. Negroes, 1960-67 (Jensen and Miller, 1971).

$$dP/dt = 3k_2 S_2 t^2 \exp(-k_2 t^3) \tag{5}$$

in which $n = 1$ and $r = 3$, entailing 3 somatic mutations, rather than the 2 postulated by Knudson.

Figure 6 shows that the age-pattern of most "familial" cases, unilateral and bilateral, is described by equation (4), although the statistics are inadequate to rule out a contribution from equation (5). Indeed, the average age at onset of unilateral familial cases (Matsunaga, 1976; Bonaïti-Pellie *et al.*, 1976), which is significantly higher than that of bilateral familial cases, suggests that the age-dependence of some unilateral cases follows equation (5). I shall call the cases in which $n = 1$, $r = 1$ — they are initiated by a single somatic mutation — Type 1 cases.

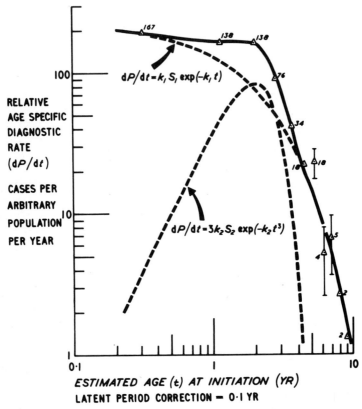

Fig.5 Relative age-specific diagnostic rate, sexes combined, for retinoblastoma in France, 1959 to 1975, in relation to estimated age at initiation (Bonaïti-Pellie *et al.*, 1976). Otherwise, as for Fig.2.

Figures 2 to 5 show that the age-pattern of the remaining cases fit the *three-mutation* equation (5), with $n = 1$ and $r = 3$. I shall call these Type 2 cases. My theory requires that these Type 2 cases should also be confined to a distinctive genotype, which constitutes a fraction, S_2, of the general population. If we postulate that predisposition to Type 2 cases is polygenic, then we can readily account for the discovery of some remarkable families with distantly-related affected members (Macklin, 1960; Bonaïti-Pellie *et al.*, 1976) in which the heredity of retinoblastoma cannot be explained in terms of an autosomal dominant gene of high penetrance. Most Type 2 cases appear to be "sporadic" because the chance of multiple members of a given family inheriting the requisite combination of predisposing genes is generally small.

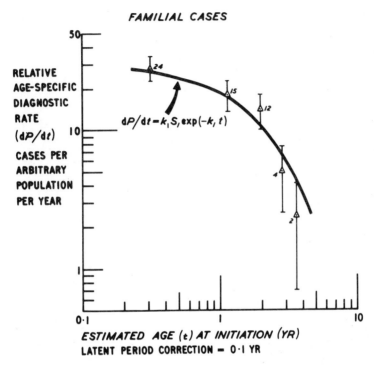

Fig.6 Relative age-specific diagnostic rate, sexes combined, for "familial" cases of retinoblastoma in the series of Bonaïti-Pellie *et al.* (1976), in relation to estimated age at initiation.

From the reproducible features of the age-pattern of retinoblastomas, and the observations of Bonaïti-Pellie *et al.* concerning laterality and the apparent distinction between "familial" and "sporadic" cases, I draw the following conclusions:

(1) Type 1 cases, unilateral and bilateral, are initiated by a single random somatic mutation: the age-pattern of initiation is described by equation (4) with the parameters listed in Table I.

(2) So-called "sporadic" cases may be of Type 1 (unilateral or bilateral) resulting from a "new" germ cell mutation or, as shown recently (see, for example, Knudson *et al.*, 1976), by the deletion in a parental germ cell of the long arm of chromosome 13; or Type 2, in which the disease (largely or wholly unilateral) is initiated by three somatic mutations according to equation (5) with parameters listed in Table I.

(3) Predisposition to Type 1 cases entails an autosomal dominant gene or a $13q^-$ chromosome. However, the phenomenon of the occasional "skipped generation" suggests, among other possibilities (Burch, 1976), that another gene (or genes) of high frequency in the general population might be necessary to

Table I. Parameters characterizing Type 1 and 2 retinoblastomas. Sexes combined.

Population	Type 1 $(n = 1, r = 1)$ k_1 (yr^{-1})	Type 2 $(n = 1, r = 3)$ k_2 (yr^{-3})	S_1/S_2*
England and Wales (mainly 1953-65) Oxford Survey (Stewart, 1969)	0.52	5.7×10^{-2}	1.6
U.S. Whites, hospital series 1960-67 (Jensen and Miller, 1971)	0.52	6.8×10^{-2}	2.7
U.S. Negroes, hospital series 1960-67 (Jensen and Miller, 1971)	0.42	6.4×10^{-2}	1.3
France (1959-1975) (Bonaïti-Pellie et al., 1976)	0.54	8.2×10^{-2}	2.55

* These data do not enable absolute values of S_1 and S_2 to be calculated.

complete the Type 1 genotype.

(4) The anatomical distribution of target cells is determined by genetic factors. The majority of Type 1 bilateral cases (18 out of 20 in Leelawongs and Regan's series) are bilateral at the first examination. This rules out the possibility that *most* bilateral cases are initiated by independent random events, one in each eye. It implies that the single initiating event occurs in a *central* location and that both eyes are affected almost simultaneously. This does not preclude the possibility that, in some instances, bilateral cases do result from two independent initiating events, and two forbidden clones, one specific for each eye. Indeed, when the interval between the onset in the two eyes is large — many months or years — such an explanation cannot be avoided.

(5) Biases of ascertainment (Bonaïti-Pellie et al., 1976) and small numbers in some categories, preclude accurate estimates of the proportion of Type 1 or Type 2 retinoblastomas among sporadic cases. However, sporadic bilateral cases in the series of Bonaïti-Pellie et al. belong mostly, but not necessarily exclusively, to Type 1. Among sporadic unilateral cases in the French series, about 53% appear to belong to Type 1, arising from "new" autosomal dominant mutations, and the remainder (47%) to Type 2, with polygenic predisposition.

IV. IMPLICATIONS FOR AETIOLOGY AND GROWTH CONTROL

Figures 2 to 5 leave little doubt about the reproducibility of the chronogenetics of "familial" and "sporadic" retinoblastomas in different populations and the conformity of their age-distributions to the statistical criteria described in Section II. According to these criteria, retinoblastoma, regardless of type, should be an autoaggressive disease. Particularly important is the demonstration of *two* genotypes, one (autosomal dominant) requiring only a single somatic mutation for the initiation of unilateral or bilateral retinoblastoma, and the other (presumably polygenic) requiring *three* somatic mutations for the initiation of (usually unilateral) Type 2 retinoblastoma. The age-patterns in Figs 2 to 5

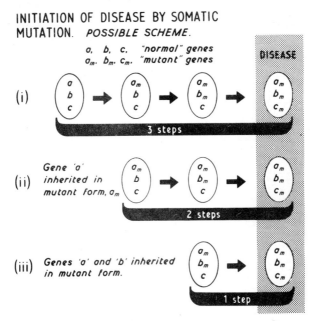

Fig. 7 A possible simple scheme for the initiation of age-dependent disease through the process of somatic gene mutation. The emergence of disease requires a cell to be formed with the three mutant genes a_m, b_m and c_m. In (i), the mutant cell emerges through a 3-step random process of somatic mutation. In (ii), gene a is inherited in the mutant form, a_m, and only 2 random-steps are needed to produce a cell with mutant genes a_m, b_m, and c_m. In (iii), the same end-result is accomplished through a single random step of somatic mutation.

give no hint of a genotype requiring two somatic mutations for the initiation of the tumour. An essentially similar situation is found with many other diseases which involve two or more distinctive predisposing genotypes (Burch, 1968, 1976). Such findings imply that any simple scheme of disease, of the kind illustrated in Fig. 7, is untenable.

If the initiation of disease depends solely on the formation of a cell with, say, the three mutant genes, a_m, b_m, and c_m, and if one genotype — genotype (i) in Fig. 7 — entails a 3-step random process of initiation from a, b, c to a_m, b_m, c_m, then we should find in the population two other genotypes — genotype (ii) and genotype (iii) — requiring two steps and one step, of initiation respectively. In genotype (ii), one gene is inherited in its mutant form and in genotype (iii), two mutant genes are inherited. This type of relationship, instead of being universal, is only rarely encountered (Burch, 1968, 1976).

The difficulty with this simple scheme, which is fundamental, can be overcome by postulating that predisposing genes perform a *dual function*: they code for recognition polypeptides in "central" and "target" cells (Fig. 8). In this scheme, autoaggressive disease is determined by a specific steric *relationship* between the macromolecule, for example $a_m b c_m d_m$, incorporating the mutant polypeptides a_m, c_m, and d_m and the target macromolecule a b c d. The number of somatic mutations required to produce a pathogenic relationship will depend on the overall structure of the recognition macromolecules, normal

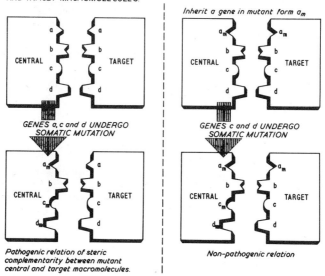

Fig.8 A scheme for the initiation of disease by somatic gene mutation that avoids the difficulties of the scheme illustrated in Fig.7. Predisposing genes perform a dual function: they code for recognition macromolecules in "central" and in "target" cells. Disease depends on a specific complementary relationship between the mutant macromolecule $a_m b c_m d_m$ and the target macromolecule a b c d. If gene a is inherited in its mutant form a_m, as in the right-hand panel, then the target macromolecule contains the polypeptide a_m. Two somatic mutations (or one) in central cells are unable to establish the complementary relationship characteristic of disease, shown in the left-hand panel.

and mutant. Similarly, the pathological consequences are determined by the nature of the interaction.

By hypothesis, a highly specific complementary steric relationship characterizes every autoaggressive disease. This suggests that the type of somatic mutation that initiates autoaggressive disease should itself involve a very special steric relationship between the mutant and nonmutant genes and between their respective polypeptide products. I proposed that initiating somatic mutations entail DNA strand-switching: that is, a changeover in transcription from the normal strand of DNA to the complementary base-paired antiparallel strand (Burch, 1968). This hypothesis makes several predictions of which some, but not all, have been verified (Burch, 1976).

It follows from the general theory that, for every target tissue element that is subject to autoaggressive disease, an identity relation exists between central and target macromolecules (coded by identical genes) before the occurrence of initiating somatic mutations. This unique relationship, which probably applies to every mosaic element of every target tissue in the body, is unlikely to be without profound biological significance. We proposed that this identity relation forms the basis of tissue recognition, at both ends of the feedback loop, in normal growth-control (Burch and Burwell, 1965 — see Fig.9).

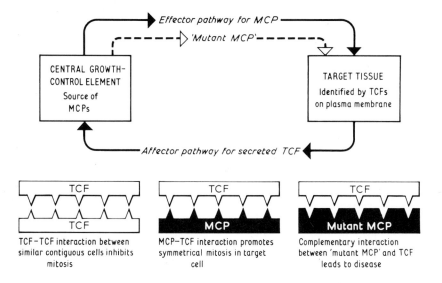

OUTLINE OF NEGATIVE-FEEDBACK CONTROL OF GROWTH
The size of each distinctive tissue is controlled by its own homoeostat

Fig.9 Recognition relations between central and target macromolecules in the normal homoeostatic control of growth and autoaggressive, including neoplastic, disease. We call effector molecules of symmetrical mitosis, *mitotic control proteins* (MCPs), and affector molecules, *tissue coding factors* (TCFs). Target cells are identified by TCFs on the cytoplasmic membrane, and receptor cells in the central control by MCPs. Humoral MCPs and TCFs have to be distinguished: the identical components of homologous macromolecules are supplemented by nonidentical protein and/or nonprotein (lipid, carbohydrate) components.

An independent argument has since been put forward which shows that if growth is centrally regulated — and a wide range of experimental evidence implies that it is — then no other genetic and molecular basis of the specificity of recognition in growth-control is possible (Burch, 1968, 1976). By definitiion, an alternative system of recognition would have to be based on one set of genes coding for recognition molecules in the central system and a different set coding for recognition molecules in target cells (Fig. 10). If perfect matching were present in generation 1, then, from known rates of maturation of germ cells and the large number of genes involved in recognition, many mismatches would be present, on the average, in an individual of generation 2 (Burch, 1968, 1976). The normal growth of such mismatched tissue elements would fail. Both the individual and the species would succumb. However, when the same set of genes is used to code for central and cognate target recognition macromolecules, a germ-cell mutation of a recognition gene leads to changes in both central and cognate target macromolecules and the all-important identity relation between the two is preserved (Fig.11). Mutual recognition, at both ends of the negative feedback control loop (Fig.9), is maintained through London-Van der Waal's self-recognition forces.

I conclude, therefore, that normal growth and neoplastic growth are intimately related. The specificity of the regulation of normal growth depends on a relationship of identity between central control and target recognition macro-

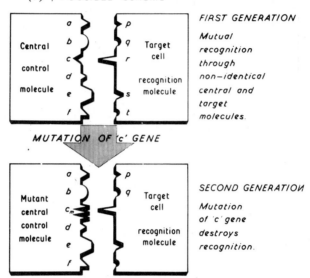

Fig.10 A hypothetical scheme for recognition in growth-control using one set of genes to code for central molecules and a different set to code for target-cell recognition molecules. Perfect matching and recognition is assumed in generation 1. However, germ-cell mutations of either central- or target-cell recognition genes will, in general, destroy the mutual recognition relation for many tissue elements in generation 2.

molecules. In autoaggressive disease, the somatic mutation of one or more genes in stem cells of the central control system changes the identity relation into an equally specific relation of complementarity. The resulting interaction is stronger and pathogenic. In neoplastic disease, which is a particular form of autoaggressive disease, the complementary interaction between central and target recognition macromolecules promotes the neoplastic growth of target cells. The Gompertzian growth curve that many tumours follow resembles that of the whole organism and in both phenomena a limiting size is attained.

REFERENCES

Bonaïti-Pellie, C., Briard-Guillemot, M.L., Feingold, J. and Frezal, J. (1976). *J. Natl. Cancer Inst.* 57, 269-276.
Brand, K.G. (1976). *J. Natl. Cancer Inst.* 57, 973-976.
Burch, P.R.J. (1962). *Nature (London)* 196, 241-243.
Burch, P.R.J. (1968). "An Inquiry Concerning Growth, Disease and Ageing." Oliver and Boyd, Edinburgh; (1969) University of Toronto Press, Buffalo, N.Y.
Burch, P.R.J. (1976). "The Biology of Cancer: A New Approach", Medical and Technical Publishing, Lancaster. University Park Press, Baltimore.
Burch, P.R.J. and Burwell, R.G. (1965). *Q. Rev. Biol.* 40, 252-279.
Burnet, F.M. (1959). "The Clonal Selection Theory of Acquired Immunity", Cambridge University Press, London.
Burwell, R.G. (1963). *Lancet* 2, 69-74.
Czeizel, A. and Gárdonyi, J. (1974). *Humangenetik* 22, 153-158.

Normal Growth and Neoplastic Growth: Are They Connected?

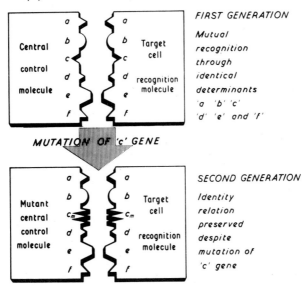

Fig. 11 The probable scheme for recognition in growth-control, in which the same set of genes codes for recognition molecules in the central and target cells of a given growth-control element. Mutation in the germ cell line, such as c to c_m, does not destroy the identity relation between central and target recognition molecules. Evolutionary change has depended in part on the entry of mutant genes into the gene pool and any viable system of growth-control has to be able to incorporate newly mutant genes.

Jensen, R.D. and Miller, R.W. (1971). *N. Engl. J. Med.* **285**, 307-311.
Knudson, A.G. (1971). *Proc. natn. Acad. Sci. U.S.A.* **68**, 820-823.
Knudson, A.G., Meadows, A.T., Nichols, W.W. and Hill, R.C. (1976). *N. Engl. J. Med.* **295**, 1120-1123.
Laird, A.K. (1964). *Br. J. Cancer* **18**, 490-502.
Laird, A.K. (1965). *Br. J. Cancer* **19**, 278-291.
Leelawongs, N. and Regan, C.D.J. (1968). *Am. J. Ophthalmol.* **66**, 1050-1060.
McCredie, J.A., Inch, W.R., Kruuv, J. and Watson, T.A. (1965). *Growth* **29**, 331-347.
Macklin, M.T. (1960). *Am. J. Hum. Genet.* **12**, 1-43
Matsunaga, E. (1976). *Hum. Genet.* **33**, 1-15.
Matsunaga, E. and Ogyu, H. (1976). *Jap. J. Ophthalmol.* **20**, 266-282.
Neilsen, M. and Goldschmidt, E. (1968). *Acta Ophthalmol.* **46**, 736-741.
Schappert-Kimmijser, J., Hemmes, G.D. and Nijland, R. (1964). *Ophthalmologica* **151**, 197-213.
Stewart, A. (1969). Private communication to author.
Sullivan, P.W. and Salmon, S.E. (1972). *J. clin. Invest.* **51**, 1697-1708.

INTERCORRELATIONS OF GROWTH AND MATURITY CHARACTERS IN THE MALE SEX: A FACTOR ANALYSIS

R. Fricke and R. Knussmann

Department of Anthropology, University of Hamburg, Von-Melle-Park 10 2000 Hamburg 10, German Federal Republic

SUMMARY

Based on an unassorted random sample in Hamburg of 1,110 19-year-olds liable for military service, 35 metric, 13 descriptive and 5 physiological parameters were taken and intercorrelated. Using factor analysis — method of principal axes and orthogonal rotation — five factors were extracted: a girth factor, a length factor, a circulation factor, a masculinity factor, and a maturity factor.

I. INTRODUCTION

The aim of the present study was to discover to what extent intercorrelations exist in the male sex between the metric and morphognostic characters to be measured during the study and dealt with in more detail below. This was done with a view to establishing the existence of factors which influence the moulding of a number of characters. The factors, which are to be ascertained by factor analysis, can be interpreted as biological growth tendencies. The type of each factor is classified according to the way it affects those characters upon which it exerts particularly strong influence.

From the historical point of view, one can differentiate between three theoretical approaches: whereas Spearman (1905) presumed that the assumption of one factor was sufficient, because, in comparison, the influence of other factors was negligible, Burt (1911) postulated the differentiation of group factors in addition to a general factor. Finally, Thomson (1919) argued that the variability could be explained solely by the overlapping of such group factors, discording with the assumption of a general factor. These differentiations, which at first only related to psychic traits, can also be applied to the factor analysis of body structure characters.

The assumption of only one single general factor proved insufficient (Spearman, 1927; Thomson, 1939), whereas the assumption of one or several group factors alongside a general size factor has been confirmed by numerous analyses: Carter and Krause (1936), mat. from Bakwin and Bakwin (1931, 1934); Burt (1938, 1943, 1944, 1974); Cohen (1938, 1939, 1940a, 1940b); Mullen (1939, 1940); McCloy (1940); Waltrop (1940); Holzinger and Harman (1941), mat. from Mullen (1939); Hammond (1942, 1953a), mat. from Adcok et al. (1947) and O'Brien et al. (1941); Hammond (1957), mat. from Hammond (1955, 1953b), Carter and Krause (1936), and Low (1952); Rees (1945, 1949, 1950, 1960, 1973); Rees and Eysenck (1945); Banks (1946); Thurstone (1946), mat. from Hammond (1942); Thurstone (1947), mat. from Rees and Eysenck (1945); Burt and Banks (1947), mat. from Banks (1946); Moore and Hsü (1946), mat. from Connolly (1939); Sills (1950); Howells (1951), mat. from Howells (1948, 1949); Petersen (1961).

Results of this kind can, with respect to the first factor, be well paralleled (Knussmann, 1968, 1970) with a typological variation series from macrosomia to microsomia (Knussmann, 1961, 1966), while the pycnomorphous-leptomorphous variation series is visible in the group factors (Kretschmer, 1921; McDonough, 1932). Several group factors are differentiated without the assumption of a general factor by McDonough (1932); McCloy (1935); Sterett (1935); Marshall (1936); Coleman (1940); Hofstätter (1943/44); Carpenter (1941a, 1941b); Cohen (1941); Heath (1952), mat. from O'Brien and Shelton (1941); Howells (1952), mat. from Dupertuis (1950); Schick (1953); Lorr and Fields (1954); Tanner et al. in Tanner (1964); Klein (1970); Knussmann and Klein (1970); Jackson and Pollock (1976).

However, the analyses of many of these authors are based either on biased character sets or on samples which were selected with a particular problem in mind. The selection of the parameters and the composition of the sample influence the type of the resultant factors. Deliberate selection of characters can, in this way, practically provoke the outcome of certain factors. The comparability of these analyses is therefore limited. This also applies to a comparison with the present study, which is based on a general set of characters and, at the same time, includes sex-specific parameters.

II. MATERIAL AND METHODS

Altogether 1,110 young men liable for military service were available for the study. They represent an unassorted random sample, since all members of the 1957 age-group mustered by a Hamburg examination board between June and December 1976 were included in the study. Scholars and students may be underrepresented because the majority of them had already been mustered during the first months of the year for organizational reasons. There were thus fewer represented over the period of study than is usually the case. Persons shown to have definite pathological defects in their locomotor apparatus were excluded from the investigation because they could not be expected to give anthropologically useful results.

Table I.

Head Measurements

1.	Maximal head length	4.	Bizygomatic diameter
2.	Maximal head width	5.	Bigonial diameter
3.	Minimal frontal width	6.	Morphological face height

Body Measurements

7.	Stature	22.	Chest girth at normal respiration
8.	Sitting height	23.	Chest girth at maximum inspiration
9.	Gnathion height	24.	Chest girth at maximum expiration
10.	Suprasternal height	25.	Waist girth
11.	Symphysis height	26.	Hip girth
12.	Length of upper arm	27.	Forearm girth
13.	Length of forearm	28.	Thigh girth
14.	Tibial height	29.	Radioulnar width
15.	Biacromial diameter	30.	Hand width
16.	Chest width	31.	Hand girth
17.	Chest depth	32.	Forearm fat (skinfold)
18.	Bicristal diameter	33.	Umbilical fat (skinfold)
19.	Distance of tubera ossis ischii	34.	Gluteal fat (skinfold)
20.	Girth above larynx	35.	Weight
21.	Girth below larynx		

Physiological Parameters

36. Grip strength
37. Blood pressure (syst. and diast., before and after physical exercise)
38. Pulse (before and after physical exercise)

Descriptive Features of the Head

39.	Frontal inclination	43.	Prominence of the chin
40.	Prominence of arcus superciliares	44.	Prominence of the larynx
41.	Prominence of the nose	45.	Intensity of the beard
42.	Inclination of nasal base		

Descriptive Characters of the Body

46.	Extension of the pubes	49.	Penis girth
47.	Size of the testes	50.	Lumbar lordosis
48.	Penis length	51.	Transversal angle of the elbow

The study covered the 51 characters listed in Table I. Each of these characters was correlated with every other character, producing 1,378 correlations, whereby the descriptive characters were specified according to a morphognostic schedule. The product-moment correlation was the method employed, and the correlation matrix was subjected to principal axes factor analysis. Rotation was orthogonal in accordance with the varimax criterion. The number of factors was restricted by the requirement eigenvalue > 1.

III. RESULTS AND DISCUSSION

The above conditions allowed 5 factors to be extracted which are interpreted below.

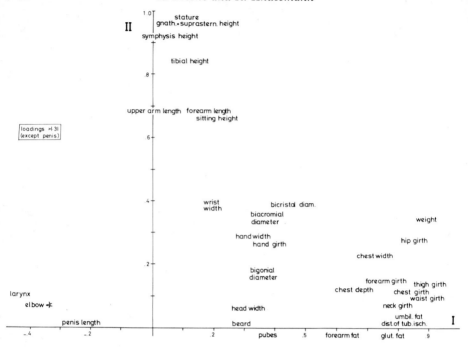

Fig. 1 Distribution of the loadings of factors 1 and 2 in the coordinate system.

A. First Factor

A large number of measurements are strongly influenced by the first factor (Fig.1): these are measurements of girth, viz., largely those influenced by the development of the soft tissue. This coincides well with the fact that the measurements of fat are also determined by this factor. Weight, too, ranks among these characters influenced by factor 1. The whole character complex calls for denoting factor 1 the girth-fat factor.

The distance between the tubera ossis ischii, too, belongs to the girth and fat measurements group. Due to the overlying soft tissue it was not possible to determine this distance directly. Consequently, the measurements taken had to be corrected by adding the corresponding (i.e., gluteal) fat measurement. However, this gave rise to the observation that the adjusted measurement always occurred in close context with the fat measurements, while the nonadjusted one did not noticeably correlate with them. Hence, it must be taken into account that the variability is caused by the corrective and not by the measurement itself. It is therefore not intended to offer an independent interpretation here of the distance of tubera ossis ischii.

It is also interesting to note that in the case of intensity of the beard and extension of the pubes the loadings act in the same direction as the other parameters determined by this factor, although they are lower, while the loadings of penis length and prominence of the larynx act in the opposite direction. It must be concluded that the characters of terminal hair are subject to the same influence as girth and fat development, while the typically male characters — prominence of the larynx and penis development — are subject to a growth tendency contrary to girth and fat development.

The transversal angle of the elbow as the lateral flexure of the forearm is affected by the first factor in the same direction as prominence of the larynx and penis length, but in the opposite direction to girth and fat development. This is surprising because, on average, men have a smaller flexure than women. However, since in the case of factor 1 the transversal angle of the elbow runs counter to the fat measurements, it could be that a strong development of the fat renders the transversal angle of the elbow less prominent.

B. Second Factor

In the case of the second factor, stature, gnathion and suprasternal height, symphysis height, and tibial height, all manifest high loadings; sitting height and length of upper arm and forearm reveal a somewhat lower loading. Excluding, first of all, sitting height from the discussion, it can be observed that the loadings, decrease with the magnitude of the measurements, i.e., factor 2 exerts a greater influence on larger distances than on smaller ones. This corresponds to the degree of heritability, which, according to the findings of investigations on twins, diminishes with decreasing distance (Knussmann, 1966). At the same time, it can readily be assumed that factor 2 is a length factor which measures the genetic basis of length growth. With regard to its loading, sitting height certainly ranks too low for its magnitude, but this is due to the difficulty involved in measuring it. For experience has shown that, despite the application of stringent criteria, small inaccuracies cannot be avoided because exact comparison of sitting positions is too difficult.

Combining factors 1 and 2, a group of characters emerges that is affected by both factors only to an average or even modest extent. These characters include the trunk width measurements, as well as the width and girth measurements of the distal section of the upper extremity, i.e., all measurements relating to the horizontal development of the skeletal structure. The intermediate position of this group of characters, between the complex of characters influenced predominantly by factor 1 and the complex affected predominantly by factor 2, arises from the fact that factor 1 acts as a girth factor upon the horizontal dimension, while factor 2 influences the length and distance development of the skeleton. Moreover, it can be gathered from the latter aspect that hip girth, which is affected chiefly by factor 1, is also clearly influenced by factor 2. Factor 2 must find expression in weight, too, since greater stature is accompanied by greater weight.

The group of measurements falling between the two factors is joined by head width measurements, which are affected more by horizontal factor 1 than by factor 2. The morphological face height, on the other hand, proves to be determined by length factor 2 (0.24), but hardly by factor 1 (0.11).

C. Third Factor

In the case of the third factor (Fig. 2), taking all loadings > 0.2 into consideration, blood pressure, pulse and, to a lesser extent, lumbar lordosis and fat measurements, including the adjusted distance of tubera ossis ischii, all show equidirectional positive loadings, while the loading for penis length bears a negative sign. The opposite loadings of fat measurements and penis length have already been observed in the case of the first factor and they are confirmed again here. It

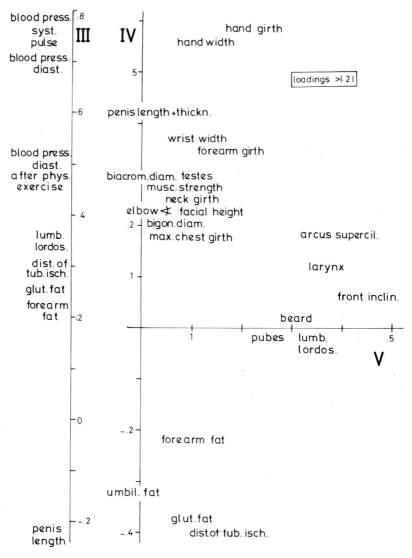

Fig. 2 Distribution of the loadings of factor 3 (on the left) and factors 4 and 5 (on the right) in the coordinate system.

should be pointed out, however, that the gluteal and forearm measurements exceed the 0.2 limit, while the umbilical fat is not still within this limit with its 0.18. This can be interpreted in the sense that, in the male sex, the constitutional obesity for blood pressure and pulse behaviour is of relative high significance, and this also applies to the specifically female fat-formation when this is based on a constitutional adiposis. The nutritively determined and, hence, temporary adiposis in the umbilical fat measurement is, in contrast, less significant. The fact that the thickness of gluteal fat coincides with a stronger lumbar lordosis, as expressed by the loadings, is understandable, in that both indicate a tendency towards female habit. The relationship between gluteal fat measurement and adjusted distance of tubera ossis ischii has already been dealt with above.

With regard to the blood pressure and pulse loadings, it is interesting to note that the systolic measurements display higher loadings than the pulse measurements and that the latter have higher loadings than the diastolic measurements. The loadings for blood pressure measurements are higher before physical exercise than after, although the difference in the case of systolic blood pressure is only small. According to these findings, then, measurements before physical exercise provide more constitutional information than measurements after physical exercise, the systolic blood pressure measurements more than the pulse measurements, while the diastolic blood pressure measurements give the least information. This is readily understandable, in so far as the systolic measurement in healthy individuals is generally highly informative, whereas, on the other hand, after physical exercise it is the systolic rather than the diastolic blood pressure that permits conclusions about a person's physical fitness. Not only blood pressure values but also pulse behaviour enables an assessment of a person's circulation to be made here, too, rather before than after physical exercise. For, as in the case of blood pressure, pulse measurements before physical exercise are constitutionally more informative than measurements taken after exercise. This is due to the fact that, provided the sample is unassorted, the varying states of physical fitness must give rise to greater fluctuations after physical exercise.

D. Fourth Factor

The fourth factor influences the primary sex characters: penis length, penis girth and size of testes. Likewise, it affects the characters of robustness, such as hand width, hand girth, and radioulnar width. Moreover, the muscular strength and appertaining girth measurements are dependent on this factor: forearm girth and girths both above and below the larynx. In addition to these, there are measurements influenced by this factor, which also are an expression of sex-specificity: morphological face height, bigonial diameter and biacromial diameter. The fat measurements, including the adjusted distance of tubera ossis ischii (see above), have loadings with an opposite sign to the above characters.

Judging by the combination of parameters, the fourth factor is one of sex-specificity for the male sex, a fact which is expressed in the factor's influence on the primary sex characters, characters of robustness and muscular strength, including the girth measurements which depend on it. It is significant that, in the case of this factor, hand girth and penis length do not have a loading with an opposite sign, as was the case with the first factor. Instead, the fat measurements in connection with the fourth factor bear an opposite sign to hand width and hand girth. The explanation for this is that the first factor encourages horizontal growth, particularly fat development, and, to a lesser degree, bone girths. However, penis development clearly runs counter to fat development, so that, in the case of factor 1, a contrast must necessarily emerge between penis development and girth measurements. Factor 4, however, is not a factor of girth growth, but a sex-specific factor, so that it influences the sexual organs and the likewise sex-specific characters of robustness in the same direction. In the case of the character of robustness, bigonial diameter, it is also conceivable that the musculature influences the magnitude of the measurement, for a strong musculature is beneficial to bone growth. It is probably due to the musculature, too, that the girth measurements are also affected by factor 4.

There is difficulty to be noted in interpreting the transversal angle of the elbow. However, as has already been noted in connection with the first factor, it should be pointed out that, at least in the male sex, the transversal angle of the elbow is all the more prominent, the less the contours are smoothed out by fat layers. Accordingly, with this factor, transversal angle of the elbow and fat measurements have opposite loadings, too.

E. Fifth Factor

Interpretation of the fifth factor would not appear to be without its problems, either, because here the form of the lumbar bordosis as a female characteristic shows a loading in the same direction as a number of male parameters, such as frontal inclination, prominence of arcus superciliares, prominence of the larynx, intensity of the beard, extension of the pubes, and hand girth, but this is only seemingly a contradiction. All these parameters are actually an expression of maturity and, with increasing maturity, the lumbar lordosis in the male sex, too, increases. The fifth factor can thus be called the maturity factor.

All five factors arising at eigenvalues > 1 have now been interpreted; they are the girth factor, length factor, circulation factor, masculinity factor, and maturity factor. The parameters covered by this study can be explained by the interaction of these five factors.

REFERENCES

Adcok, E.W., Hammond, W.H. and Magee, H.E. (1947). *J. Hyg.* **45**, 65-69.
Banks, C. (1946). "Factor Analysis Applied to Current Problems Based on Data from H.M. Forces", Thesis, University of London Library, London.
Bakwin, H. and Bakwin, R.M. (1931). *J. clin. Invest.* **10**, 369-376.
Bakwin, H. and Bakwin, R.M. (1934). *Hum. Biol.* **6**, 612-626.
Burt, C. (1911). *Child Study* **4**, 1-44; 92-101.
Burt, C. (1938). *Br. J. med. Psychol.* **17**, 158-188.
Burt, C. (1943). *Nature* **152**, 75.
Burt, C. (1944). *Man* **44**, 82-86.
Burt, C. (1947). *Psychometrika* **12**, 171-188.
Burt, C. and Banks, C. (1947). *Ann. Eugen.* **13**, 238-256.
Carpenter, A. (1941a). *Res. Q. Am. Assoc. Phys. Educ. Recreat.* **12**, 34-39.
Carpenter, A. (1941b). *Res. Q. Am. Assoc. Phys. Educ. Recreat.* **12**, 714.
Carter, H.D. and Krause, R.H. (1936). *Child Dev.* **7**, 60-68.
Cohen, J. (1938). *J. Ment. Sci.* **84**, 495-512.
Cohen, J. (1939). *Nature* **144**, 944.
Cohen, J. (1940a). *Eugen. Rev.* **32**, 81-84.
Cohen, J. (1940b). *Proc. natn. Acad. Sci. U.S.A.* **26**, 524-526.
Cohen, J. (1941). *Br. J. med. Psychol.* **18**, 323-337.
Coleman, J.W. (1940). *Res. Q. Am. Assoc. Phys. Educ. Recreat.* **11**, 47-53.
Connolly, C.J. (1939). "Physique in Relation to Psychosis", Stud. Psychol. Psychiatr. **4**, Catholic Univ. Amer. Press.
Dupertuis, C.W. (1950). *Am. J. Phys. Anthropol. N.S.* **8**, 367-385.
Hammond, W.H. (1942). *Man* **42**, 4-11.
Hammond, W.H. (1953a). *Hum. Biol.* **25**, 65-80.
Hammond, W.H. (1953b). *Br. J. Prevent. Soc. Med.* **7**, 231-239.
Hammond, W.H. (1955). *Br. J. Prevent. Soc. Med.* **9**, 152-158.

Hammond, W.H. (1957). *Hum. Biol.* **29**, 40-61.
Heath, H. (1952). *Psychometrika* **17**, 87-95.
Hofstätter, P.R. (1943/44). *Z. Menschl. Vererb. - u. Konstit.-Lehre* **27**, 579-602.
Holzinger, K.J. and Harman, H.H. (1941). "Factor Analysis. A Synthesis of Factorial Methods", Univ. of Chicago Press, Chicago.
Howells, W.W. (1948). *Am. J. Phys. Anthropol. N.S.* **6**, 449-460.
Howells, W.W. (1949). *Am. J. Phys. Anthropol. N.S.* **7**, 101-108.
Howells, W.W. (1951). *Am. J. Phys. Anthropol. N.S.* **9**, 159-192.
Howells, W.W. (1952). *Am. J. Phys. Anthropol. N,S.* **10**, 91-118.
Jackson, A.S. and Pollock, M.L. (1976). *Med. Sci. Sports* **8**, 196-203.
Klein, K. (1970). *Z. Morphol. Anthropol.* **62**, 121-165.
Knussmann, R. (1961). *Homo* **12**, 1-16.
Knussmann, R. (1966). *In* "Humangenetik" (P.E. Becker, ed), Vol. I, pp.197-279, 280-437. Georg Thieme Verlag, Stuttgart.
Knussmann, R. (1968). *Biom. Z.* **10**, 199-218.
Knussmann, R. (1970). *Ringelh. Biol. Umsch.* **25**, 193-208.
Knussmann, R. and Klein, K. (1970). *Homo* **21**, 133-137.
Kretschmer, E. (1921). "Körperbau und Charakter". Springer Verlag, Berlin.
Lorr, M. and Fields, V. (1954). *J. clin. Psychol.* **10**, 182-185.
Low, A. (1952). "Growth of Children". Aberdeen Univ. Press, Aberdeen.
Marshall, E.L. (1936). *J. exper. Educ.* **5**, 212-228.
McCloy, E. (1935). *Res. Q. Am. Assoc. Phys. Educ. Recreat.* **Suppl. 6**, 114-121.
McCloy, C.H. (1940). *Child Dev.* **11**, 249-278.
McDonough, M.R., Sr. (1932). "Aspects of the New School". Brenzinger Brothers, New York.
Moore, T.V. and Hsü, E.H. (1946). *Hum. Biol.* **18**, 133-157.
Mullen, F.A. (1939). "Factors in the Growth of Girls Seven to Seventeen Years of Age". Doctoral Dissertation, Department of Education, University of Chicago.
Mullen, F.A. (1940). *Child Dev.* **11**, 27-41.
O'Brien, R., Girshick, M.A. and Hunt, E.P. (1941). Miscellaneous Publication No.366, U.S. Government Printing Office, Washington.
O'Brien, R. and Shelton, W.C. (1941). Miscellaneous Publication No.454, U.S. Government Printing Office, Washington.
Petersen, K. (1961). *Homo* **12**, 26-33.
Rees, W.L. (1945). *Eugen. Rev.* **37**, 23-27.
Rees, L. (1949). *J. Ment. Sci.* **95**, 171-179.
Rees, L. (1950). *J. Ment. Sci.* **96**, 619-632.
Rees, L. (1960). *In* "Handbook of Abnormal Psychology" (H.J. Eysenck, ed), pp.344-392. Pitman Medical, London.
Rees, L. (1973). *In* "Handbook of Abnormal Psychology" (H.J. Eysenck, ed), pp.487-540. Pitman Medical, London.
Rees, L. and Eysenck, J.H. (1945). *J. Ment. Sci.* **91**, 8-21.
Schick, C. (1953). *Z. Menschl. Vererb. - u. Konstit. - Lehre* **32**, 1-31.
Sills, F.D. (1950). *Res. Anat. Ass. Hlth. Phys. Educ.* **21**, 424-437.
Spearman, C. (1905). *Am. J. Psych.* **15**, 201-293.
Spearman, C. (1927). "Abilities of Man". Macmillan, London.
Sterett, J.E. (1935). *Res. Q. Am. Assoc. Phys. Educ. Recreat.* **7**, 112.
Tanner, J.M. (1964). *In* "Human Biology" (G.A. Harrison, J.S. Weiner, J.M. Tanner and N.A. Barnicot, eds) pp.297-397. Clarendon Press, Oxford.
Thomson, G. (1919). *Br. J. Psychol.* **10**, 319-326.
Thomson, G. (1939). "Factorial Analysis of Human Ability". Univ. of London Press, London.
Thurstone, L.L. (1946). *Psychometrika* **11**, 15-21.
Thurstone, L.L. (1947). *Am. J. Phys. Anthropol. N.S.* **5**, 15-28.
Vandenberg, S.G. (1968). *Human. Biol.* **40**, 295-313.
Waltrop, R.S. (1940). *Psychol. Bull.* **37**, 578.

GENETIC CONTROL OF HUMAN GROWTH AND ECOSENSITIVITY

N. Wolański

Department of Human Ecology, Polish Academy of Sciences
Nowy Swiat 72, 00-330 Warsaw, Poland

SUMMARY

The genetic control of human growth and development is discussed both in general and with respect to the interplay of genetic predisposition with the surrounding ecologic factors acting in the course of development and with the individual ecosensitivity. Numerous unexplained phenomena are found with respect to the influence of genetic and ecologic factors on prenatal and postnatal growth and development. Apart from the mechanisms of genetic control and paragenetic influences, and from the modifications induced by ecologic factors and way of living, the mechanisms of self-regulation and various feedbacks appear to be of essential significance for the course of both ontogenesis and postnatal growth and development

I. INTRODUCTION

The purpose of this paper is to discuss several problems of essential importance for the understanding of ontogenetic development of man. The data collected as yet do not allow any final explanation, but the complex of problems outlined may encourage more profound studies in this field. Particularly numerous unexplained mechanisms are found in the connection between the influence of genetic and ecological factors, on the one hand, and the course of postnatal development in connection with the character of prenatal development on the other. Apart from the mechanisms of genetic control and paragenetic influences, as well as the modification role of ecological factors and way of life, the mechanisms of self-regulation and various feedbacks seem to be of essential significance for the course of ontogenetic development.

II. INITIATING IMPULSES

A very important problem is the nature of the impulses initiating the development. In the whole course of ontogenesis, this mechanism is active only twice: for the first time, in the embryonal stage of life, and then in the stage called puberty. In both these life periods, the phenomenon is identical, although it differs in absolute values, particularly in relation to the initial values. The impulse appears and causes initially a very great acceleration of development of an exponentially positive character; after some time, the positive acceleration decreases and changes into a negative increasing acceleration, which in turn decreases after some time. In the end-stage, acceleration wears down. The first stage beings in the early embryonal phase and continues to the age of 3-4 years; the second stage begins in the early prepubertal period of life and continues until disappearance of body growth. Both these stages have been isolated on the basis of an analysis of changes occurring in body height (length). As is known, this is also true of other features. Investigations on intelligence have demonstrated, however, that similar processes take place at puberty (Wolański, 1970a, b).

The cause of the first developmental acceleration seems to be the change in the mass of the zygote in relation to its surface. As a result of the division of the fertilized ovum into 2, 4, 8, etc., cells, the size and mass of the embryo increase precipitously. The initiating factor may be a biophysical process.

The cause of the second developmental acceleration may be the maturation of the organism, in which the principal mechanism is the neurohormonal stimulation of the hypophysis by the hypothalamus. The initiating factor may be a biochemical process.

In the first case, the initiating factor causes an exponential increase in the number of cells (proliferation); in the second case, the initiating factor causes an exponential increase in the mass of cells and tissues. The end result of the first process is achievement of a relative individual maturity; the end result of the second process is species-specific maturity.

Explanation of the mechanisms of the stimulation and the secondary attenuation of developmental acceleration described here may provide us, in times to come, with the possibility of stimulating or inhibiting (which is as yet much more difficult) the processes of development in a controlled way.

III. SEX DETERMINATION

Another process calling for extensive studies is the problem of primary (at fertilization) and secondary (at birth) sex ratio. The views on the primary sex ratio differ greatly (from 140 to 400 male per 100 female fertilized ova). The secondary sex ratio is about 106-108. The relationship of male to female embryos and fetuses with regard to various features demonstrable at a given developmental stage would be particularly interesting. This problem arises in connection with the hypothesis that the size and mass of the body show sexual dimorphism between male and female newborns due to more frequent death of male fetuses with poor development. The surviving male infants are, thus, on the average, bigger than the more viable females, as has been also confirmed in investigations of stillborn fetuses (Wolański, 1970a,b). It would be extremely

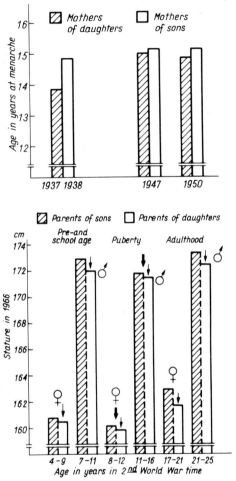

Fig.1 Age at menarche (above) and adult stature (below) of future mothers and fathers of sons and of daughters.

important to establish to what extent selection, rather than simple genetic determination, may be responsible for the presence of many similar somatic features exhibiting sexual dimorphism.

The problem of the prevalence of daughters in certain families and of sons in other families, has been frequently discussed in the literature. This problem, as well as the one of alternating births of sons and daughters, has been extensively documented for about one hundred years (e.g., Geissler in 1889), but it is still far from being solved with respect to its underlying factors. In recent years, certain observations have, however, been made, which may serve as an interesting starting point for further investigations. The main one was the observation of certain differences between the mothers of sons and those of daughters. It has been found that women maturing later (Fig.1, above), and, because of that, growing longer and being taller (Fig.1, below), are more frequently mothers of sons, while women maturing earlier and being shorter more frequently give birth to daughters. Sons are born also more frequently to women not reaching puberty in winter (Wolański, 1973), as well as to those with shorter menstrual cycle (Bozilow et al., 1973).

The problem may arise, in this connection, whether the menstruation rhythm is associated with biochemical or physiological features that determine a greater susceptibility to fertilization by a spermatozoon with X or Y chromosome. The set of observations discussed is, generally speaking, a closed chain of causes and effects. A woman with shorter menstrual cycle would reach puberty later only if there were a negative feedback mechanism between these facts, but longer growth is a direct consequence of later maturation with consequently greater height. We do not know the relationship of more frequent maturation in the summer season to the above phenomena. The relationship between age and menarche season of the year was studied frequently, but the results were often controversial and the problem must be regarded as unsolved as yet (Kowalska et al., 1963; Bojlén and Bentzon, 1971, and others).

IV. PARAGENETIC FACTORS

Another extensive field for investigations, particularly from the standpoint of fetology, is the role of the paragenetic and maternal factors in the control of intrauterine development and in their relation to environmental influences (Fig.2). This is also of great importance for the understanding of postnatal development. The role of paragenetic factors may be seen in the influence of the

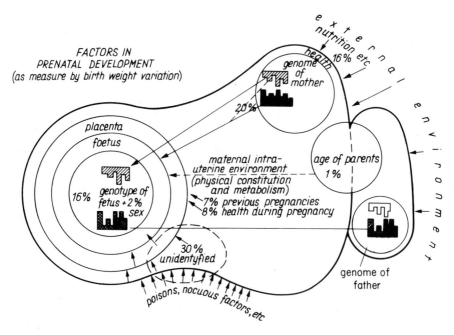

Fig.2 Factors affecting fetal development, and birth weight variation connected with different genetic and environmental factors (data from Penrose).

protoplasm of the ovum, and that of maternal factors, in the effect of the genome part which has not been transmitted by the mother to the fetus, but which is, however, present in the maternal tissues where the embryo is developing. Also of significant importance are the biochemical features of maternal tissues, which also depend on way of life, nutrition, and other not yet identified factors. This "maternal" aspect of "heredity" has already been extensively studied in zoology (Walton and Hammond, 1938; and others) and in human biology (Ohno, 1968; Ounsted and Ounsted, 1968; Wolański, 1970a,b; and others).

We recently studied this problem using a set of 65 somatic, physiological, biochemical, and psychomotor traits (Wolański, 1975). The correlation coefficients between father and son, father and daughter, mother and son, and mother and daughter, were analyzed in 22 age groups for each trait (in a part of the traits, the correlations were studied in offspring in the age range from 18 months to 60 years). No traits were found in which the similarity of the offspring to the father prevailed over that to the mother. On the other hand, in many traits the similarity to the mother prevailed significantly over that to the father; this was in height, trunk length, height of face, breadth of forehead and mandible, ear shape, strength of shoulder muscles, and endurance fitness. Though to a lesser extent, a more pronounced similarity to the mother than to the father was also found in such traits as: ear size, foot breadth, diastolic arterial pressure, and haptoglobin level. With regard to the remaining traits, the similarity to the father was greater in some age groups, and that to the mother, in other age groups.

The importance of maternal factors has been confirmed by the results of studies on the effects of heterosis (Chrzastek-Spruch, 1977). When the father is short and the mother tall (for their respective sex), the offspring is taller than when the mother is short and the father tall.

The results obtained in these investigations have demonstrated that, in the somatic, physiological (cardiovascular) and biochemical (enzyme blood levels) traits, the similarity of offspring to the mother prevailed over that to the father. With regard to psychomotor traits, the prevalence of similarity to one parent was sporadic. The importance of this observation lies in the fact that the observed similarity may be due to the personality-forming influence of the mother, particularly on the psychic features and behaviour of the child, since the contact of children with their mother is closer and more frequent than with their father. If the similarity to the mother is less evident than that to the father, or if there is no prevalence of similarity, it may be assumed that only genetic factors take part in the control of the trait. However, if paragenetic and maternal factors play a role, we can observe a rather greater similarity of the offspring to the mother. The study of the mechanism of these influences is a very important task for the future, especially for fetology and neonatology.

V. ASSORTATIVE MATING

The considerable role of the influence of ecological factors on the physiological, biochemical, and psychomotor traits, is suggested by studies on assortative mating. It is known that a positive assortative mating is observed with regard to most of the above traits, for which spouses are similar (Spuhler, 1968; Wolański, 1973). What is even more interesting is that the spouses are also similar in such traits as arterial blood pressure and blood level of various enzymes. It is

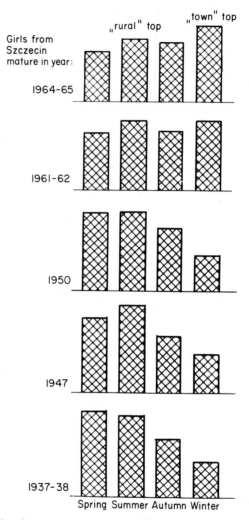

Fig. 3 Age at menarche of women coming to Szczecin after puberty (menarche in 1937-38), maturated immediately after coming to Szczecin (1947-50), and 10-14 years after coming to Szczecin (1961-62), and of those whose conception and birth had occurred in Szczecin (1966).

possible that this depends on certain phenotype properties, but this assumption has not been proved. Therefore, it would be more reasonable to assume that the similarity in many physiological, biochemical, and even psychomotor traits, is a result, rather than a cause, of elective selection, i.e., of living in the same household (identical nutrition), of similar way of life, similar experiences, etc.. Should this hypothesis hold true, one might consider the possibility that the traits of the fetus might be influenced through the maternal organism, so far as this would be allowed by the permeability of the placental barrier and the tolerance of the fetus. This is a very encouraging direction of investigation.

VI. BIORHYTHMS AND MATURATION

Mothers and daughters have been shown to reach menarche at a similar age, in a similar season of the year, and to have a similar length of menstrual cycle, etc. (Kowalska, 1966, 1968). A correlation also seems to exist between the season of birth and the season of menarche (Stukovsky and Valsik, 1968). This may suggest that the cyclic patterns of the mother are inherited by the daughter. Perhaps, there is a definite monthly cyclic pattern, which is transmitted during pregnancy to the daughters, who thus reach puberty in a strictly defined time (age and season).

Some light on this problem has been shed by our investigations in Szczecin 1962-1966 (Wolański, 1970b; Wolański et al., 1970). Interestingly, we found that women coming to Szczecin had menarche usually in summer in their previous place of residence (Fig.3). In fact, in women from rural populations menarche more frequently occurs in the summer season, in connection with undernutrition in the spring, heavy work in the fields, etc. (Kowalska et al., 1963). However, in women from urban populations, menarche more frequently occurs in winter, in connection with vitamin deficiency, increased loss of heat, etc.. Women from rural areas coming to Szczecin at the age of 10-14 years and maturing in this city had menarche still more frequently in summer. Only in the years 1961-1962 both peaks — winter and summer — became equal. This was the time when about half the girls at the age of puberty had been born in Szczecin and the other half came to this city already after birth. Only when most girls studied by us (in 1966) had been conceived and born in Szczecin, a shift of maximum menarche frequency to the winter season was observed, which is a typical pattern for the urban population. This fact may indicate the existence of some deeper process which may be steering the fetal biorhythms and this rhythm is realized by the organism in further life. In that case, the rhythm would only change if the mother of a given girl had already adapted, in some degree, to the conditions of life in the city before conception, with a consequent change of her seasonal, or even monthly, rhythm.

VII. GENES AND/OR ENVIRONMENT

An important problem to which recent investigations have contributed new data is the similarity of relatives. The main question is, which factors play the predominant role, genes or environment?

The genetic similarity between parent and offspring, randomly selected siblings, or dizygotic twins, amounts to 50% (for same-sex subjects). However, the environment in which two twins grow is nearly identical; that in which two ordinary siblings grow (if close enough in age) is usually very similar; but the environment of parents and offspring at developmental age differs considerably. Because of that, the correlation coefficients between traits of dizygotic twins overestimate genetic similarity. They are sometimes twice as high as those between a random pair of ordinary siblings, and the latter may be much higher than those between parent and child (Fig.4).

Two populations not differing in gene frequency (Wolański and Jarosz, 1969) — an urban population with a high standard of living and nutrition, and a rural one from poor villages — showed differences in the similarity of traits

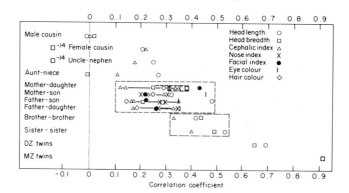

Fig. 4 Correlations of some cephalometric traits, hair and eye colour, between different relatives (data from Pearson and Lee, 1903; Newman *et al.*, 1937; Ikeda, 1953; Wolański and Charzewska, 1967; Wiercińska, 1969; Wierciński, 1969; Czesnis, 1971).

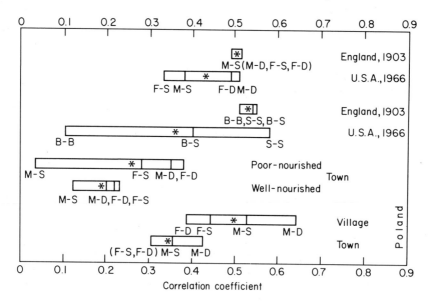

Fig. 5 Correlations for stature in Poland: town and villages (Charzewska and Wolański, 1964), poor- and well-nourished town families (Bielicki and Welon, 1966); in U.S.A. (Garn and Rohmann, 1966) and England (Pearson and Lee, 1903). F = father, S = son, M = mother, D = daughter, B = brother, S-S or B-S = correlation with sister.

between offspring and parents (Fig. 5): higher correlation coefficients were observed in families with lower standard of living (Charzewska and Wolánski, 1964). Two years later, the results of these investigations were confirmed by Bainbridge and Roberts (1966) and Bielicki and Welon (1966). The latter obtained similar results studying poorly nourished groups living in urban areas. The same result is obtained comparing the data reported by Garn and Rohmann (1966) for optimal life standard in the U.S.A. with data of Pearson and Lee (1903) for groups with a low living standard in England, in both parent-child and sib-sib relations (Fig.5).

In connection with the above views, Bielicki and Welon (1966) and Tanner et al. (1970) suggested that correlation coefficients decrease due to a change in the participation of the environmental component in phenotypic variability. That means that the similarity of the living standard of children to that of their parents in childhood has diminished since the parents moved to higher social strata. The probability is greater in families with a good socioeconomic standard.

In fact, in the urban population, with a higher standard of living, the dispersion of traits is greater than in the rural population with a lower standard and more similar conditions of life (Wolański and Lasota, 1964). As is known, however, with a greater variability of the trait, the correlation coefficients overestimate the actual parent-child similarity. On the other hand, a homogeneous material masks the distinctiveness of correlation which could be obtained under optimal conditions for heritability (Plochinskij, 1964). Furthermore, animal experimentation has shown that, with unification of the environmental component, the individuals with favorable breeding traits are unable to develop to an optimum, while, under better conditions, the phenotypic differences between individuals decrease. Under good economic conditions, more or less enterprising men can equally well earn their daily bread. The earnings above an indispensable minimum are without significance for the mode of nutrition. This shows that, under worse conditions of life, genetic traits manifest themselves not less well than under conditions of optimal feeding and living (Johansson, 1953).

It seems that it is not the general economic level, but the degree of differentiation of nutrition and the way of life of a population that are of decisive importance for phenotypic differentiation. Because of that, we have put forward another hypothesis for explaining the fact that the similarity of traits in the parent-child and sib-sib relations is greater when nutrition is poorer and lesser when nutrition is better and more differentiated (Wolański and Charzewska, 1967b; Wolański, 1970a). We believe that the lower similarity of children to their parents under conditions of good nutrition is due to wider possibilities in using different metabolic pathways in view of the greater amount of the various nutritional components. On the other hand, when the offspring ascend to higher socioeconomic strata, their traits adapt to better feeding and achieve a higher level of development, which is not necessarily connected with a change in their position in relation to their parents, and because of that, the correlation coefficient remains unchanged (Wolański, 1974a).

Higher correlation coefficients between relatives may be also due to quite different causes. It is known that positive assortative mating and consanguinity exert an effect inflating heritability (Fisher, 1918; Susanne, 1976; see also Fig. 15).

The hypothesis of an effect of social status on parent-child correlation has been recently refuted by its authors (Bielicki and Charzewski, 1975), who have shown parent-child correlation coefficients for stature to be identical in socially rising families and in families who had not risen socially. This was to be expected, however, since remarkable environmental changes have taken place in a period of two generations (in Poland at least) with regard to both types of families.

VIII. GENETIC CONTROL AND ECOSENSITIVITY

The last problem is the genetic determination of the organism's sensitivity to environmental factors, the so-called ecosensitivity. The investigations were up to now concerned with the degree of heterozygosity, sexual difference, developmental period, and various traits.

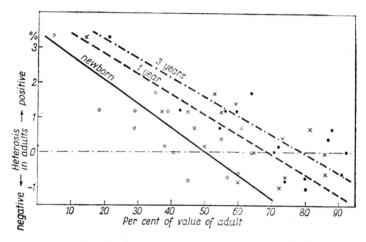

Fig.6 Positive heterosis (increased susceptibility in heterozygotes) of traits developing mainly in postnatal period, and negative heterosis (reduced susceptibility in heterozygotes) of traits developing mainly in prenatal period.

The investigations we have so far carried out (Fig.6; Wolański, 1974b) showed that these features, which in the period of prenatal development reach over 60% of the value of the adult individual, are lower in individuals from exogamous than in those from endogamous families. This phenomenon, the so-called negative expressivity, evidences that traits developing mostly in the prenatal period show a lower susceptibility to environmental factors (which will be discussed below). These traits include nasal breadth, breadth of face and head, head shape, etc., that is, the traits recognized for a long time by the anthropologists as diagnostic in racial typology, owing to their susceptibility to stimuli from the external environment.

The same diagram (Fig.6) suggests, however, that traits developing to less than 45% in the fetal period (in relation to their value in adulthood) are greater in heterozygotic individuals (from exogamous families) than in homozygotic ones (from endogamous families). This occurs, however, only when nutrition is adequate.

Under conditions of good nutrition, heterozygotic offspring are taller than homozygotic ones (Wolanski, 1974c; Fig.7). Similar results had been obtained earlier by Garn (1962a, b) and Malina *et al.* (1970), but they were not interpreted

Genetic Control of Human Growth and Ecosensitivity 43

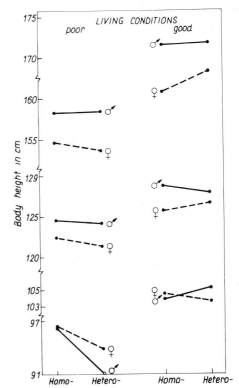

Fig. 7 Stature in children of stature-homogeneous parents (positive assortative mating) and in children of stature-heterogeneous parents (negative assortative mating) in poor and good living conditions.

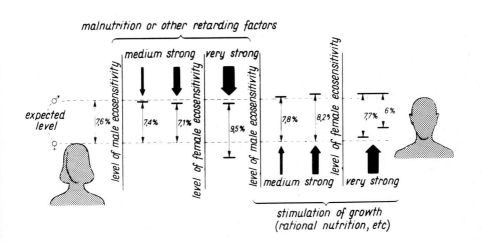

Fig. 8 Per cent differences in stature between males and females under various living conditions.

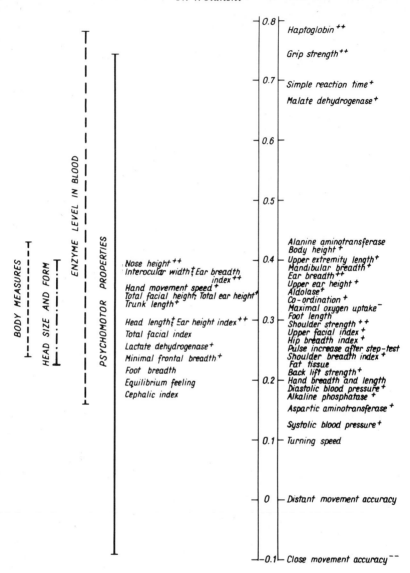

Fig. 9 Midparent-offspring correlation coefficients for different traits in Polish rural population: ++ strong positive, + moderate positive, − moderate negative, −− strong negative, assortative mating of parents (all statistically significant).

then in this way. On the other hand, under unfavourable conditions (moderate malnutrition), heterozygotic offspring are shorter than homozygotic ones. Heterozygotic individuals are much more sensitive to environmental factors (ecosensitive) than homozygotic ones. Males react with a shorter stature to moderate undernourishment, while females respond in this way only to high-grade disturbances of living conditions (Fig.8; Wolański, 1975). When the conditions of life improve (e.g., realimentation), men are first to respond with in-

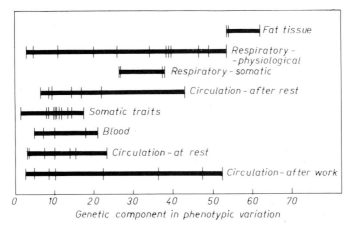

Fig.10 Genetic component (sex-differences) in phenotypic variation (SD) for some somatic and physiological traits in three Polish populations in different living and climatic conditions.

creasing stature, while women respond later (Fig.8). This confirms the hypothesis that men are more ecosensitive. In women, it is more difficult to change the actual pathway of development, but when this change has taken place it is more persistent and less likely to return to its previous pathway. This is probably due to a stronger development determination in the females, who have two X chromosomes (perhaps the inactivation of one of them is not complete). The XY heterochromosomes may therefore underlie the greater proneness of males to environmental adaptation.

IX. GENETIC COMPONENT IN PHENOTYPIC VARIATION

On the basis of parent-child correlation, we tried to assess, in a different way from the twin method, the heritability of somatic, psychomotor, and physiological traits. The traits showing the highest degree of genetic determination include (Fig.9) haptoglobin level, grip strength, reaction time and malate dehydrogenase blood level. An average degree of genetic determination is shown by most enzyme blood levels, most psychomotor properties, body measures, head size and shape, and blood pressure. Among the somatic traits, head dimensions are more similar between parent and child than the dimensions of the hand and foot. Movement accuracy shows only very small parent-child similarity.

The genetic component was assessed by comparison of sex-differences with environmental (between populations: Wolański and Jarosz, 1969), or with phenotypic variability (common standard deviation: Fig.10). The data thus obtained suggest that the thickness of adipose tissue, and respiratory traits such as chest structure, are strongly genetically determined. The traits of the cardiovascular system, somatic and blood characters, are less genetically controlled. The adipose tissue shows significant sex-dependent differences, as well as interpopulation, environmentally-dependent differences. In other words, the characters of the adipose tissue, and, to a lower extent, those of the respiratory system, are at the same time strongly genetically determined and highly influenced by nutrition and physical activity, as well as by climatic factors. This may explain

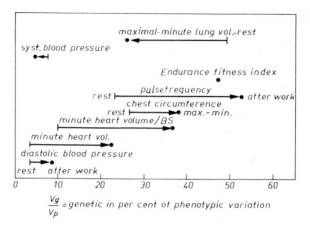

$\frac{V_g}{V_p}$ = genetic in per cent of phenotypic variation

Fig.11 Genetic component in phenotypic variation in three Polish populations in rest and in time of submaximal work load.

the different assessment of the genetic component of these traits by various authors using different methods.

The submaximal or maximal values are usually much more strongly controlled genetically than resting ones (Fig.11). During rest, the organism thus shows a much greater adaptation to environmental factors, while, during exercise, genetic differences become more evident.

X. CLOSING REMARKS

In view of these problems, a host of research projects arises, concerning the mechanisms of development of various congenital developmental anomalies as well as their possible prevention. It has been found that disturbances of traits developing mainly in the prenatal period cause excessive growth (e.g., the size of the head) while disturbances of traits developing to only a small degree in the prenatal period tend to result in development inhibition (e.g., the length of lower extremities). A theoretical model of this phenomenon is shown in Fig.12, and it will be the aim of future investigations to check the validity of this model in various developmental disturbances, and also to elaborate a proper method of management, already during pregnancy, for reducing the negative effects of these disturbances.

REFERENCES

Bainbridge, D.R. and Roberts, S.F. (1966). *Hum. Biol.* **38**, 251-278.
Bielicki, T. and Charzewski, J. (1975). Personal communication.
Bielicki, T. and Welon, Z. (1966). *Hum. Biol.* **38**, 167-174.
Bojlén, K. and Bentzon, M.W. (1971). *Hum. Biol.* **43**, 493-501.
Bozilow, W., Sawicki, K. and Scheller, J. (1973). Materiały XX-lecia Zakładu Antropologii PAN, Wrocław.
Charzewska, J. and Wolański, N. (1964). *Pr. Mater. Nauk. Inst. Matki Dziecka* **3**, 9-42.
Chrzastek-Spruch, H. (1975). Ph.D. Thesis, Akademia Medyczna, Lublin.
Chrzastek-Spruch, H. (1977). Paper presented at 1st Int. Congr. of Auxology, Rome.
Czesnis, C. (1971). *Vopr. Antropol.* **37**, 92-99.

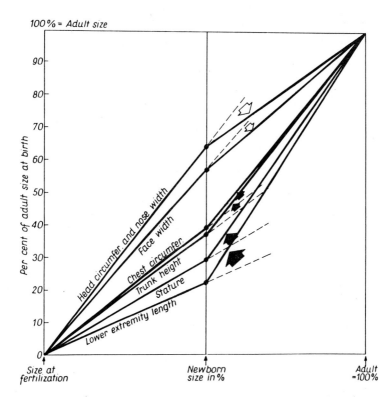

Fig.12 Theoretical model of development of traits reached in newborn infants (per cent of adult values assumed as 100%) and deviations (broken lines) in postnatal development. The direction of deviation has a different value in traits developing mainly in pre- or in postnatal life, because of the absence of retardation or acceleration of growth in the postnatal period.

Fisher, R.A. (1918). *Trans. Roy. Soc. Edinb.* 52, 399-433.
Garn, S.M. (1962a). *In* "De Genetica Medica" (L. Gedda, ed), Vol.2, pp.415-434. Istituto Gregorio Mendel, Rome.
Garn, S.M. (1962b). *Mod. Probl. Paediatrics* 7, 50-54.
Garn, S.M. and Rohmann, C.G. (1966). *Pediatr. Clin. North Am.* 13, 353-379.
Ikeda, J. (1953). *Acta Med. Biol. (Niigata)* 1, 181-189.
Johansson, I. (1953). *Acta Genet. Statis. Med.* 4, 221-231.
Kowalska, I. (1966). *Pr. Mater. Nauk. Inst. Matki Dziecka* 8, 69-77,
Kowalska, I. (1968). *Przegl. Antropol.* 34, 317-323.
Kowalska, I., Valšik, J.A. and Wolański, N. (1963). *Pr. Mater. Nauk. Inst. Matki Dziecka* 1, 81-108.
Malina, R.M., Harper, A.B. and Holman, J.D. (1970). *Res. Q. Am. Assoc. Health Phys. Educ. Recreat.* 41, 503-509.
Newman, H.H., Freeman, F.N. and Holzinger, K.H. (1937). "Twins. A Study of Heredity and Environment". Univ. Chicago Press, Chicago.
Ohno, S. (1968). *In* "Human Genetics", Birth Defects Orig. Artic. Series 4, pp.45-50, National Foundation March of Dimes, New York.
Ounsted, M. and Ounsted, C. (1968). *Nature (London)* 220 (5167), 599-600.
Pearson, K. and Lee, A. (1903). *Biometrica* 2, 357-462.
Płochinskij, N.A. (1964). Nasledujemost. Akademia Nauk, Nowosibirsk.
Spuhler, J.N. (1968). *Eugen. Q.* 15, 128-140.

Štukovsky, R. and Valšik, J.A. (1968). *Przegl. Antropol.* **34**, 123-129.
Susanne, C. (1976). *Glas. Antropol. Drustva Jugosl.* **13**, 11-20.
Tanner, J.M., Goldstein, H. and Whitehouse, R.H. (1970). *Arch. Dis. Child.* **45**, 755-762.
Walton, A. and Hammond, J. (1938). *Proc. R. Soc. London Series B* **125**, 311.
Wiercińska, A. (1969). *Genet. Pol.* **10**, 277-280.
Wierciński, A. (1969). *Genet. Pol.* **10**, 269-275.
Wolański, N. (1970a). *Hum. Biol.* **42**, 349-368.
Wolański, N. (1970b). *Przegl. Zachodniopomorski* **14**, 5-39.
Wolański, N. (1973). *Studies in Human Ecology* **1**, 182-188.
Wolański, N. (1974a). *In* "Nutrition and Malnutrition " (A.F. Roche and F. Falkner, eds) pp. 231-269. Plenum Press, New York, London.
Wolański, N. (1974b). *In* "Biology of Human Populations" (W. Bernhard and A. Kangler, eds) pp.16-30. Fischer Verlag, Stuttgart.
Wolański, N. (1974c). *Studies in Human Ecology* **2**, 7-75.
Wolański, N. (1975a). Škola Bioloska Antropologie, Predavanije, Zagreb.
Wolański, N. (1975b). *Anthropologiai Kozlemenyek* **19**, 207-211.
Wolański, N. (1976). Druga Škola Bioloska Antropologie, Predavanije, Zagreb.
Wolański, N. and Charzewska, J. (1967a). *Acta Genet. Statis. Med.* **17**, 365-381.
Wolański, N. and Charzewska, J. (1967b). *In* "Trudy 7 Miezdunarodnogo Kongresa Etnograficzeskich i Antropologiczeskich Nauk", Vol.2, pp.159-163.
Wolański, N. and Jarosz, E. (1969). *Acta Med. Auxol.* **1**, 122-130.
Wolański, N., Jarosz, E. and Pyzuk, M. (1970). *Soc. Biol.* **17**, 1-16.
Wolański, N. and Lasota, A. (1964). *Z. Morphol. Anthropol.* **54**, 272-292.

GROWTH, NUTRITIONAL STATUS AND FOOD CONSUMPTION OF ROMAN SCHOOLCHILDREN

A. Mariani, B. Lancia, P.A. Migliaccio and D. Sorrentino

National Institute of Nutrition, Via Lancisi 29, Rome, Italy

SUMMARY

A longitudinal study was started in 1969 on 673 Roman schoolchildren of both sexes, belonging to different socioeconomic levels. The investigation, started at the age of 6 years, included: clinical examination; nutritional-status assessment; measurement of height, weight, skinfolds, several diameters and circumferences, skeletal age; analysis of some biochemical indices of the nutritional status and food intake. The results show that, for the whole sample, food intake as well as clinical and nutritional status are satisfactory. Worthy of consideration, however, is the existence of a socioeconomic gradient which influences the energy intake and the quality of the diet. Moreover, for each age class in the underprivileged groups, the same socioeconomic gradient goes with a growth retardation, as shown by a lower height increment, a retarded skeletal maturation, and the HOP index. It remains to be established to what extent other non-nutritional factors do intervene in determining the observed facts, such as, first of all, the environmental hygienic conditions. In view also of the psychological and behavioural effects, it has to be kept in mind that an unsatisfactory nutrition, both qualitatively and quantitatively, is part of a social context that only global interventions can modify.

In the interaction between genotype and ecosystem, several factors enter, promoting and/or interfering with health and physical development. It is generally acknowledged that, among these, the socioeconomic factors play a major role, as they can influence, to a potentially determinant degree, the nurture as well as other conditions essential to life, e.g., hygienic environmental conditions.

In spite of this, documentary evidence of the influence which the diversity of socioeconomic conditions may have on growth rate and well-being in the critical periods of development is scarce in western industrialized societies.

Situations of poorer nutrition in relation to lower socioeconomic status had already been described also in affluent societies, such as in metropolitan areas of New York (Baker et al., 1967; Christakis et al., 1967; Ziffer et al., 1967) and in urban areas of Great Britain (Tanner et al., 1966).

What is mainly needed is adequate information about those countries where inequalities or unbalances in the economic development may be responsible for nutritional repercussions, possibly marginal, to take place on the most vulnerable groups of population. Such is the case of Italy, where the chronic scarcity of food, which afflicted up to perhaps some twenty years ago large sections of the rural population, particularly in the "Mezzogiorno" (The South), has disappeared. But the transformation of the economy and the social growth have taken place in Italy in a tumultuous and disorderly fashion. Particularly in the main industrial cities and in the metropolitan areas of Rome, population expansion and urbanization have created new problems. Unsafe and unhygienic dwellings, unemployment and underemployment, have created situations which paradoxically, for a distorted development, make believe that the problems of nutritional inadequacy and/or unbalance in our country are being transferred, at least in part, from rural to urban areas.

With the aim of contributing to fill this information gap, a longitudinal study has been carried out on the relationship between socioeconomic level of the family and growth, clinical and nutritional status and diet of boys and girls, 6 to 13 years old, living in Rome.

In the first five years of the present study, which began in 1969, the socioeconomic conditions of the sample families have undoubtedly improved; however, even taking this general improvement into consideration, the profession of the family head, and, consequently, the family income, have maintained the original differences.

During the first stages of the investigation, 673 boys and girls were examined. After five years, 432 children (219 males and 213 females) were still under control; of these, 262 belonged to an underprivileged socioeconomic class (subgroup 1) and 170 to a privileged class (subgroup 2).

The food consumption (Table I), assessed according to the method of Burke (1947), modified and adapted and integrated with the use of visual aids, appears to differ ($p < 0.01$), for each age class, only between the sexes. However, examining the subgroups (the two sexes are combined for ease of presentation) an interesting trend is revealed. Up to the age of 7 years, subgroup 1 shows a lower energy intake. At 8 years, the difference in energy intake between the two subgroups diminishes. At 9 years, dietary control intervenes in the case of overweight children in families of the privileged socioeconomic level, and the situation changes. For all subjects, the adequacy of the energy intake reaches, or more often exceeds, the recommended level.

The history of food choices, at any rate, reveals substantial differences for the two groups. Figure 1 demonstrates the structural characteristics of the diet over the five years of the investigation. At each age class, total protein contributes 13% and 14% of the total energy for subgroups 1 and 2, respectively. Moreover, small but interesting differences in the percentage of total energy intake

Table I. Energy intake (Kcal).

Age (yr)	Boys	Girls	Subgroup 1	Subgroup 2
6.5	1875 ± 350	1725 ± 350	1750 ± 375	1800 ± 300
7.5	2000 ± 350	1850 ± 350	1900 ± 400	2000 ± 350
8.5	2050 ± 475	1900 ± 375	1975 ± 450	1990 ± 400
9.5	2250 ± 400	2000 ± 300	2175 ± 400	2125 ± 375
10.5	2400 ± 475	2175 ± 425	2325 ± 450	2325 ± 500

derived from animal and vegetable protein exist between the two subgroups, i.e., 8% v. 5% (for subgroup 1) and 9% v. 5% (subgroup 2).

Even more striking (also in a dynamic sense), are the differences in the percentage of total energy derived from fat. In subgroup 1, the percentage of the total energy intake derived from vegetable fats is always higher than that from animal fat. A change may be observed in subgroup 2 in the relationship between percentage contributed by vegetable and animal fats; animal fat exceeds vegetable fat, at the expense, however, of the percentage of total energy intake derived from carbohydrates, which diminished in the period under consideration for those children belonging to families of a higher socioeconomic level.

The different caloric percentage derived from animal and vegetable food sources reflects a qualitative difference in the diets of the two subgroups. This is confirmed by the evidence provided by various nutritional indicators, such as, besides the already indicated lower intake of animal protein per kg of bodyweight, the lower intake of Fe, Ca and vitamin A (Lancia et al., in preparation).

It is known that anthropometry constitutes, particularly in Western industrialized societies, the most valid method for identifying differences in growth and development of nutritional interest.

For both sexes, as far as height and weight are concerned (Fig.2), growth curves (P_{50}) are significantly higher than those of British children (Tanner et al., 1966) and those reported by Ferro Luzzi and Sofia (1967) for the whole of Italy.

When comparing the growth curves of the two subgroups (Fig.3), a significant difference in yearly increments of height may be observed for both sexes. As regards weight, on the other hand, there is a significant difference only for males. For females, the pattern is different; at 9 years as already mentioned, dietary control is introduced and the difference is reduced.

Table II shows that, for both sexes in the two subgroups, height and weight have a similar range of variation, as reflected by the differences between P_{10} and P_{90}.

A socioeconomic gradient has also been evidenced in the sample when analysing other anthropometric indices. Table III shows that, for males, the thickness (P_{50}) of the triceps skinfold, which constitutes among the various skinfolds the most reliable index of the state of energy nutrition (Arroyave et al., 1964), is decidedly lower in subgroup 1 in respect of subgroup 2. The situation changes with time with respect to the data reported by Tanner and Whitehouse (1975) for Great Britain and by Ferro Luzzi and Sofia (1967) for the whole of Italy in the sixties. This supports the thesis, as already reported in other countries (Whitelaw, 1971, for Great Britain) that also in Italy (as shown by Sorrentino et al., in preparation) a higher deposition of fat is found in the subcutaneous tissue

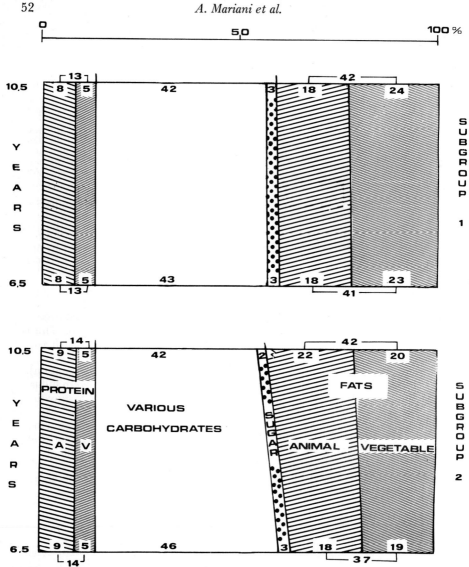

Fig.1 Calories derived from fats, carbohydrates, and proteins, as percentage of total calories (according to age and socioeconomic status).

for subjects belonging to higher socioeconomic conditions.

The improvement of diet for all classes in recent years makes our values (even for subgroup 1, after 9 years) higher than those observed in Great Britain, as well as those found in preceding years in our own country.

Similar considerations can be made for female subjects. Moreover, the observation already made in relation to weight, regarding the effect of diet control in subgroup 2, is confirmed.

The auxologic comparison between the two subgroups closes, as far as boys are concerned, with Fig.4, which shows a slight but significant retardation in

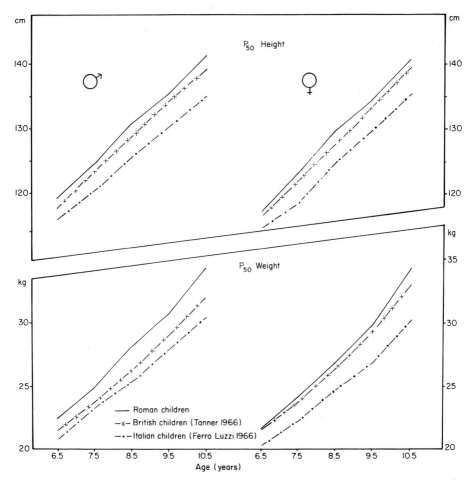

Fig. 2 Height and weight (P_{50}) of Roman children compared with British (Tanner *et al.*, 1966) and Italian children (Ferro Luzzi and Sofia, 1967).

skeleton maturation as evaluated according to the method of Greulich and Pyle (1959) following Garn recommendations (Garn *et al.*, 1964).

The growth retardation of subgroup 1 in respect to subgroup 2 is confirmed (Fig. 4) for male subjects by the HOP index, the variations in which, despite the reservations advanced after it was proposed by Whitehead (1965), can be taken as a valid expression of the growth rate. As regards females (Fig. 5), similar observations can be made for the first two years. Although the differences disappear between 8 and 9 years, the analysis of the data collected in successive years will confirm whether the differences, which become significant again after the age of 10, are the expression of a more precocious growth at the onset of puberty.

Obviously, once the existence of a growth variation in respect to socioeconomic conditions is recognized, the extent to which the various factors are responsible in determining this phenomenon remains to be established.

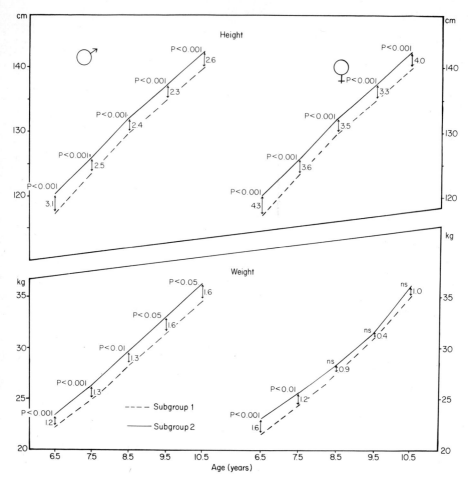

Fig. 3 Mean values of height and weight of Roman children of different socioeconomic status.

If, as suggested by Young (1970), the growth potential does not differ greatly for at least 90% of the entire world population, we may then presume that the differences in growth between groups of differing socioeconomic conditions in our sample may not be genetically determined. Even if, in Italy, different patterns in growth have been revealed from region to region (Ferro Luzzi and Sofia, 1967), it is beyond any reasonable doubt that the lowest levels of growth in our country were always found wherever the most underprivileged living conditions existed ("Mezzogiorno" and the Islands). In the future development of this study, we shall investigate the relationships between height of parents and of children, also in relation to the region of origin of the population of a metropolitan area such as Rome, although (Roche et al., 1975) the height of the parents does not constitute a valid index for predicting the height of the children, as demonstrated by the correlation coefficients calculated by Welch (1970).

As regards the specific Italian situation, however, it does not seem that what geneticists call "hybrid vigour" can explain why the highest statural increments

Table II.

HEIGHT (cm)

Age (yr)	Boys P_{10} Sub-group 1	Boys P_{10} Sub-group 2	Boys P_{50} Sub-group 1	Boys P_{50} Sub-group 2	Boys P_{90} Sub-group 1	Boys P_{90} Sub-group 2	Girls P_{10} Sub-group 1	Girls P_{10} Sub-group 2	Girls P_{50} Sub-group 1	Girls P_{50} Sub-group 2	Girls P_{90} Sub-group 1	Girls P_{90} Sub-group 2
6.5	111.9	114.2	117.3	120.6	123.4	126.2	107.8	113.3	115.0	119.5	122.0	125.0
7.5	117.7	120.6	123.2	125.9	130.1	132.1	113.8	119.1	121.7	125.6	128.2	130.5
8.5	123.9	127.1	129.4	131.7	137.2	138.3	118.9	124.4	127.0	131.9	135.1	135.6
9.5	128.8	131.9	134.7	136.7	143.0	143.2	124.1	129.2	132.9	136.4	140.7	143.2
10.5	133.7	135.9	139.8	142.2	147.0	149.6	129.4	134.2	138.3	143.0	147.7	151.7

WEIGHT (kg)

Age (yr)	Boys P_{10} Sub-group 1	Boys P_{10} Sub-group 2	Boys P_{50} Sub-group 1	Boys P_{50} Sub-group 2	Boys P_{90} Sub-group 1	Boys P_{90} Sub-group 2	Girls P_{10} Sub-group 1	Girls P_{10} Sub-group 2	Girls P_{50} Sub-group 1	Girls P_{50} Sub-group 2	Girls P_{90} Sub-group 1	Girls P_{90} Sub-group 2
6.5	18.3	19.8	22.0	23.0	26.3	27.8	17.4	19.0	21.5	22.3	26.7	29.1
7.5	21.2	21.8	23.9	25.6	31.1	30.8	18.7	21.0	23.6	24.5	30.3	31.6
8.5	22.6	25.3	27.5	28.8	35.5	36.0	21.4	23.7	26.3	27.9	35.2	34.6
9.5	25.2	27.5	30.3	32.3	40.1	41.5	23.6	25.3	29.4	29.8	40.9	40.5
10.5	27.9	29.5	33.8	35.5	44.0	45.8	26.3	28.4	34.0	34.7	46.0	46.9

Table III. Comparison of triceps and subscapular skinfolds of Roman children, British reference population and average Italian population.

TRICEPS SKINFOLD (P_{50})

Age (yr)	Boys						Girls				
	6.5	7.5	8.5	9.5	10.5		6.5	7.5	8.5	9.5	10.5
Subgroup 1	6.5	7.0	8.0	10.0	10.0		8.0	9.0	11.0	13.0	14.0
Subgroup 2	8.0	8.5	10.0	12.0	11.5		9.5	10.5	12.0	13.5	14.0
Tanner (1975)[a]	8.2	8.0	8.1	8.3	8.6		9.7	9.8	10.4	10.9	11.3
Ferro Luzzi (1967)	7.3	7.2	7.2	7.4	7.7		8.7	8.8	9.2	9.6	10.0

SUBSCAPULAR SKINFOLD (P_{50})

Age (yr)	Boys						Girls				
	6.5	7.5	8.5	9.5	10.5		6.5	7.5	8.5	9.5	10.5
Subgroup 1	5.0	5.0	5.5	6.5	6.5		5.0	5.5	7.0	8.5	9.5
Subgroup 2	5.0	5.0	6.0	7.0	7.0		6.0	6.0	7.0	8.5	10.0
Tanner (1975)[a]	5.2	5.2	5.3	5.5	5.8		6.0	6.1	6.4	6.8	7.4
Ferro Luzzi (1967)	4.4	4.4	4.4	5.1	4.9		5.0	5.1	5.3	5.7	6.1

[a] Derived by Standards: British children, 1970.

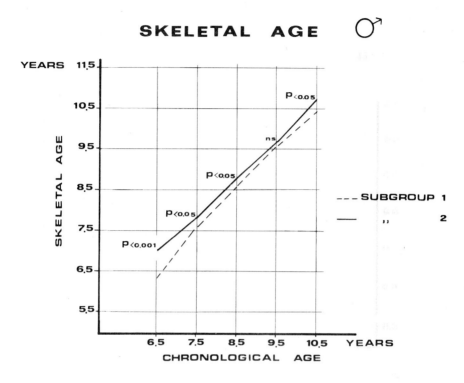
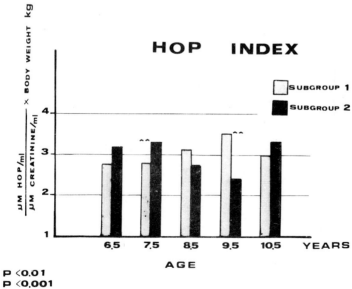

Fig. 4 HOP index and comparison of the relation between skeletal age and chronological age of Roman schoolboys of different socioeconomic status.

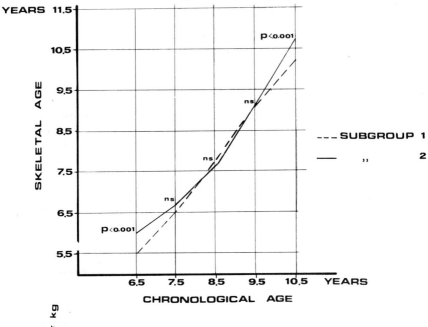

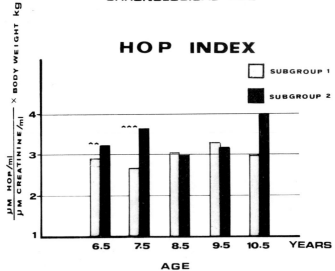

Fig.5 HOP index and comparison of the relation between skeletal age and chronological age of Roman schoolgirls of different socioeconomic status.

Table IV. Percent distribution of clinical data[a].

Age (yr)	Subgroup 1						Subgroup 2				
	6.5	7.5	8.5	9.5	10.5		6.5	7.5	8.5	9.5	10.5
Clinical impression											
good	60	54	57	56	61		79	77	74	76	74
fair	30	33	23	21	20		19	20	21	17	14
poor	10	13	20	23	19		2	3	5	7	12
Tonsil											
hypertrophic	34	34	42	26	24		35	45	39	33	27
tonsillectomy	31	38	35	39	39		17	22	27	28	28
Lymphadenopathy											
present	69	80	78	71	51		62	69	64	57	43
Hepatomegaly											
present	14	14	8	17	15		10	7	5	10	8
Caries											
present 1-4	38	46	48	54	64		37	42	50	46	54
> 5	21	28	38	33	19		11	23	21	27	18

[a] Data are relative to the whole sample, sexes combined.

of the last 50 years have been observed in the regions where population movements have been minimal, as in Marche (+ 10 cm), Abruzzi (+ 9 cm), Basilicata (+ 9 cm), Sardegna (+ 9 cm), while lower increases have been seen in regions where movements and possibilities of genetic interchange have been maximal, as in Lazio (+ 6m) and Lombardia (+ 7 cm) (ISTAT, 1975).

The attention, thus, is to be directed towards the intervention of major environmental factors: nutrition, on the one hand, hygienic conditions and infection on the other. The fact that the differences encountered can be evidenced mainly by anthropometric indices, supports the hypothesis of the role played by environmental factors, which act in the long run. Anthropometric determinations express, in fact, the whole story, namely, the integration of all the preceding phases of growth of the individual. They register the effects of critical periods in growth, of temporary deficiencies or unbalance in the diet, as a result of persistent qualitative defects of the diet, or of the intervention of other factors.

In the case of our sample, where the disadvantageous situation in subgroup 1, confirmed by the dietary history, existed already at the beginning of the study, it would be unrealistic to attempt to distinguish the weight which different environmental factors bear. It is not a matter of opposing nutrition to infection, but rather of acknowledging that, in situations of disadvantage, a major exposure to illness and infections is related to a nutrition which is, at least qualitatively, less rich, or marginal.

The clinical picture (Table IV) obtained from the examination of the subjects (two sexes combined) is less favourable in the case of the children of subgroup 1. In this subgroup, one may also observe a more frequent micropathology and the appearance of negative clinical signs associated with situations which interfere with, or condition, nutrition (as, for example, a higher frequency of lymphadenopathy, hepatomegaly, tooth decay).

In the description of the results, a socioeconomic gradient has been evidenced, but it does not reach alarming dimensions, either on the biological or social plane, except for the fact that the study has revealed other factors which are worthy of consideration. Representative groups of the sample have been subjected to measures of the intellectual level, and a correlation between performance on tests, food intake and anthropometric development, in relation to socioeconomic situations, could be established. Since in an industrialized society the influence of nutrition on the development of the central nervous system is not a common condition, we have proceeded to analyse the subjects using suitable techniques to establish correlations between food intake, physical development, and psychological self-perception and perception of the relationships with "significant others" (Caprara et al., 1977).

This analysis has given rise to the hypothesis which identifies in self-perception a determinant factor in the relationship that can be established between quality of the diet and anthropometric development, on the one side, and mental efficiency on the other. In other words, the hypothesis is advanced that a better-nourished child, being also more beautiful and healthy, and consequently in the physical conditions of being and feeling better accepted, is in the psychological condition of being and feeling more appraised, even in terms of his intellectual performance.

In conclusion, even if this hypothesis must be further substantiated, our multiple approach provides the opportunity to extend the analysis to identify all the variables which can actually mediate or codetermine the relationship between nutrition and psychological and physical development.

It is beyond doubt, above all in view of programs of intervention, that a nutrition which is qualitatively and quantitatively less satisfactory is part of a social context associating, to a less adequate nutrition, other precariousness factors at the level of life conditions, such as type of housing, of schooling, of social life. Only global interventions may induce measurable improvements.

ACKNOWLEDGEMENTS

The authors express their appreciation to Drs. M.R. Bollea, L. Cetorelli, M.A. Frasca, and V. Pennetti, for their collaboration in the course of the research; to G.D. Addesa, G. Maiale, and S. Uliscia, for their technical assistance in the course of the project development; and to all the students for the contribution offered in the course of their theses. Particular thanks are expressed to Drs. A. Ferro Luzzi and C. Lintas for valuable comments and suggestions in the preparation of the manuscript.

REFERENCES

Arroyave, G., Flores, M. and Behar, M. (1964). *Am. J. clin. Nutr.* **15**, 331-340.
Baker, H., Frank, O., Feingold, S., Christakis, C. and Ziffer, H. (1967). *Am. J. clin. Nutr.* **20**, 850-857.
Burke, B.S. (1947). *Am. J. Diet. Ass.* **23**, 1041-1046.
Caprara, G.V., D'Alessio, M., Ercolani, A.P. and Mariani, A. (1977). *Nutr. Rep. Int. (in press).*
Christakis, G., Miridjanian, A., Neth, L., Khurana, H.S., Lowell, C., Archer, M., Frank, O., Ziffer, H., Baker, H. and George, I. (1967). *Am. J. clin. Nutr.* **21**, 107-126.
Ferro Luzzi, G. and Sofia, E. (1967). *Quad. Nutr.* **17**, 269-292.
Garn, S.M., Silverman, F.N. and Rohmann, C. (1964). *Ann. Radiol.* **7**, 297-307.
Greulich, W.W. and Pyle, S.I. (1959). "Radiographic Atlas of Skeletal Development of the Hand and Wrist", 2nd ed. Stanford University Press, Stanford, California.
ISTAT (1975). "Annuario Statistico Italiano". Istituto Centrale di Statistica, Rome.
Lancia, B., Carnovale, E. and Miuccio, C.F. *(in preparation).*
Roche, A.F., Wainer, H. and Thissen, D. (1975). *In* "Predicting Adult Stature for Individuals" (A.F. Roche, H. Wainer and D. Thissen, eds), Monographs in Paediatrics, **Vol.3**. Karger, Basel.
Sorrentino, D., Ferro Luzzi, A., Migliaccio, P.A. and Mariani, A. *(in preparation).*
Tanner, J.M., Whitehouse, R.H. and Takaishi, M. (1966). *Arch. Dis. Child.* **41**, 454-471; 613-635.
Tanner, J. and Whitehouse, R.H, (1975). *Arch. Dis. Child.* **50**, 142-145.
Welch, Q.B. (1970). *Growth* **74**, 293-312.
Whitehead, R.G. (1965). *Lancet* **2**, 567-570.
Whitelaw, A.G.L. (1971). *Hum. Biol.* **23**, 414-434.
Young, H.B. (1970). *In* "Malnutrition is a Problem of Ecology" (P. Gyorgy and O.L. Kline, eds), Biblioteca Nutritio et Dieta **No.14**, pp.43-46. Karger, Basel—New York.
Ziffer, H., Frank, O., Christakis, G., Talkington, L. and Baker, L. (1967). *Am. J. clin. Nutr.* **20**, 858-865.

2. THE TIMES OF AUXOLOGY
Prenatal, Perinatal, Early Postnatal, and Pubertal Growth

FETOSCOPY: DIRECT VISUALIZATION AND SAMPLING OF THE FETUS: AN OVERVIEW

J.M. Phillips

Section of Gynecological Endoscopy, Department of Obstetrics and Gynecology University of California, Irvine, California, USA

SUMMARY

The fetus has successfully avoided direct observation during the nine most critical months of human development. Techniques now available for detection of fetal disease or distress include amniocentesis, radiography of skeletal and soft tissues (amniography, fetography), ultrasonography, electrocardiography, biopsy of fetus, placenta and membranes, and maternal blood and urine evaluation. Needle puncture of the pregnant uterus for amniocentesis is a relatively safe procedure during the second trimester; therefore, it has been projected that a needle slightly larger with lens and light source should be tolerated by the fetus and the uterus. This would make possible under direct vision: (1) observation of the fetus and detection of anatomic defects, (2) amniocentesis, (3) fetal skin biopsy, (4) blood sampling, (5) blood transfusions, and (6) intrauterine fetal surgery. However, to be useful, the morbidity and mortality risk to the fetus and the mother must be less than the congenital disorder under investigation.

The fetus has successfully avoided direct observation during the nine most critical months of human development. Techniques now available for detection of fetal disease or distress include amniocentesis, radiography of skeletal and soft tissues (amniography, fetography), ultrasonography, electrocardiography, biopsy of fetus, placenta, and membranes, and maternal blood and urine evaluation (Emery, 1973; Burton *et al.*, 1974; Holmes, 1974).
Needle puncture of the pregnant uterus for amniocentesis is a relatively safe procedure during the second trimester (Milunski *et al.*, 1970; Burton *et al.*,

1974); therefore, it has been projected that a slightly larger needle with lens and light source should be tolerated by the fetus and the uterus. This would make possible under direct vision: (1) observation of the fetus and detection of anatomic defects, (2) amniocentesis, (3) fetal skin biopsy, (4) blood sampling, (5) blood transfusions, and (6) intrauterine fetal surgery. However, to be useful, the morbidity and mortality risk to the fetus and the mother must be less than the congenital disorder under investigation (Wheeless, 1974).

Fetoscopy must be differentiated from amnioscopy, a procedure used in late at risk pregnancies, in which the amniotic fluid is not entered, but is observed vaginally for color, often predicting fetal distress. Amnioscopy also facilitates scalp blood sampling for fetal monitoring during labor. It is important that these terms not be used interchangeably, that article titles contain the appropriate word, and that indexers be familiarized with the new word, so that upcoming advances may be listed in the literature. A recent computer search to the National Library of Medicine detected only 5 of 13 references to this new technique, and included 18 references to the less useful technique of amnioscopy. *Fetoscopy* should be defined as direct visualization of the fetus through an abdominal and uterine entry, using a fiberoptic endoscope. *Amnioscopy* is the visualization of the amniotic fluid through a vaginal approach with instruments and light source that allow color detection, or membrane rupture for fetal scalp blood sampling.

The endoscopes for fetoscopy are still in the process of development (Prescott, 1974). Early fetoscopy was done with modifications of a pediatric cytoscope (Valenti, 1972), endoscope designed for orthopedic intraarticular diagnosis (Patrick *et al.*, 1974) and a fiberoptic bronchoscope (MacKenzie, 1974). Other endoscopes specifically designed for fetoscopy are in experimental use in various laboratories (Scrimgeour, 1973; Wheeless, 1974). Rigid and flexible instruments with diameters ranging from 2 to 16 mm have been used (Laurence *et al.*, 1974). Wheeless (1974) demonstrated a direct relationship between morbidity and the size of the needle and most research groups prefer the smaller instruments because of less risk of abortion and fetal death. However, the smaller the instrument, the more the field of vision, the depth of focus, and illumination are diminished, and magnification of fetal parts is greatly enhanced and distorted. Amniotic fluid conducts light easily, unless contaminated with blood, meconium or bilirubin.

Gynecologic laparoscopy is now being utilized in many medical centers internationally for diagnosis and treatment of disorders of the female reproductive system (Phillips and Keith, 1974). Skill in the use of endoscopic investigation of the abdomen can be applied to endoscopy of the fetus. Specific techniques for fetoscopy are summarized in Table I from major investigators (Scrimgeour, 1973; Valenti, 1973; Chang *et al.*, 1974; Hobbins and Mahoney, 1974b; Hobbins *et al.*, 1974; Laurence *et al.*, 1974; Patrick *et al.*, 1974; Wheeless, 1974).

A very recent variation is transcervical extraamniotic fetoscopy, in which a flexible-tip bronchofiberscope was used for observation in 28 patients just prior to abortion (MacKenzie, 1974). The obvious advantage is that this technique avoids penetration of membranes. It also allows observation at an earlier gestation, from 8 to 20 weeks; termination could then be done earlier in an affected fetus. The risk of infection would be higher, although it did not occur in this series. Visualization would be less complete.

The optimal time for fetoscopy is from approximately the 16th to the 20th

Table I. Summary of specific techniques for fetoscopy.

1. Ultrasonic confirmation of fetal age, localization of placenta, fetal parts, and distances between.
2. Anesthesia — general or local.
3. Incision — needle puncture percutanoues, or abdominal incision, or mini-laparotomy.
4. "A mode transducer", with oscilloscope for insertion of introducer needle, enables operator to know where needle is at all times.
5. Endoscope and light source inserted through introducer needle or cannula.
6. Skin biopsies taken with biopsy forceps.
7. Amniotic fluid sample taken.
8. Blood sampling via placental plate or umbilical cord.

week of gestation. If gestation is less than 15 weeks, entry into the amniotic sac is difficult, and there is less development of the abnormality under investigation. By 21 weeks, the size of the fetus makes it difficult to move and visualize.

The history of antenatal diagnosis by direct visualization of the fetus begins with Westin in 1954. He described a technique of hysterophotography performed between 16 and 20 weeks gestation, with an instrument introduced through the cervical canal in patients about to undergo therapeutic abortions. He observed fetal limb movements and swallowing when local anesthesia was used, but neither if general anesthesia was used. In 1970, Valenti and Emery each reported the use of small endoscopes introduced through the abdominal wall directly into the amniotic cavity for visualization of the fetus *in utero*.

References to 227 fetoscopies performed during the second trimester of pregnancy were found up to November 1974 (Scrimgeour, 1973; Laurence et al., 1974) (Table II). A few (7) were done for diagnostic purposes in pregnancies at risk for severe malformations (Table III). The rest (220) were performed on patients undergoing second trimester abortion. Pregnancies were usually terminated at the close of the fetoscopy procedure, thereby giving good information on immediate effects to the fetus, but no information on the long-term outcome of the pregnancy. Wheeless (1974) has reported postponement of the therapeutic abortion procedure for a week in seven patients, at which time all fetuses were dead. One patient went into spontaneous labor three days after fetoscopy.

A few live births following fetoscopy have been reported. Scrimgeour (1973) has reported six patients monitored with fetoscopy, and Laurence et al. (1974) followed one patient to term following fetoscopy. All patients had two previous children with severe malformation. Table III summarizes the results of diagnostic fetoscopy.

The feasible uses of information gained at fetoscopy are rapidly increasing with advances in early fetal biochemical detection of genetic disease. As infectious diseases decline in frequency, and environmental conditions improve, genetic disease is becoming a more important cause of mortality and morbidity (Smith,

Table II. Fetoscopies reported – 1972-1974.

Reporter	Number of fetoscopies	Week of gestation	Size	Maternal complications
Aladjem, 1974	14			None
Benzie and Doran, 1974	65	16-18	2 mm	1 – cystotomy (fetoscopy stopped)
Hobbins and Mahoney, 1974	51	15-20	2.2 mm	1 – severe uterine cramping
Laurence et al., 1974	13	14-18	9 mm	1 – leakage of amniotic fluid later in pregnancy
Levine et al., 1974	12		2 mm	1 – infection
Patrick et al., 1974	27	16-20	2.1 mm	None
Scrimgeour, 1973	25	15-21	2.2 mm	None
Valenti, 1973	11	14-18	18 Fr	None
Wheeless, 1974	9	16-18	5 mm	1 – hematoma, small tear in marginal sinus of placenta
Total	227	14-21	2-9 mm	5

Table III. Diagnostic fetoscopy.

Reporter: Patient No.	Week	Fetoscopy results	Complications	Baby at birth
Scrimgeour, 1973				
1	18	No defects seen	None	Normal
2	15	No visualization	Blood-covered lens	Pregnancy terminated
3	20	No defects seen	None	Spina bifida, died
4	18	No defects seen	None	Not yet reported
5	18	No defects seen	None	Not yet reported
6	18	No defects seen	None	Not yet reported
Laurence et al., 1974	18	No defects seen	Blood-stained amniotic fluid leakage after 26 weeks, required hospitalization, early C-section	Normal

1973). Congenital abnormalities have increased as a cause of infant mortality from about 5 to 20% in the last 70 years (Carter, 1963), and in one survey, 42% of hospital deaths of children were genetic or partly genetic in origin (Roberts et al., 1970). As much as 3.5% of newborns have genetic disease of some kind (Stevenson, 1961); 2% of newborns have a serious malformation (Holmes, 1974), and malformations constitute the most common cause for hospitalization of children in North America (Day and Holmes, 1973; Scriver et al., 1973; Holmes, 1974). Detection of defects early in pregnancy could lead to prevention or elimination of severe deformities and inheritable disease, to intrauterine therapy of treatable disorders, or to medical and parental preparedness for the affected baby.

Congenital anomalies that could be easily visualized are meningomyelocele, spina bifida, anencephalus, hydrocephalus, omphalocele or gastroschisis, skeletal defects, and harelip.

Amniocentesis is our most valuable tool for diagnosing genetic disease (Burton et al., 1974). When it is done under direct vision, it would avoid the most frustrating complications of this useful procedure — injury to the fetus during a blind puncture. Fetal cells obtained by amniocentesis in early pregnancy have two limitations, however; they are only 80 to 90% viable in cell culture, which takes at least three weeks for adequate growth, and the cells tested are shed dead cells of uncertain tissue origin (Milunsky et al., 1970). Direct biopsy with visualization of the biopsy site provides an increased yield of live cells that have an almost 100% take in cell culture in three to ten days.

Blood drawn directly from the fetus can be used for early detection of hemoglobinopathies such as sickle-cell anemia, thalassemia, other blood dyscrasias, and clotting abnormalities (Kan et al., 1972; Kazazian, 1972; Chang et al., 1974). Blood has been obtained during fetoscopy from the cord and from a fetal vein on the placenta (Valenti, 1973; Benzie and Doran, 1974; Hobbins and Mahoney, 1974a,b; Patrick et al., 1974). When an anterior placenta makes blood sampling impossible under direct vision, aspiration through the placenta has been done without complications (Kan et al., 1974). With blood sampling possible, fetal transfusion will also be possible to aid the fetus in distress with erythroblastosis fetalis. Other forms of drug therapy and replacement therapy will be possible.

And for the dreamers among us, the future might see intrauterine fetal surgery. Possibly, some life-threatening developmental abnormalities such as hydrocephalus or omphalocele could be corrected or palliated before severe disproportions occurred.

However, much work must precede these dreams. The risks and limitations of fetoscopy at the moment are many (Astedt et al., 1974; Benzie and Doran, 1974; Goodlin, 1974; Hobbins and Mahoney, 1974a,b; Hobbins et al., 1974; Laurence et al., 1974; Levine et al., 1974; MacKenzie, 1974; Patrick et al., 1974; Prescott, 1974; Valenti, 1974; Wheeless, 1974; White and Gordon, 1974) and are listed in Tables IV, V and VI.

We must await results from laboratories using animal models to determine the effect of fetoscopy on the uterus, on the duration of pregnancy, and on the developing fetus. Pregnant monkeys (Valenti, 1973) and sheep are presently being used for this purpose. Animal pregnancies do not completely parallel human ones; so, many more procedures will have to be done on preabortion fetuses, lengthening the time between fetoscopy and abortion.

For full development of the potential of fetoscopy, we need a forum for

Table IV. Limitations and risks of fetoscopy — to the fetoscopist.

1. Inadequate knowledge about uterine and fetal physiology; i.e., why does the fetus die, and why does the uterus go into premature labor? The animal uterus allows experimentation — how are the human uterine and fetal balances different, why, and what methods can be used to control premature labor and other effects of instrumentation.
2. Location of placenta limits the approach; an anterior placenta makes entry difficult, and blood sampling impossible.
3. Because of loose attachment of membranes, uterine penetration is difficult.
4. Cord pulsation and movement of fetus make drawing blood and visualization difficult.
5. Bleeding into amniotic fluid obscures vision.
6. Rigid instrument limits use.
7. Magnification and narrow field of vision reduces reliability.
8. Position of fetus and area of entry limit visualization.
9. Much experience is necessary for easy recognition of fetal parts, which are only visualized at random. Sex determination is difficult.
10. Direct relationship exists between morbidity and the size of the instrument. Trade-off of survival for visibility.
11. The larger/older the fetus, the less room for visibility; the younger the fetus, the less development of abnormality.
12. Biochemical techniques for unequivocal diagnosis are not yet available.
13. Patients will ask for anything that is publicized (legal risk).

Table V. Limitations and risks of fetoscopy — to the mother.

1. Fetomaternal transfusion with maternal sensitization.
2. Spontaneous abortion.
3. Leakage of amniotic fluid during later months of pregnancy.
4. Uterine bleeding.
5. Scarring of uterus.
6. Infection.
7. Puncture of other organs.
8. Discomfort.
9. Psychiatric disturbance depending on outcome.

continuing dialog, because the year lag before research results reach the literature and indexes makes it impossible to keep up with the accelerating pace of information. It is essential that there be total and aggressive open communication

Table VI. Limitations and risks of fetoscopy — to the fetus.

Same as in amniocentesis:
1. Fetal death.
2. Fetal hemotoma.
3. Scarring.
4. Puncture or other injury.
5. Prematurity.
6. Anesthesia carries unknown risk.
7. Intense light source may damage developing eye.
8. Blood loss from sampling may be damaging to development.

between obstetricians, fetologists, pediatricians, neonatologists, hematologists, pediatric and adult endocrinologists, pediatric surgeons, neurosurgeons, geneticists, and medical instrumentation engineers. Sociologists, psychiatrists, and clergy should be included as we discuss the implication of our work with fetoscopy on the whole subject of antenatal diagnosis of genetic disease.

Fetal experimentation is the subject of much legislative and judicial debate, and decisions now being made by research committees and organizations granting financial support will affect our research potentialities in the future. The medical profession's immediate and long-term goals in genetic engineering must be more adequately defined, so the patient's specific needs can be met. The concerned women first asks her obstetrician for advice. Criteria and standards for the use of fetoscopy must be established at the same time techniques for its use are being developed.

REFERENCES

Aladjem, S. (1974). Unpublished abstract, Fetoscopy Symposium, Birth Defect Conference, Newport Beach, California.
Astedt, B., Gennser, G., Grennert, L., Liedholm, P. and Ohrlander, S. (1974). *Lancet* **1**, 941-942.
Benzie, R.J. and Doran, T.A. (1974). *Ob. Gyn. News* **9**, 15.
Burton, B.K., Gerbie, A.B. and Nadler, H.L. (1974). *Am. J. Obstet. Gynec.* **118**, 718-746.
Carter, C.O. (1963). *In* "Congenital Malformations" (M. Fishbein, ed), p.306. International Medical Congress, New York.
Chang, H., Hobbins, J.C., Civadalli, G., Frigoletto, F.D., Mahoney, M.J., Kan, Y.W. and Nathan, D.G. (1974). *N. Engl. J. Med.* **290**, 1067-1068.
Day, N. and Holmes, L.B. (1973). *Am. J. Hum. Genet.* **25**, 237-246.
Emery, A.E.H. (1970). *Fogarty International Center Proceedings, Milan* **6**, 31
Emery, A.E.H. (1973). *In* "Antenatal Diagnosis of Genetic Disease" (A.E.H. Emery, ed), pp.1-3. Williams and Wilkins, Co., Baltimore.
Goodlin, R.C. (1974). *Lancet* **1**, 357.
Hobbins, J.C. and Mahoney, M.J. (1974a). Unpublished abstract, Fetoscopy Symposium, Birth Defects Conference, Newport Beach, California.
Hobbins, J.C. and Mahoney, M.J. (1974b). *N. Engl. J. Med.* **290**, 1065-1067.
Hobbins, J.C., Mahoney, M.J. and Goldstein, L.A. (1974). *Am. J. Obstet. Gynec.* **118**, 1069-1072.
Holmes, L.B. (1974). *N. Engl. J. Med.* **291**, 763-773.
Kan, Y.W., Dozy, A.M., Alter, B.P., Frigoletto, F.D. and Nathan, D.G. (1972). *N. Engl. J. Med.* **287**, 1-5.

Kan, Y.W., Valenti, C., Guidotti, R., Carnazza, V. and Rieder, R.F. (1974). *Lancet* 1, 79-80.
Kazazian, H.H., Jr. (1972). *N. Engl. J. Med.* 287, 41-42.
Laurence, K.M., Pearson, J.F., Prosser, R., Richards, C. and Rocker, I. (1974). *Lancet* 1, 1120-1121.
Levine, M.D., McNeil, D.E., Kaback, M.M., Frazer, R.E., Okada, D. and Hobel, C.J. (1974). Unpublished abstract, Fetoscopy Symposium, Birth Defects Conference, Newport Beach, California.
MacKenzie, I.Z. (1974). *Lancet* 2, 346-347.
Milunsky, A., Littlefield, J.W., Kanfer, J.N., Kolodny, E.H., Shih, V.E. and Atkins, L. (1970). *N. Engl. J. Med.* 283, 1370-1381; 1441-1447; 1498-1504.
Patrick, J.E., Perry, T.B. and Kinch, R.A.H. (1974). *Am. J. Obstet. Gynec.* 119, 539-542.
Phillips, J.M. and Keith, L., eds (1974). "Gynecological Laparoscopy: Principles and Techniques". Symposia Specialists, Florida.
Prescott, R. (1974). Unpublished abstract, Fetoscopy Symposium, Birth Defects Conference, Newport Beach, California.
Roberts, D.F., Chavez, J. and Court, S.D.M. (1970). *Arch. Dis. Child.* 45, 33-38.
Scriver, C.R., Neal, J.L., Saginur, R. and Clow, A. (1973). *Can. Med. Assoc. J.* 108, 1111-1115.
Scrimgeour, J.B. (1973). In "Antenatal Diagnosis of Genetic Disease" (A.E.H. Emery, ed), pp.49-52. Williams and Wilkins Co., Baltimore.
Smith, C. (1973). In "Antenatal Diagnosis of Genetic Disease" (A.E.H. Emery, ed), pp.137-155. Williams and Wilkins Co., Baltimore.
Stevenson, A.C. (1961). *Br. Med. Bull.* 17, 254-259.
Valenti, C. (1970). *Fogarty International Center Proceedings, Milan* 6, 34.
Valenti, C. (1972). *Am. J. Obstet. Gynec.* 114, 561-564.
Valenti, C. (1973). *Am. J. Obstet. Gynec.* 115, 851-853.
Valenti, C. (1974). *Ob. Gyn. News* 9, 1.
Westin, B. (1954). *Lancet* 2, 872.
Wheeless, C.R. (1974). *Am. J. Obstet. Gynec.* 119, 844-848.
White, J.M. and Gordon, H. (1974). *Lancet* 1, 352.

THE RELATIONSHIP BETWEEN MATERNAL URINARY ESTROGENS AND RESPIRATORY INSTABILITY IN THE HUMAN NEONATE

A. Steinschneider, R.W. Abdul-Karim, S.N. Beydoun and M. Pavy

*Departments of Pediatrics and Obstetrics and Gynecology
State University of New York, Upstate Medical Center
750 East Adams Street, Syracuse, New York 13210, USA*

SUMMARY

Maternal 24-hour urinary total estrogens, expressed as a ratio of the amount of creatinine in the urine (E/C), were determined serially on 47 normal pregnant women from the 13th week of pregnancy till term. All ratios were within the normal range. Subsequent analysis employed the individually calculated slope of the relationship between E/C ratio and gestational age, as well as the E/C ratio at 40 weeks. Respiratory instability in the newborn was assessed as the risk status for prolonged sleep apnea during the 1st and 4th week of life. A previous study had resulted in the development of a measure (PSA_4), obtained during a single nap, which discriminated infants with prolonged sleep apnea from those who were normal. Based on the PSA_4 score at a given age, infant-mother pairs were classified into two risk groups (high and low). In general, infants who demonstrated respiratory instability in either the 1st or 4th week of life were born from mothers who had a smaller slope and lower E/C ratio at 40 weeks of gestational age. The latter determination (E/C ratio at 40 weeks) was statistically significant. These findings were independent of race, weight, or sex of the neonate. The results suggest that maternal urinary estrogens, within the "normal range", are related to sleep respiratory instability in the neonatal period.

I. INTRODUCTION

It is with some trepidation that we present to this august body our findings on the relationship between maternal urinary estrogens and respiratory instability in the newborn since they are based on an exploratory study. The justification for presenting the data at this time is the hope of provoking interest and further research efforts.

It is now well established that many aspects of postnatal development reflect prenatal influences. The general problems facing perinatology include: (1) elucidating those aspects of postnatal development which reflect prenatal influences and, (2) identifying prenatal indices that prognosticate certain features of infant and child development. The current research touches on both of these aspects and stems from two previously independent lines of investigation; one related to estrogens and fetal development, and the other to respiratory instability in the newborn.

The value of serial maternal urinary estrogen determination in the assessment of fetal well being is well recognized (Ostergard, 1973) and graphs depicting the mean and range of normal levels are readily available (Shelley et al., 1970). It should be recognized, however, that these limits of normalcy are arbitrary statistical standards. Hence, differences in estrogen levels within these limits might have important consequences on both fetal and postnatal development.

An insufficiently studied aspect of estrogens in pregnancy is the possible influence of these steroids on fetal development. In a series of studies on pregnant rabbits (Abdul-Karim et al., 1968, 1970a,b, 1972, 1976; Abdul-Karim and Marshall, 1969; Abdul-Karim and Prior, 1969; Abdul-Karim and Bruce, 1972; Beydoun et al., 1974) it was found that estrogens regulate fetal and placental growth. Possibly of greater significance to the present investigation were their findings that estrogens influenced acetylcholinesterase activity (and hence, presumably, that of acetylcholine) in fetal brain (Abdul-Karim et al., 1970a). In a separate study, Drillien (1970) reported that 10 of 17 infants from mothers with low urinary estriol levels had a suspected mental or neurologic abnormality, compared to only 1 in 11 infants whose mothers had normal urinary estriols. This led Abdul-Karim et al. (1970a) to conclude that, despite the fact that many factors may be invoked to explain this finding, "the role of estrogens in the development of the human fetal brain and the possibility of supplemental estrogen therapy are worthy of consideration."

Other investigators have demonstrated an effect of estrogens on the central nervous system, although such studies were carried out postnatally. Curry and Heim (1966) have shown tht estradiol 17 B accelerates brain myelination in neonatal rats. Heim (1966) and Heim and Timiras (1963) found that estradiol diprorionate given to newborn rats hastened brain maturation as assessed by maximal electroshock seizure response. Casper et al. (1967) demonstrated that estradiol-treated rats had a higher amount of cerebrosides in the spinal cord and cerebrum. More recently, Cavallotti and Bisanti (1972) found that the administration of estradiol 17 B to neonatal rats increased the cerebrosides, sulfatids and protein levels in the brain as well as the activities of enzymes involved in glycolysis and nerve cell function, whereas ovariectomy produced the opposite effects. In another study, Bisanti and Cavallotti (1972) showed that ovariectomy retarded, and estradiol 17 B accelerated, the electrophysiological maturation of the rat brain. Finally, Westley and Salaman (1974) have evidence for the presence of a specific estradiol-binding protein in the neonatal rat brain.

Concomitant with research on the influence of estrogens on fetal development, studies were being conducted to assess and evaluate the significance of neonatal respiratory instability. These latter studies are part of a more extensive examination of the hypothesis that prolonged sleep apnea is part of the pathophysiologic process resulting in the sudden infant death syndrome (Steinschneider, 1972). In one such investigation, sleep recordings were obtained from a group of infants with documented episodes of sleep apnea in excess of twenty seconds in duration (Steinschneider, 1977). The results demonstrated that infants with prolonged sleep apnea, in comparison to controls, also had more frequent and, on the average, longer brief apneic pauses. This same study resulted in the development of a composite measure, obtained from a single nap and referred to as the PSA_4 score, which discriminated infants with prolonged sleep apnea from controls. Thus, infants with prolonged sleep apnea had PSA_4 scores $\geqslant -0.04$ whereas control infants had more negative values. A parametric study (Steinschneider, 1975) revealed that the degree of respiratory instability was inversely proportional to birth weight. Of additional relevance to this report is the observation by Henning et al. (1977) that full-term, apparently healthy, newborns with increased respiratory instability ($PSA_4 \geqslant -0.04$) in the 1st week of life had mental and psychomotor developmental delays at nine months of age when compared to infants whose PSA_4 values were less than -0.04. In view of these findings, it seemed reasonable to inquire into the possible relationship between prenatal factors and the degree of neonatal respiratory instability.

The specific direction of this investigation was based on the increasing evidence demonstrating a relationship between central nervous development and estrogens. Consequently, this study was designed, primarily, to examine the relationship between maternal urinary estrogens obtained from normal pregnant women with no complications and neonatal sleep respiratory instability.

II. SUBJECTS AND METHODS

Women attending the Low Risk Pregnancy Clinic were invited to participate in this study and written informed consents obtained at this time. They were instructed that 24-hour urines were to be collected at home at regular intervals during the remainder of the pregnancy and that their newborn would be studied in a sleep laboratory within the 1st and again in the 4th week of life. Results from the urine and neonatal measurements were assessed separately and independently, so that a true "double blind" approach was achieved. Forty-seven mother-infant pairs participated. All the mothers had accurate gestational ages and were free of medical or obstetrical complications. They received no drugs, other than iron and prenatal vitamins.

All patients had normal pregnancies and deliveries. Accurate 24-hour urine collections were obtained at repeated intervals between the 13th and 42nd weeks of gestation and analyzed for their estrogen (Rourke et al., 1968) and creatinine (Folin, 1914) contents, and the estrogen/creatinine (E/C) ratio was calculated. The mean gestational age at the initial and last collections were 23.9 and 38.4 weeks respectively. The number of collections per patient ranged between 3 and

17 with a mean of 9. Employing the method of least squares, a liner regression analysis was performed for each pregnancy separately, and the slope of the line obtained which described the relationship between the E/C ratio and gestational age (E/C · slope). In addition, the individually derived linear regression was employed to calculate the estimated E/C ratio at 40 weeks gestation (E/C · 40).

All infants were Caucasian (except four), born at term, had an Apgar score greater than 7 at 5 min, and weighed between 2552 and 4423 g at birth. The group consisted of 25 males and 22 females. Twenty-three infants were breast-fed, while the rest received a preparatory formula. The initial sleep study was conducted between ages 2 and 11 days postnatally (40 of the 47 infants had their first study performed within the 1st week of life). A repeat sleep study was obtained when the infants were approximately 4 weeks old (between 23 and 30 days). At each of these time periods, the infant was studied during a complete nap (1–3 hours long) in a temperature controlled (32.2 ± 1.1°C) room. Subsequent to being brought into the test room, surface electrodes were applied for the continuous recording of both respiratory activity and extraocular movements, and the infant fed. The sleep recording was initiated shortly thereafter, when the infant visually was observed to be asleep. Respiratory activity was measured by means of a mercury strain gauge positioned across the lower thorax and a nasal thermister taped below one nostril. Disc electrodes taped to the nasion and both outer canthi were employed to record extraocular movements. The recording was made on a Grass 7 polygraph (Grass Instrument Co., Quincy, Mass.) at a paper speed of 10 mm/sec, and an amplifier gain set at 50 uv/cm for the eye movement channels.

Each 15-sec sleep period was identified as either a Rapid Eye Movement (REM) or Nonrapid Eye Movement (NREM) epoch on the basis of the occurrence, within the epoch, of rapid eye movements. In addition, all apneic pauses of at least 2 sec in duration were measured to a tenth of a second. This information was employed to obtain four apnea measures: Mean Duration (Total), Percent Apnea REM, Percent Apnea NREM and A/D Percent. Table I contains a list and definition of each of these apnea measures. The four measures derived for each infant and sleep session were utilized in the calculation of a measure of respiratory instability: PSA_4 (see Table I). Infants were categorized as having low respiratory instability (Low Group) or high respiratory instability (High Group) if the PSA_4 score was < -0.04 or ≥ -0.04 respectively. This categorization was performed separately from the initial and repeat sleep study.

Parametric tests of statistical significance were conducted to determine if the maternal E/C · slope and E/C · 40 differed depending upon the neonatal respiratory instability group. A t test, employing a pooled error term, was used to evaluate the statistical significance of observed mean differences. In addition, a multiple linear regression analysis was performed, which included birth weight, race, sex, and method of feeding as background variables. Since these two approaches resulted in the same conclusions, only the results employing the simpler method (t test) will be presented.

III. RESULTS

There were 29 infant-mother pairs in the Low Group 1 and 18 in the High Group 1 based on the PSA_4 values obtained from the initial sleep trace. The

Table I. Definitions of laboratory apnea measures (Apneic pauses ≥ sec).

Apnea measure		Definition
Mean Duration: Total (MT)		Average duration (sec) of apneic pauses
Percent Apnea	REM (AR)	Percentage of REM epochs during which at least one apneic pause was initiated
	NREM (AN)	Percentage of NREM epochs during which at least one apneic pause was initiated
A/D Percent		$\dfrac{\text{Duration of all apneic pauses}}{\text{Sleep duration}} \times 100$

$$PSA_4 = -2.695 + 0.607 \text{ (MT)} + 0.023 \text{ (AR)} + 0.042 \text{ (AN)} - 0.143 \text{ (A/D\%)}$$

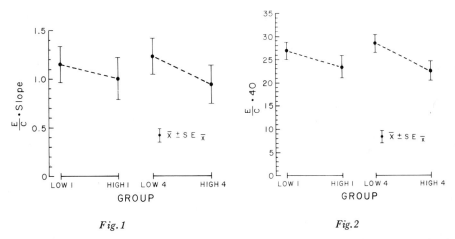

Fig. 1

Fig. 2

Fig. 1 The average slope (± 1 SE) of the linear regression between the E/C ratio and gestational age in mothers whose neonates had different degrees of sleep respiratory instability when tested within, approximately, the 1st (left half) and 4th (right half) week of life.

Fig. 2 The average estimated E/C ratio (± 1 SE) at 40 weeks gestation in mothers whose neonates had different degrees of sleep respiratory instability when tested within, approximately, the 1st (left half) and 4th (right half) week of life.

sample size of the Low Group decreased to 24 (Low Group 4) and increased to 23 for the High Group (High Group 4) when employing the results from the 4th week sleep study.

A. *Slope of the E/C by Gestational Age Regression (E/C · slope)*

The left portion of Fig. 1 contains the average slope for Low Group 1 and High Group 1. Although the average slope of Low Group 1 (1.15) was greater than that of High Group 1 (1.00), the t test failed to demonstrate statistical significance ($t = 0.524$, df = 45, $p > 0.05$). Comparable data for Low Group 4 and High Group 4 are presented in the right portion of Fig.1. Once again, the mothers in the Low Group had a greater average slope than that of mothers in the High Group (1.23 v. 0.94), though the difference failed to achieve statistical significance ($t = 1.058$, df = 45, $p > 0.05$).

B. Estimated E/C at 40 Weeks Gestational Age (E/C · 40)

Mothers who delivered newborns with relatively stable respirations during sleep (Low Group 1) had, on the average, higher E/C · 40 values when compared to mothers in High Group 1 (see Fig.2). The difference between the groups (26.9 v. 23.3) was not statistically significant ($t = 1.17$, df = 45, $p > 0.05$). However, when group categorization was based on the infants' 4th week sleep study, not only did the Low Group 4 have a greater average E/C · 40 score (28.4) than the High Group 4 mothers (22.4), but the observed difference was statistically significant at less than the 0.05 level ($t = 2.083$, df = 45).

IV. DISCUSSION

These results demonstrate that sleep respiratory instability within the fourth week of life is associated with significantly lower total maternal urinary estrogens. Although the group differences failed to achieve statistical significance, the similarity of trend would suggest that the same conclusion probably applies, as well, to the first postnatal week. These observations assume added importance when it is recognized that all E/C ratios obtained in this study were within the accepted normal range for the stage of pregnancy. It should also be noted that there was a tendency for neonatal respiratory instability to be associated with a decrease in the slope of the line describing the relationship between the E/C ratio and gestational age.

For the present, one can only speculate on the implications of finding lower E/C values in mothers of neonates having increased respiratory instability during sleep. It is possible that the observed relationship is a reflection of an intrauterine disturbance manifested both in the lowering of estrogen levels and an increased tendency towards respiratory instability in the newborn. This hypothesis implies no direct causal link between estrogens and respiratory instability. Conceptually, it is similar to the other better-known associations between low maternal urinary estrogens and fetal development (e.g., fetal size and fetal distress).

An alternative hypothesis would attribute an intrautero influence of estrogens on respiratory function. A thorough search of the literature has failed to reveal information directly relevant to this latter suggestion. For the most part, research has focused on the role of estrogens on the sexual (including the neuroendocrine) differentiation of the brain. However, ancillary evidence raises the possibility that estrogens may influence other aspects of CNS function.

Thus, studies have demonstrated: (1) the presence of radioactive material in several areas of the mammalia brain and spinal cord following radioactive estradiol administration (Stumpf et al., 1975); (2) the in situ formation of estrogens in the amygdala, hippocampus and limbic system (Kato, 1975; Naftolin and Ryan, 1975); (3) the presence of 17 B oxireductase and sulfatase activities in the hypothalamus and cerebral cortex of fetal lambs (Jenkins et al., 1975); (4) that limbic and cerebral cortex tissues from mid-trimester human fetuses are capable of forming catecholestrogens (Fishman et al., 1976) and contain a macromolecular substance that binds estradiol with high affinity (Davies et al., 1975); and (5) the presence of high levels of cytoplasmic and nuclear binding of estradiol in the parietal cortex of neonatal, but not adult, rats (Maclusky et al., 1976). As the latter authors state, "the presence of high levels of putative estrogen receptors in the neonatal cerebral cortex suggests that cortical cells may be transiently responsive to estrogen during early postnatal life."

In summary, the results obtained from this study are sufficiently suggestive to justify the tentative conclusion that estrogens are related to neonatal sleep respiratory instability. As such, it is hoped that they will provide further impetus and direction for more in-depth investigations.

ACKNOWLEDGEMENT

This work was supported in part by a research grant from the National Institute of Child Health and Human Development (RO1-HD-07460).

REFERENCES

Abdul-Karim, R.W. and Bruce, N.W. (1972). *J. Reprod. Fertil.* **30**, 477-480.
Abdul-Karim, R.W. and Marshall, L.D. (1969). *Toxicol. Pharmacol.* **15**, 185-188.
Abdul-Karim, R.W. and Prior, J.T. (1969). *J. Reprod. Med.* **2**, 140-146.
Abdul-Karim, R.W., Prior, J.T. and Marshall, L.D. (1968). *J. Reprod. Med.* **1**, 397-405.
Abdul-Karim, R.W., Drucker, M. and Rizk, P. (1970a). *Obstet. Gynec.* **36**, 719-721.
Abdul-Karim, R.W., Marshall, L.D. and Nesbitt, R.E.L. (1970b). *Am. J. Obstet. Gynec.* **107**, 641-644.
Abdul-Karim, R.W., Nesbitt, R.E.L., Drucker, M.H. and Rizk, P.T. (1971). *Am. J. Obstet. Gynec.* **109**, 656-661.
Abdul-Karim, R.W., Haviland, M.E. and Beydoun, S.N. (1972). *Oxford Med. Sch. Gazette* **24**, 17-20.
Abdul-Karim, R.W., Pavy, M., Beydoun, S.N. and Haviland, M.E. (1967). *Biol. Neonate* **29**, 89-95.
Beydoun, S.N., Abdul-Karim, R.W. and Haviland, M.E. (1974). *Am. J. Obstet. Gynec.* **120**, 918-921.
Bisanti, L. and Cavallotti, C. (1972). *Prog. Brain Res.* **38**, 319-327.
Casper, R., Vernadakis, A. and Timiras, P.S. (1967). *Brain Res.* **5**, 524-526.
Cavallotti, C. and Bisanti, L. (1972). *Prog. Brain Res.* **38**, 69-83.
Curry, III, J.J. and Heim, L.M. (1966). *Nature (London)* **209**, 915-916.
Davies, J., Naftolin, F., Ryan, K.J. and Siu, J. (1975). *J. clin. Endocr. Metab.* **40**, 909-912.
Drillien, C.M. (1970). *Pediatr. Clin. North Am.* **17**, 9-24.
Fishman, J., Naftolin, F., Davies, I.J., Ryan, K.J. and Petro, Z. (1976). *J. clin. Endocr. Metab.* **42**, 177-180.
Folin, O. (1914). *J. biol. Chem.* **17**, 469-481.
Heim, L.M. (1966). *Endocrinology* **78**, 1130-1134.
Heim, L.M. and Timiras, P.S. (1963). *Endocrinology* **72**, 598-606.
Henning, L., Steinschneider, A. and Sheehe, P. (1977). *Pediatrics (submitted for publication).*
Jenkin, G., Henville, A. and Heap, R.B. (1975). *J. Endocrinol.* **64**, 22pp.
Kato, J. (1975). *J. Steroid Biochem.* **6**, 979-987.
Maclusky, N.J., Chaptal, C., Lieberburg, I. and McEwen, B.S. (1976). *Brain Res.* **114**, 158-165.
Naftolin, F. and Ryan, K.J. (1975). *J. Steroid Biochim.* **6**, 993-997.
Ostergard, D.R. (1973). *Obstet. Gynec. Surv.* **28**, 215-231.
Rourke, J.E., Marshall, L.D. and Shelley, T.F. (1968). *Am. J. Obstet. Gynec.* **100**, 331-335.
Shelley, T.F., Cummings, R.V., Rourke, J.F. and Marshall, L.D. (1970). *Am. J. Obstet. Gynec.* **35**, 184-190.

Steinschneider, A. (1972). *Pediatrics* **50**, 646-654.
Steinschneider, A. (1975). *In* "Minnesota Symposia on Child Psychology" (A.D. Pick, ed), **Vol.9**, pp.106-134. The University of Minnesota Press, Minneapolis.
Steinschneider, A. (1977). *Pediatrics (in press)*.
Stumpf, W.E., Sar, M. and Keefer, D.A. (1975). *Adv. Biosci.* **15**, 77-88.
Westley, B.R. and Salaman, D.F. (1974). *J. Endocrinol.* **63**, 54-55.

EARLY POSTNATAL GROWTH EVALUATION IN FULL-TERM, PRETERM, AND SMALL-FOR-DATES INFANTS

F. Falkner

Department of Pediatrics, University of Cincinnati College of Medicine
and
Fels Research Institute, Yellow Springs, Ohio 45387, USA

SUMMARY

Compared to full-term infants, the growth patterns of preterm and small-for-date infants is different in the early postnatal period, and related to the more important outcome, as regards growth, of these infants. These patterns are presented together with a newly suggested ratio concerning growth per unit of body size. Among full-term infants, there appears to be differing patterns also as to whether the infant is breast- or formula-fed. Since standards in use in the developed countries are largely based upon samples of formula-fed infants, supplementation of breast-fed infants who are in the lower percentiles may in fact not be needed or desirable.

I. INTRODUCTION

It is not all that long ago that a newborn infant weighing less than 2.5 kg at birth was labelled "a premature". Now science, helpfully and progressively, labels this infant as Infant of Low Birth Weight (ILB). For, in the last decade progress, study and interest have revealed the importance of gestational age; the prematurely born or appropriate-for-gestational age infant (PTI); the necessary high-standard specialized neonatal care; the small-for-dates infants (SFDI); the concept of intrauterine growth, both normal and abnormal, with the multifactorial influences upon both; and, importantly, auxology in the perinatal period.

One stimulus to our interest was while following the growth of a pair of monozygous (MZ) twins in the Louisville Twin Study. Both twins were born near 40 weeks gestation. The smaller twin weighed approximately half the birth weight of the larger twin — 1.6 kg $v.$ 3.3 kg. Thus, in past times and terminology, the smaller MZ twin born at full term was "a premature".

Parenthetically, but we think importantly, their monochorionic placenta, when divided along the vascular equator, showed the smaller twin having about one-third the placental mass of that of his brother, as his share of placental supply.

The small twin grew much more rapidly than his twin in all anthropometric measures, including head circumference, for the first nine months. Then he grew at approximately the same rate, and now, at 14 years, as from three years, there is a notable size difference with the smaller twin not having achieved "complete" catch-up, and, for example, being 4 kg lighter and 5 cm shorter.

Our interest, accentuated by the possible placental nutritional factor, was heightened by the sudden realization that the smaller MZ twin's growth pattern was exactly that of a SFDI.

Of all ILB in the developed countries, very approximately one-third are SFDI. In the developing countries, not only are the total numbers of ILB much higher, but also that proportion is reversed and two-thirds are SFDI. Why?

We considered that careful assessment of growth in the perinatal period was needed — in view of the multifactorial nature of potential influences. Compared to full-term infants (FTI), the growth patterns of PTI and SFDI are different, and are related to the more important *outcome*, as regards growth, of these infants. This led to a re-evaluation, and a newly suggested ratio concerning growth per unit of body size.

Postnatal growth of weight, length/height, and head circumference, is generally assessed by comparing an individual to reference data that give a mean value, with ranges, for the measure at a certain age-distance curve. In the evaluation of growth of a PTI for a SFDI, it is likely, using this approach, that such infants with birth weights under the 10th percentile will remain below for several months. This is in spite of their growth velocity, that distance curves do not evaluate. Especially when studying growth patterns of different categories of infants, the increment of a measurement over time is a valuable addition to the distance value ascertained. But, of course, reference data and distributions are needed here too.

It is, for example, inappropriate to expect the same growth velocity in two infants of the same age between whom there is a body weight difference of 2 kg. We would like to present growth patterns of early postnatal growth for weight, length/height, and head circumference, in FTI, PTI, and SFDI, using distance and velocity data together with a new concept of growth per unit of body weight.

II. METHODS

A. Sample

This was 112 healthy Caucasian infants of similar socioeconomic background in the City of Montevideo, with birth dates 1972-74. In the neonatal period, they were cared for under a continued and progressive care system (essentially, this is care by a neonatalogist and trained neonatology nursing services), operative in the obstetric and neonatal services. All the infants were recruited into a follow-up clinic. Here, we are concerned with the birth-to-two years age range.

FTI were above 37 weeks gestational age and over 2500 g birth weight. PTI were defined as having a gestational age of less than 37 weeks (259 days) (Treloar et al., 1967). SFDI were those above 37 weeks gestation and below the 10th percentile for birth weight on the Lubchenco curves (Lubchenco et al., 1963).

B. Gestational Age

This was determined by estimation from the last menstrual period; and a clinical method developed by Dubovitz and modified by Capurro (1973). All infants in the study differed by less than 14 days in the estimate from both methods.

C. Body Measurements

Body measurements were made at the following ages and limits:
- At birth, and monthly to 6 months, ± 3 days of correct date.
- At 6, 8, 10 and 12 months, ± 6 days of correct date.
- At 18 and 24 months, ± 2 weeks of correct date.

When an individual was measured outside the above limits, a parabolic interpretation was made using the previous and following values:
- Body weight was determined as nude weight on regularly checked infant scales.
- Length and head circumference were measured following the technique and apparatus described by Falkner (1961).

D. Growth Velocity

In each of the three groups, median growth velocity (MGV) and median growth velocity per unit of body size (MGVU) were determined by the formulae:

$$MGV = \frac{\text{Measurement value at each age} \; minus \; \text{Measurement value at previous age}}{\text{Time (in days) of period between the two ages}}$$

$$MGVU = \frac{MGV}{\text{Measurement value at previous age}}$$

Thus, MGVU represents the daily increase in body weight in grams, per kilogram of body weight at the previous measurement age; in length and head circumference it is the daily increase in centimeters, per centimeter of length or head circumference at the previous measurement age.

III. RESULTS

A. MGV

Reference curves for MGV were constructed, and these show that in the first 6 months of life all groups (FTI, PTI, SFDI) grew most rapidly in all measures. And PTI and SFDI grew more rapidly than FTI. From this point, although the shape of the curves for body weight are similar, the PTI and SFDI, as was to be expected, grew faster in order to achieve catch-up. Growth patterns in length after 6 months were similar to weight, though the SFDI grew more slowly and suddenly slower than FTI at 12 months. For head circumference, after 6 months, while PTI remained more rapid than FTI, SFDI dropped and remained below the FTI curve up to 2 years of age.

B. *MGVU in Relationship to a Previous Measurement*

A distribution of MGVU according to the value of a measurement at the previous age would hopefully contribute to the evaluation of an individual's growth progress.

In order to produce appropriate curves, the following procedure was carried out: To give one example of MGVU against body weights at a previous measurement between 8 and 12 kg, in order to obtain a useful distribution of MGVU for a previous body weight value, the mean previous body weight of all infants having a previous body weight between 8 and 12 kg was determined. In this example it was 9245 g.

A regression line for all the individual MGVUs was constructed, and then each individual MGVU was transferred to a new abscissa at the mean value previously described — 9254 g — and parallel to the regression line. This statistical procedure is based on a method used by Wingerd et al. (1971). Next, a distribution of MGVU was obtained from all the points on the new abscissa and the 90th, 50th, and 10th percentiles calculated for all points.

Table I. Previous weight interval, its median and percentiles for MGVU in g/day/kg.

Weight interval (g)	Median of interval (g)	P_{10}	P_{50}	P_{90}
750 – 2,000	1,521	5.35	14.32	23.90
2,001 – 4,000	3,061	4.82	9.97	15.66
4,001 – 6,000	4,986	2.73	4.90	7.77
6,001 – 8,000	6,911	0.99	2.29	4.20
8,001 – 12,000	9,254	0.52	1.14	1.96
12,001 – 16,000	12,460	0.06	0.48	0.64

The above procedure was repeated for five other previous weight groups between 0 and 16 kg (Table I). It was then possible to construct the curves shown in Fig.1. Knowing an individual infant's previous body weight and using these curves, will show the infant's MGVU status and growth prospects, while also indicating the expected MGVU distribution.

It was previously determined that there was no statistical difference ($p > 0.05$) in MGVU distributions for previous weight about 2 kg in the three groups, FTI, PTI, and SFDI, and therefore, for this indicator, all infants are included regardless of those groups.

Finally, Figs 2 and 3 show the same distributions for length and head circumference.

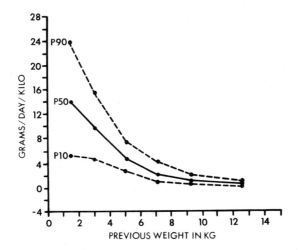

Fig. 1 MGVU for body weight: 10th, 50th, and 90th percentiles.

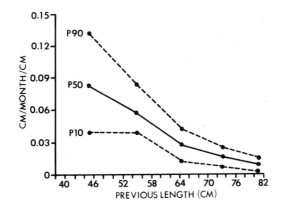

Fig. 2 MGVU for length: 10th, 50th, and 90th percentiles.

IV. DISCUSSION

A. MGV

Our findings are in agreement with others that the largest values for MGV for all three groups (FTI, PTI, and SFDI) for the three measures studied are in the first 6 months of life (e.g., Cruise, 1973; Karlberg et al., 1968).

Recently, since outcome for any individual infant is the crux, attention has been given to physical growth outcome for PTI and SFDI compared to FTI. To oversimplify, generally speaking, given good care and in the absence of serious pathological states, the PTI have caught up with the FTI by 3 years of age, whereas SFDI have not done so. More recently, however, Brandt (1977) has provided strong evidence that SFDI may well be divided into two subgroups:

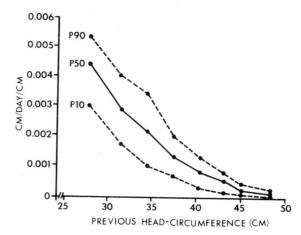

Fig.3 MGVU for head circumference: 10th, 50th, and 90th percentiles.

those, particularly in head circumference, who do in fact catch-up by 3 years; and those who do not. She suggests that the degree of intrauterine adverse factors upon growth may be the key to eventual separation of the two groups.

Our data not only may support this view, they also exemplify a notorious obstacle, and pitfall in longitudinal studies. Was the marked drop in length MGV at 12 months of the SFDI due to the fact that a greater proportion of the less-favored SFDI were measured at this particular age (due to missed visits of the better-favored) than at other ages? In head circumference MGV after 10 months, the SFDI fall below the FTI. Is there in our whole SFDI sample a high proportion of the "non-catch-up" SFDI subsample? In a future publication we shall examine these questions (Martell and Falkner).

B. MGVU

The concept of MGVU presents a new approach to growth evaluation. The energy required for body metabolism changes its relationship to body size as growth proceeds. If indeed one factor causing SFDI is intrauterine undernutrition, then prenatal growth rate is slowed, and postnatal catch-up may occur with the aim of restoring appropriate weight for length. It therefore seems beneficial to estimate, in the case of weight, the increase in grams per day for each kilogram of body weight, and this will give a more usual indication of general body weight changes than MGV. Using MGVU, infants (and children) of greatly differing body weights may be compared.

Mean MGVU is clearly higher when body weight is lower, and length and head circumferences smaller.

For a particular body weight, length, and head circumference, it appears that daily MGVU is independent of gestational age and of birth weight. Using this approach, the picture for clinical use is broadened; and in following an individual infant, distance standards and absolute values are relegated as MGV and MGVU are used.

The MGVU/previous measurement curves have of course been constructed from analysis of the raw data in this longitudinal study. A word on the use of the day as a time unit — as for example in Fig.1 and Table I. In a longitudinal study where measures are made close together in time, use of days allows greater accuracy, especially when, say, a planned measurement is 10 days after the appointed and correct age. This is so for body weight, though clearly time intervals using length and head circumference should not be used unless they are more than 30 days.

It is appropriate to give two examples of how the MGVU curves may be used.

(a) To evaluate the growth of an infant for whom no previous weight (for example) is known, the weight is recorded at the first examination. After, say, 5 days, the infant's weight is again recorded. The MGVU now calculated is placed on the MGVU distribution curves for weight at the previous weight of 5 days ago.

(b) With body weight it is necessary to have negative values for MGVU gm/day/kilo (see Fig.1). Body weight may be lost (i.e., malnutrition) over a time interval, and hence the MGVU will have a negative value — that, in itself, is a valuable indicator and marker for future progress or deterioration.

In conclusion: *(a)* Growth velocity from a biological viewpoint is a function of body mass and composition; *(b)* independently of birth weight and gestational age, with two measures separated by time a useful evaluation of growth is made; *(c)* expressing growth per day per unit relates well to daily nutritional and other requirements.

Related to these discussions is the need to examine our use of other reference data, especially concerning the *Breast-fed Infant*. There is currently an international controversy concerning the infant formula industry and the proponents of breast feeding. While we could not believe more strongly that breast feeding for up to four months is the ideal regimen for infants, the proponents in a variety of areas are inclined to rely on emotion rather than hard data to support their cause. And the industry involved is inclined to over-react.

One area in need of reference data concerns early growth patterns, and, again, the all-important outcome. Reference growth data, curves and distributions in the developed countries are based on data from cohorts of largely formula-fed infants. Healthy breast-fed infants do not grow, on average, as fast as formula-fed babies. The former, when plotted on commonly used reference data may well follow a lower centile curve. There is evidence that in certain communities these infants will be placed on a supplement formula regimen in order to "move the infant to a higher centile". If breast-fed infant growth reference data were available, such supplementation could be shown to be neither desirable nor necessary.

We hope to stimulate interest in this dilemma and gap when the results are available from our present study.

ACKNOWLEDGEMENTS

I gratefully acknowledge that much of this work has been done by colleagues and friends in our collaborative study at the P.A.H.O. Centro Latino-Americano Perinatologica in Montevideo; and I am particularly grateful to Dr. Miguel Martell.

REFERENCES

Brandt, I. (1977). *In* "Human Growth. A Comprehensive Treatise" (F. Falkner and J.M. Tanner, eds) **Vol.1**. Plenum Publishing Corporation, New York *(in press)*.
Capurro, H. (1973). Tesis de doctorado. Montevideo, Uruguay.
Cruise, M.O. (1973). *Pediatrics* **51**, 620-628.
Falkner, F. (1961). *Pediatr. Clin. North Am.* **8**, 13-18.
Karlberg, P., Klackenberg, G., Klackenberg-Larsson, I. et al. (1968). *Acta Pediatr. Scand.* **Suppl. 187**, 9-27.
Lubchenco, L.O., Hansman, C., Dressler, M. and Boyd, E. (1963). *Pediatrics* **32**, 793-800.
Martell, M. and Falkner, F. "Outcome for infants of low birth weight" *(in preparation)*.
Treloar, A.E., Behn, B.G. and Cowan, D.W. (1967). *Am. J. Obstet. Gynec.* **99**, 34-45.
Wingerd, J., Schoen, E.J. and Solomon, I.L. (1971). *Pediatrics* **47**, 818-825.

HORMONAL CONTROL OF PUBERTAL DEVELOPMENT

D. Gupta

*Department of Diagnostic Endocrinology, University Children's Hospital
7400 Tübingen, German Federal Republic*

SUMMARY

The mechanisms that lead to the onset of the process of sexual maturation are still imperfectly understood. A vast amount of new data is now available in man, which supports the concept that the CNS, and not the pituitary gland or gonads, is responsible for the activation of the hypothalamic-pituitary-gonadal axis. Accordingly, the major event that precipitates puberty is an increase in the threshold (or set point) of the inhibitory feedback receptors, which respond to changes in gonadotropin output. This paper discusses two of the control mechanisms which are related to the onset of pubertal manifestation: *(a)* the existence of a negative feedback control of secretion of gonadotropins before puberty, which exhibits a change in sensitivity as the subject approaches toward puberty; and *(b)* the quantitative and qualitative changes in the pituitary responsiveness to the gonadotropin-releasing hormone (LH-RH) with the onset of puberty.

I. INTRODUCTION

The morphologic events which take place during puberty, as much as one can say, are neither subtle nor elusive. Yet, a definitive explanation of the various events taking place at this time escapes us, and what initiates puberty still remains a biological mystery. The difficulties in understanding the underlying processes arise from the fact that reproductive maturity does not have a single determinant, but many determinants, each influencing the other in a complex way. The problem is, therefore, to understand how many different events taking place at various levels, sequentially or simultaneously, interact in the temporal fashion to produce the fully sexually mature member of a species.

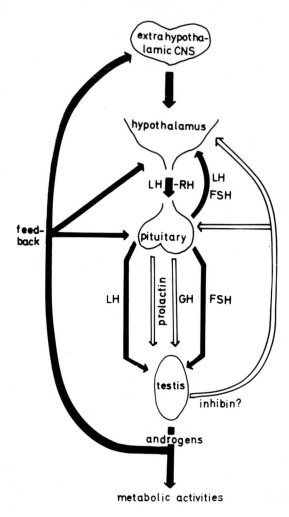

Fig.1 A schematic representation of the hypothalamic-pituitary-gonadal axis.

Figure 1 depicts a physiologist's-type schematic diagram and, therefore, essentially shows a simple relationship within the hypothalamic-pituitary-gonadal axis with the negative inhibiting effect of the sex steroids on the gonadotropins. The dominant change that triggers puberty is presumed to be happening in the hypothalamus. But this does not infer that the hypothalamus is master of itself and functions in a physiological vacuum.

The hypothalamus is singled out to define a hypothetical crucial period in sexual maturation, the time before the onset of reproductive maturity. At that time, the powerhouses of the steroid hormones, i.e., the testes and ovaries, are all competent and ready to their respective tropic hormones, but none is released in sufficient quantity to trigger puberty until some event occurs in the hypothalamic-pituitary system.

II. THE CURRENT CONCEPT

In this respect, a key concept in current thinking is one that we call the hypothesis of differential sensitivity. According to this hypothesis, the major event that precipitates puberty is an increase in the threshold (or set point) of the inhibitory feedback receptors which respond to changes in circulating steroid levels by inducing reciprocal changes in gonadotropin output. The onset of puberty is associated with an increase in the set point of the hypothalamic negative feedback receptors. Consequently, the low concentrations of sex steroids can no longer effectively suppress the gonadotropin secretion. This results in increased release of hypothalamic LH-RH, and therefore also the gonadotropins and increased stimulation of gonadal steroid production. Finally, the system attains an adult set point of gonadostat.

This hypothesis has the virtue of explaining a variety of phenomena without resorting to unattractive concepts like the reversal of hypothalamic function from inhibitory to facilatory at puberty. Moreover, that changes in the sensitivity of the hypothalamic-pituitary axis may be a critical one in reproductive events, is indicated by various experiments, including several in our laboratory (Donovan and van der Werff ten Bosch, 1965; Ramirez, 1972; Gupta, 1974; Gupta et al., 1975). Of many aspects we have selected only two of the control mechanisms which are related to the onset of puberty:

(1) The presence of a negative feedback control of the secretion of gonadotropins before puberty, which exhibits a change in sensitivity (or set point) with the onset of puberty.

(2) The change in pituitary sensitivity to LH-RH at puberty and the qualitative change in the LH response.

From experimental evidence, including the data from our laboratory, it seems that puberty really does not represent the sudden activation of the previously dormant system, but an increasing function of a system that has been continuously active from the very early state of development.

III. EXPERIMENTAL EVIDENCE

The following figures throw some light on the question of increment in the threshold of the inhibitory feedback mechanism as the subject matures. These experiments were carried out in male laboratory rats. This stress on the male animal, however, does not signify any bias against the females as such. It was taken only because less information is available in the male animals.

Figure 2 shows the magnitude of the gonadotropin response after castration at various ages of rat. The minimum increase occurred with 15-day-old rats, and the maximum occurred at 58 days of age when spermatogenesis takes place. At 15 days of age, the LH level declined to the initial level when testosterone was applied 10 μg/100 g body wt. The FSH level also fell, but came down to the initial value only with a higher dose of testosterone.

At 28 days, testosterone at the level of 10 μg/100 g body wt was not successful to reduce both LH and FSH values to those of the initial stage. Testosterone, when given 25 μg/100 g, suppressed the levels of both gonadotropins further, nearly to the level of the initial stage. At age 58 days, the 10 μg dose was no longer successful in reducing the gonadotropin levels, indicating that the sensitivity had considerably changed with sexual maturation of the subjects.

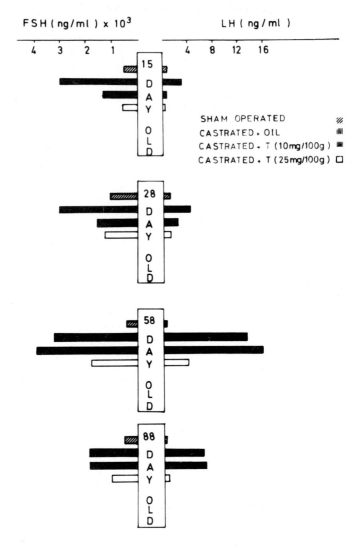

Fig. 2 The magnitude of the gonadotropin response after castration and followed by testosterone treatment at different stages of sexual maturation in male rat.

Figure 3 shows the different levels of sensitivity in the hypothalamic-pituitary axis during prepubertal, late pubertal, and adult stages, in the male rats when the endogenous source of sex steroids is removed. The left side of the figure shows the positive change in the plasma levels of LH (+ ΔLH) and negative change in testosterone level (− ΔT). The right side of the figure shows the amount of increment in plasma LH when related to the decrement of 1 ng of testosterone per 100 ml (ΔLH/ΔT). At the age of 20 days, this value was 16.6 ng/ml, but, with the onset of puberty, it decreased to 1.23 ng/ml and remained fairly constant (0.96 − 1.10 ng/ml) throughout later stages of development.

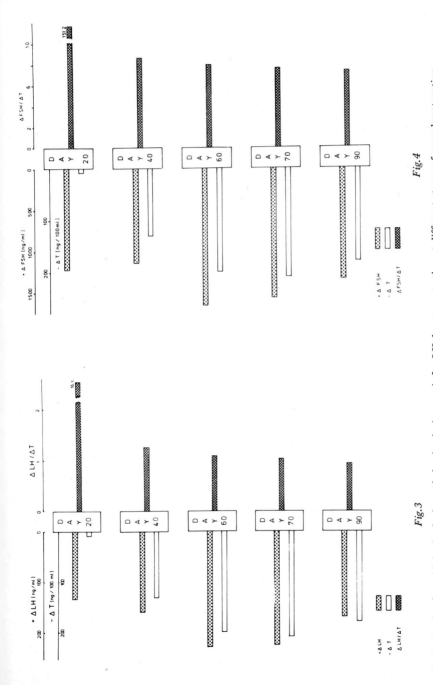

Fig. 3 Levels of sensitivity in the hypothalamic-pituitary axis for LH due to castration at different stages of sexual maturation.

Fig. 4 Levels of sensitivity in the hypothalamic-pituitary axis for FSH due to castration at different stages of sexual maturation.

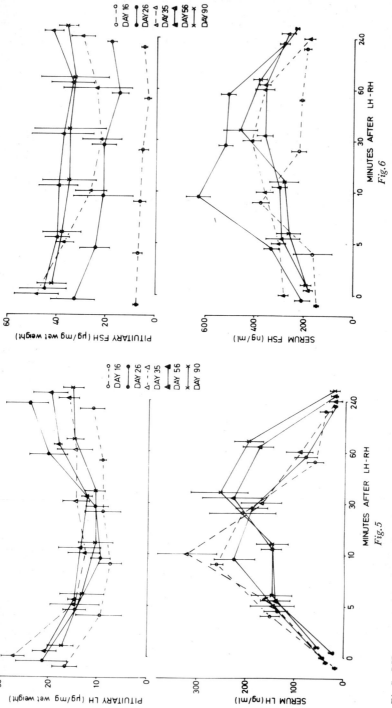

Fig. 5 LH concentrations due to the administration of synthetic LH-RH at different stages of sexual maturation in the male rat: pituitary (above); serum (below).

Fig. 6 FSH concentrations due to the administration of synthetic LH-RH at different stages of sexual maturation in the male rat: pituitary (above); serum (below).

Figure 4 illustrates the similar changes in the FSH profile with maturation due to total gonadectomy. At age 20, the increment in plasma FSH concentration (+ ΔFSH), due to fall of 1 ng/100 ml testosterone ($-\Delta$T), was 151.2 ng/ml. In the pubertal and postpubertal animals, this increment in FSH per 1 ng/100 ml decrement in T was much lower, remaining between 7.40 and 8.71 ng/ml.

IV. SYNTHESIS AND RELEASE OF GONADOTROPINS

Therefore, in the next step, we examined the influence of LH-RH on the pituitary release of gonadotropins and the age-dependence of the hypothalamic-pituitary-gonadal axis. There is not much information regarding the age-related influence of LH-RH, nor on the pituitary content of the gonadotropins under LH-RH in developing male rats. Measuring only serum concentration of a gonadotropin provides only a limited information in that such measurement reflects merely the release of the hormone. For the synthesis and release of the hormone under LH-RH, pituitary content, as well as the blood level, needs to be examined.

Figure 5 shows that serum LH levels (see bottom) in all age groups record about a 3-fold response within the first 5 min to the releasing hormone, which is statistically significant. There was, however, a qualitative difference in the time course required to reach the maximal response. The prepubertal and pubertal animals responded to the releasing hormone maximally after 10 min, while the older animals reached the peak level 30 min after administration. The decline in serum LH level after injection in the older groups was sluggish and, after 60 min, it was still high.

The releasing hormone resulted in a depletion of pituitary LH concentration in all age groups within 5 min (upper part of the picture). The 33-day-old group presented the highest initial concentration of pituitary LH and had relatively the maximal depletion, which was significantly different from other groups.

Figure 6 shows that the administration of LH-RH induced a significant increase in serum FSH levels in all age groups, but the magnitude of the response and time course varied among different groups. The maximal response was observed in the 16-day and 25-day-old groups 10 min after LH-RH administration. The peak concentrations of serum FSH in the pubertal and mature animals was, however, reached at a later stage of 30 min. Between the two immature groups of 16-day and 26-day-old animals, there was a considerable difference in the pattern of response. In the younger group, no significant response was observed after 5 min of LH-RH administration and, within 30 min, the peak level declined to the initial level.

The pituitary concentration of FSH in the 16-day-old group is not significantly depleted, whilst the 26-day-old group shows significant progressive depletion, reaching the lowest level at 30 min after the injection of releasing hormone. The depletion of pituitary FSH concentration in the older groups is sluggish and its difference does not reach statistical significance at various time-points.

One of the striking findings in this study is the large increase in serum gonadotropin accompanied by pronounced depletion of pituitary gonadotropins in the prepubertal animals, within a very short time after LH-RH injection. Thus, the present indication is that differences in neural functions and sensitivity set-points between immature and mature animals are responsible for the quantitative difference in the synthesis and release of the gonadotropins. These obser-

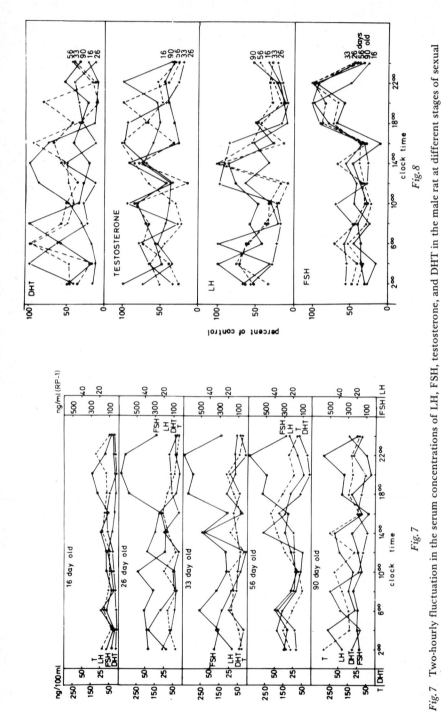

Fig. 7 Two-hourly fluctuation in the serum concentrations of LH, FSH, testosterone, and DHT in the male rat at different stages of sexual maturation.

Fig. 8 The data of Fig. 7 plotted as the percentage of maximum.

vations indicate, but do not prove, that with sexual maturation the threshold of sensitivity changes.

Studies in prepubertal and pubertal children also demonstrated a clear difference in pituitary responsiveness to synthetic LH-RH (Grumbach et al., 1974; Reiter et al., 1976; Attanasio et al., 1977). These studies suggest the existence of striking difference between the pituitary "reserve" of the FSH and LH in prepubertal and pubertal individuals, as well as a sex difference in readily releasable pituitary FSH.

V. CIRCADIAN RHYTHM IN HORMONE CONCENTRATIONS

During these studies and other manipulative investigations, we noticed a considerable variation in plasma hormone levels in the control animals when samples were taken at different time period of the day in the same age group of animals. This observation, together with the reports of episodic bursts of LH and testosterone (T) secretion in ram (Katangole et al., 1974) and fluctuations in plasma T levels in male rats (Bartke et al., 1973), prompted us to look into the 24-hour profile of plasma LH, FSH and T, and dihydrotestosterone (DHT) in the same samples, and see if there is any effect of age on the daily profile.

Figure 7 shows the plasma LH, FSH, T and DHT values in 16, 26, 33, 56 and 90-day-old male rats throughout the day when examined every 2 hours. The left side scale gives the measures of T and DHT and the right side only gives that for plasma FSH and LH.

In the 16-day-old animal, at 0600 h, there is a peak for all the hormones, except DHT. They do not show much fluctuation in the blood level, except there is a prominent rise in FSH in the evening and, at 2200 h, T also seems to rise. In the 26-day-old animals, except FSH, the other hormones do not seem to fluctuate a great deal. For FSH, the characteristic evening surge is again seen, but no prominent 14-hour peak is noticed in older animals. However, in the pubertal animals, i.e., the 33-day-old group, examinations of the patterns of LH discharge in the evening is observed in all age groups, whether prepubertal, pubertal, or adult, except that its magnitude is higher in the 26-day, 33-day, and 56-day-old animals than in the immature and mature groups.

Figure 8 shows the same results arranged on the basis of single hormones and calculated on the percent basis, taking maximum concentration as 100.

Comparatively large differences in the percent concentration between plasma samples collected at 2-hour intervals can easily be noticed. The percentage change in concentration is large and could be as much as 90% from one timepoint to the next time-point in one hormone level in a particular age group.

Although no consistent pattern of changes in plasma T and LH was observed by other investigators, we found that, at certain time-points, the hormone concentrations correlated with each other quite well. For example, the 1400 h examination reveals peaks, though minor in FSH, in all the hormones in nearly all age groups. For DHT, the 16-day-old age group has a single peak, and by 33-day the peaks are increased to 3 in number. The mature group again shows a single major peak.

The profile of LH is in sharp contrast to that of FSH, especially in the pattern of evening discharge. The immature group showed the maximum at 0600 h, which in the 26-day-old group shifted to 1200 h, and in the older animals shifted further to 1400 h.

It becomes apparent that the male rat, like ram, bull (Katangole et al., 1974) and man (Naftolin et al., 1972) secretes the pituitary gonadotropins and androgens episodically throughout the 24 hours. Here, the testosterone peaks often seem to be related to surges of LH, which was also observed in the bull and ram. The results presented here suggest that it could be the frequency, as well as the amplitude, of hormone discharge that alters during sexual maturation.

However, the meaning of the large fluctuation in basic levels of hormone is not clear, but should provide some kind of warning regarding the so-called "control" values. Evidence for this kind of episodic or pulsatile discharges of LH and FSH can also be found in human subjects (Grumbach et al., 1974). It is also possible that the episodic secretion of gonadotropin is a maturation phenomenon related to the changes in the CNS that effect pulsatile release of LH-RH.

VI. FACTORS CONTROLLING SEXUAL MATURATION

It can be seen, from the preceding discussion, that a complex interplay of various factors influences the mechanism which regulates the secretion of the pituitary gonadotropins. This complex interplay in a three-tiered hierarchy of hormonal control slowly reveals the mysterious mechanisms which are at work to uphold the temporal organization in the sexual functions.

As already discussed, specifically hypothalamic receptors mediating the inhibiting effects of gonadal steroids on hypophyseal functioning are generally thought to undergo an increase in the threshold as puberty approaches. These less-sensitive inhibitory receptors permit an increment in the gonadotropin discharge and, in turn, an increase in gonadal activity. The role of differential sensitivity in relation to the initiation of puberty should be evaluated separately in the two sexes.

A. The Male Development

First, let us take up the male sex. We have already seen that the prepubertal male, in comparison with the adult, appears to require significantly smaller doses of testosterone to inhibit the castration-induced increase of gonadotropins. However, contradictory data were recently presented by Odell's group (Swerdloff et al., 1972) who found no difference in the requirement of testosterone to suppress castration-induced high gonadotropin levels, whether in the immature or in the adult animal. This discrepancy seems to lie in the fact that testosterone replacement was started 5 days after castration in this experiment and it is possible that this delay perhaps produces an apparent change in the hypothalamic-pituitary set point. Smith and Davidson (1967), by implanting testosterone in the median eminence, also demonstrated the heightened sensitivity of the immature animal.

In the male, testosterone may not be the only testicular substance which is involved in the regulation of gonadotropin secretion. It is possible that another unidentified substance from the germinal epithelium, arbitrarily termed as *inhibin*, is also responsible for the regulation of the gonadotropins, especially FSH. This hypothesis draws its support from two observations: *(a)* the decline of plasma FSH, seen after 35 days of age, is related to the appearance of mature sperm in the seminiferous tubules; *(b)* when the spermatogenic tissue of the immature male is damaged through cryptorchidism, FSH levels remain at a significantly higher level.

But it is not only the hypothalamic-pituitary unit which shows age-related changes in the sensitivity threshold, but the testes also show changes in sensitivity with maturation. When intact males of various age groups are given LH as ng/100 g body wt, the rise in plasma testosterone is greater, the older the animal (Odell and Swerdloff, 1974). The role of FSH is complementary to LH and time-related. The older the animal, the more is the exposure to FSH and the more are the testes sensitive to LH. According to Odell et al. (1973), the increasing end-organ sensitivity to LH is a result of FSH-preconditioning, which in turn is perhaps a major determinant of puberty in the male.

B. The Female Development

Females also show age-related changes in sensitivity to the negative feedback effect of the estrogens. The studies on the feedback sensitivity in the female rat demonstrated that, before 15 days of age, administered estradiol failed to reduce castration-initiated levels of gonadotropins in plasma. Generally speaking, these data support the hypothesis that, underlying the onset of puberty, is a change in feedback interaction between the ovary and hypothalamus. The positive feedback to estradiol is present from day 21 and becomes gradually more sensitive as the animal matures (McCann and Ojeda, 1977). However, puberty does not take place, although ovulation can be induced prematurely by exogenous means, until finally a group of follicles matures, estrogen secretion is enhanced, and the preovulatory discharge of gonadotropins occurs. But, why this development takes place at a particular time, is still far from clear. There may be a subtle decrease in negative feedback and also subtle changes in positive feedback.

VII. THE "DIFFERENTIAL SENSITIVITY" HYPOTHESIS AND NEW PROBLEMS

The hypothesis of differential sensitivity is incomplete, in that it fails to explain why changes in feedback sensitivity occur at all. This hypothesis also does not account for the various forms of periodic female cycles, or for the acyclic pattern of males. The suggestion that the last step in approach to puberty may be the fine adjustment of receptors is also not without question. Also of experimental interest is the fact that 5α-androstane-3α-17β-diol is biotransformed to 3β epimer by an enzyme system that is FSH-induced (Eckstein et al., 1973), which can be related to the high levels of FSH seen early in development.

Work on pituitary biotransformation of testosterone to its 5α-reduced metabolites in rats, which has been presumed to have some bearing on the question of sexual maturation, has demonstrated that, in fact, this biotransformation is not age-dependent.

To examine this last point, we investigated the acute effect of orchidectomy on the pituitary biotransformation of testosterone to its 5α-metabolites, relating this effect to the circulating levels of LH and FSH in the same animals. We also examined whether sexual maturation in the male animal has any direct effect on this biotransformation.

Figure 9 shows that, in the intact animal, DHT constituted the main mass of metabolites. Androstanediol was also formed in large quantities. The formation of androstenedione was low. No significant age difference in biotransformation could be detected in any of the steroids considered. Orchidectomy caused an increment in the total amount of 5α-metabolites. This increase in DHT and andro-

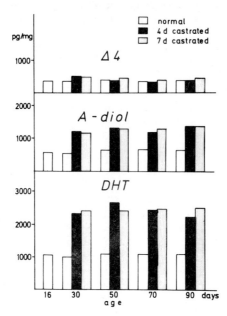

Fig. 9 Pituitary biotransformation of testosterone to 5α-metabolites and androstenedione in intact and castrated animals (for 4 days and 7 days).

stanediol could be seen in all age groups studied 4 days after castration. No further change was noted when the postsurgery period was extended to 7 days. The formation of androstenedione remained nearly constant following orchidectomy.

These results on the increase of selective steroid metabolizing enaymes, namely, 5α-reductase and not 17β-hydroxysteroid dehydrogenase, after orchidectomy, postulate contrasting views to the general belief of postcastration induction of overall growth in pituitary cells. It is more likely that LH-RH, which is increased due to castration, perhaps specifically and favourably influences the 5α-reductase system. The striking relationship between 5α-reduced metabolites of testosterone and the circulating levels of LH and FSH adds further support to this belief.

VIII. QUALITATIVE ASPECTS OF FEEDBACK CONTROL

Finally, we present another line of evidence which says that, for at least one pituitary hormone, *feedback control* is not only *quantitative* but also *qualitative*. By this, it has been meant that the target gland (in this case the gonad) in its "talk-back" to the pituitary, or hypothalamic-pituitary system, can dictate, not only the amount of tropic hormone to be secreted, but also the *kind* of tropic hormone as well.

The recent investigations of Peckham *et al.* (1973) and Bogdanove *et al.* (1975) have introduced a new dimension to the seemingly eternal question of peptide hormone economics. They have shown that, when animals are castrated, the quality of FSH alters, depending on the sex of the animal. In these experiments, it has been shown that, in the absence of gonadal steroids, the pituitary (male or female) synthesizes, stores and releases a type of FSH, that is called *neuter-FSH*. In response to androgen replacement, both stored and circulating forms of FSH change. The pituitary begins to secrete a different hormone, "andro-FSH", for which this index of discrimination, apparent molecular size and capacity to survive in the circulation, all are greater than they are for *"neuter-FSH"*. Ovarian steroids appear to have an opposite effect, producing FSH which is relatively short-lived. These are called *"gyno-FSH"*. It is not known whether this kind of pleomorphism for FSH also exists in human subjects under androgen or estrogen treatment.

IX. CONCLUDING REMARKS

During the course of experimentation one becomes slowly conscious of the extremely dynamic interplay of the total axis. Under slight pertubration of one component, the others may profoundly be affected, which may alter the function of the whole system. Therefore, the concept of differential sensitivity, which is essentially a qualitative one, should take note of other manipulative factors, such as introduction of large amounts of electricity, steel tubes, crystalline steroids or rubber pouch fixed on the shoulder for intermittent blood collection. All these can easily upset the delicate equilibrium of this system and alter the timing of puberty and influence on the pubertal growth. In the search for a definitive explanation of the events during puberty, we have to be careful not to upset the highly sensitive orchestration that still eludes us.

REFERENCES

Attanasio, A., Attanasio, A., Rohr, M., Eichner, M., Rager, R., Leitz, R. and Gupta, D. (1977). *In* "Current Topics in Hypothalamic Hormones" (D. Gupta and W. Voelter, eds). Verlag Chemie, Weinheim and New York.
Bartke, A., Steele, R.E., Musto, N. and Caldwell, B.W. (1973). *Endocrinology* **92**, 123.
Bogdanove, E.M., Nolin, J.M. and Campbell, G.T. (1975). *Rec. Prog. Horm. Res.* **31**, 567.
Donovan, B.T. and van der Werff ten Bosch, J.J. (1965). "Physiology of Puberty". Williams and Wilkins, Baltimore.
Eckstein, B., Golan, R. and Sheni, J. (1973). *Endocrinology* **92**, 941.
Grumbach, M.M., Roth, J.C., Kaplan, S.L. and Kelch, R.P. (1974). *In* "Control of the Onset of Puberty" (M.M. Grumbach, G.D. Grave and F.E. Mayer, eds). Wiley and Sons, Inc., New York.
Gupta, D. (1974). *In* "Recent Progress in Reproductive Endocrinology" (P.G. Crosignani and V.H.T. James, eds). Academic Press, London.
Gupta, D., Rager, K., Zarzycki, J. and Eichner, M. (1975). *J. Endocrinol.* **66**, 183.
Katongole, C.B., Naftolin, F. and Short, R.V. (1974). *J. Endocrinol.* **60**, 101.
McCann, S.M. and Ojeda, S.R. (1977). *In* "Current Topics in Hypothalamic Hormones" (D. Gupta and W. Voelter, eds). Verlag Chemie, Weinheim and New York.
Naftolin, F., Brown-Grant, K. and Corker, C.S. (1972). *J. Endocrinol.* **53**, 17.
Odell, W.D. and Swerdloff, R.S. (1974). *In* "Control of the Onset of Puberty" (M.M. Grumbach, G.D. Grave and F.E. Mayer, eds). Wiley and Sons, Inc., New York.
Odell, W.D., Swerdloff, R.S., Jacobs, H.S. and Hescox, M.A. (1973). *Endocrinology* **92**, 160.
Peckham, W.D., Yamaji, T., Dierschke, D.J. and Knobil, E. (1973). *Endocrinology* **92**, 1660.
Ramirez, V.D. (1972). *In* "The Use of Non-Human Primates in Research on Human Reproduction" (E. Diczfalusy and E. Stanley, eds). WHO Symposium, Sukhumi.
Reiter, E.O., Root, A.W. and Duckett, G.E. (1976). *J. clin. Endocr.* **43**, 400.
Smith, E.R. and Davidson, J.M. (1967). *Am. J. Physiol.* **212**, 1385.
Swerdloff, R.S., Jacobs, H.S. and Odell, W.D. (1972). *In* "Gonadotropins" (B.B. Saxena, C.G. Beling and H.M. Gandy, eds). Wiley Interscience, New York.

PUBERTY IN PATIENTS WITH GROWTH HORMONE DEFICIENCY

A. Pertzelan, R. Kauli, Z. Zadik, I. Blum and Z. Laron*

Institute of Paediatric and Adolescent Endocrinology
Beilinson Medical Centre, Petah Tikva and Sackler School of Medicine
Tel Aviv University, Israel

I. BASAL PLASMA LH AND FSH AND THEIR RESPONSE TO I.V. LH-RH IN PATIENTS WITH ISOLATED GROWTH HORMONE DEFICIENCY

It is generally accepted that children with isolated growth-hormone deficiency (IGHD) characterized by short stature and retarded bone age (BA), have delayed sexual maturation, as seen by the delayed appearance of secondary sex characteristics and pubertal growth spurt (Tanner and Whitehouse, 1975).

It has been demonstrated that when boys with IGHD are treated with exogenous testosterone alone, the effect of the testosterone on the acceleration of growth and induction of secondary sex characteristics is less than that observed in normal boys (Zachmann and Prader, 1970; Aynsley-Green *et al.*, 1976; Tanner *et al.*, 1976).

When patients with the same syndrome are treated with exogenous hGH alone, puberty begins at about the same stage of skeletal maturity as in normal children. Tanner and Whitehouse (1975) calculated the mean ages at which puberty is initiated in these patients: in boys, mean bone age (BA) of 11.8 years and in girls 10.9 years. However the chronological age (CA) of their patients was: in boys a mean CA 15 years and in girls 13.7 years.

For the initiation of puberty a specific degree of maturity of the hypothalamus and pituitary is required (Grumbach *et al.*, 1974). The hypothalamic-releasing hormones induce the release of the pituitary gonadotrophins which in turn stimulate the secretion of gonadal sex hormones. The gonadal sex hormones stimulate the development of secondary sex characteristics and act on bone growth and maturation.

---

* Established Investigator of the Chief Scientist's Bureau, Ministry of Health.

Table I. Isolated growth hormone deficiency — prepubertal (P_1) males.

No.	Name	CA	BA	LH		FSH	
				Basal	Peak	Basal	Peak
		Yrs: Mos			Gn Rating		
1*	M.G.	9:8	6:0	2	2	1	2
2	A.T.	11:0	7:0	3	2	1	2
3	E.I.	11:0	9:6	4	2	1	1
4	M.G.	12:0	9:6	1	4	1	1
5	N.E.	12:8	8:0	2	2	2	2
6*	S.D.	13:7	6:0	2	1	1	1

* Before treatment with hGH. CA - chronological age. BA - bone age. Yrs - years. Mos - Months. Gn - gonadotrophin.

Table II. Isolated growth hormone deficiency - prepubertal (P_1) females.

No.	Name	CA	BA	LH		FSH	
				Basal	Peak	Basal	Peak
		Yrs: Mos			Gn Rating		
1	L.I.	10:2	8:0	1	2	2	2
2	E.A.	10:11	7:0	2	2	2	2
3	H.A.	11:1	9:0	1	1	2	2
4*	H.M.	11:9	7:0	1	1	2	2
5	E.A.	11:11	9:0	1	2	5	2

* Before treatment with hGH. CA - chronological age. BA - bone age. Yrs - years. Mos - months. Gn - gonadotrophin.

The literature does not state whether the defect in the activation of the Hypothalamic-Pituitary-Gonadal axis in patients with IGHD lines in the hypothalamic-pituitary centres or in the production of the sex steroids. Recently, it has been possible to clarify this question by determining the degree of response of plasma LH and FSH to LH-releasing-hormone (LH-RH).

Subjects and Methods

Twelve patients (8 males and 4 females) with IGHD were investigated. They were classified according to the pubertal rating of Tanner (1962), with respect to pubic hair and genital development in boys, and pubic hair and breast development in girls. P_1 represents the prepubertal stage and P_5 full maturity. Some of the patients were tested on several occasions, and at different pubertal stages.

The plasma LH and FSH was measured by radioimmunoassay before and in response to i.v. administration of LH-RH (50 $\mu g/m^2$) in one bolus injection. The results were rated from P_1 (prepubertal) to P_5 (full mature) according to the standards obtained by Dickerman et al. (1976) for normal boys and girls.

Table III. Isolated growth hormone deficiency — beginning of puberty (P_2).

No.	Name	CA	BA	LH Basal	LH Peak	FSH Basal	FSH Peak
		Yrs: Mos			Gn Rating		
Males							
1	V.S.	11:11	8:0	1	2	2	2
2	E.I.	12:10	11:6	4	2	2	2
3	N.E.	13:5	11:6	2	2	3	2
4*	B.S.	14:5	11.0	4	2	2	2
5	M.T.	15:0	10:0	1	3	1	2
Female							
1	E.A.	13:6	11:0	5	5	2	4

* Before treatment with hGH. CA - chronological age. BA - bone age. Yrs - years. Mos - months. Gn - gonadotrophin.

Results

The results are summarized in Tables I, II and III.

Table I shows the degree of LH and FSH response to LH-RH in 6 prepubertal (stage P_1) boys with IGHD. The chronological ages varied from $9^{8}/12$ yrs to $13^{7}/12$ yrs; the BA from 6 to 9½ yrs. All the BA estimations were made according to the Atlas of Greulich and Pyle (1959). All the patients except for patient no.1 were at the CA at which normal children show the first signs of puberty, that is P_2. Patients no.1 and no.6 were not treated with growth-hormone (hGH) before the test, the others received hGH therapy Crescormon®, Kabi (3 mg 3 times a week) for periods varying from 6 months to 4½ years before the test. It is evident in Patients no.1 to 5 that the LH response was higher than should be expected for their clinical pubertal stage and bone maturity, in these patients the LH response corresponding to P_2 or more. The FSH response in 3 patients (no.1, 2,5) corresponded to P_2, and 3 patients had a prepubertal response. It is known that the FSH response in the male is less than that of LH at all pubertal stages (Dickerman et al., 1976). Patient no.6, who showed a prepubertal type response of LH and FSH, had the greatest retardation in BA. The presence of a rise of LH and FSH after LH-RH administration excludes total gonadotrophin deficiency. In this patient, however, partial deficiency cannot be excluded. Patient no.3 was retested at a later pubertal stage (Patient no.2, Table III).

Table II shows the degree of LH and FSH response in 4 clinically prepubertal (P_1) girls with IGHD, one girl being tested twice (no.2,5). The CA was from $10^{2}/12$ to $11^{11}/12$ yrs, the BA from 7 to 9 yrs. All showed a rise of FSH corresponding to pubertal rating 2, a finding which is compatible with the response seen at the beginning of puberty. The FSH response in girls at this stage, P_2, is higher than in fully mature girls in the proliferative phase of the cycle (Dickerman et al., 1976). The LH response in 3 of the tests (no.1,2,5) was also that seen for P_2, whereas in the remaining two, only prepubertal levels were reached. Patient

E.A. was tested at CA $10^{11}/12$ yrs (no.2) and one year later (no.5), as well as at a later stage of puberty. This is presented in Table III.

Table III shows the LH and FSH response to LH-RH in 5 boys and 1 girl with IGHD who displayed initial signs of puberty (Tanner's rating P_2). The CA of the boys was $11\ {}^{11}/12$ yrs to 15 yrs, that of the girl was 13½ yrs. The BA of the boys ranged from 8 to 11½ yrs, the girl's BA was 11½ yrs. In the boys the degree of response of LH and FSH is comparable to their clinical stage of puberty and to their BA. Patient no.5, aged 15 yrs, shows an LH response compatible with advanced puberty P_3, despite the fact that his BA is retarded by 5 yrs. Patient no.2 who was tested 2 years earlier at a clinically prepubertal stage (Table I, no.3) and who had then shown an LH response normally observed at a more advanced stage of puberty, at present revealed an LH and FSH response corresponding to his pubertal and skeletal maturation stage. The girl E.A. was at CA 13½ yrs which is normally considered as mid-puberty (P_{3-4}). Her clinical pubertal signs were that of stage P_2. She showed an excessive LH and FSH response compatible with her CA.

The few patients with IGHD tested at more advanced pubertal stages showed the expected response of LH and FSH for their clinical stage of puberty.

Conclusions

The main finding of our present study is that the gonadotrophins response to LH-RH in patients with IGHD in P_1 and P_2 is in advance of their clinical developmental stage and skeletal maturation, but corresponds to the CA. This phenomenon could be explained by the hypothesis that in the absence of growth hormone (GH) alone the hypothalamo-pituitary clock, regulating sexual development, matures within normal time limits, but lack of GH has an adverse effect on the sensitivity of the gonads with respect to hormone synthesis and secretion, as well as on the peripheral action of these hormones (Zachmann and Prader, 1970). This could explain the small testes and penis in young boys with IGHD, as previously described by us (Laron and Sarel, 1970), and the slow sexual and skeletal maturation. The presence of a normally-maturing hypothalamus and pituitary, enables the catch-up in sexual development once exogenous hGH therapy is instituted (Tanner and Whitehouse, 1975).

II. GROWTH RESPONSE TO EXOGENOUS hGH THERAPY IN PATIENTS WITH GH DEFICIENCY AGED 15 YRS OR MORE

Among other aspects of the relationship between GH and sexual development is the growth response to exogenous hGH therapy in patients at the pubertal age (CA 15 yrs or more). It is assumed that the normal growth spurt seen at puberty is the result of an additive effect of GH and androgens (Zachmann and Prader, 1970; Blizzard et al., 1974; Aynsley-Green et al., 1976; Tanner et al., 1976).

Subjects and Methods

We compared the effect of exogenous hGH administration (Crescormon® Kabi, 3 mg 3 times a week) in 3 groups of patients with GH deficiency.

Group 1: Five patients (2 males and 3 females) with isolated growth hormone deficiency (IGHD).

Table IV. Group 1. Isolated growth hormone deficiency: effect of hGH treatment in patients over 15 yrs of age.

No.	Sex	CA	BA	Height	Pre hGH	Growth On hGH	
		Yrs: Mos		cm	cm/y	cm/mos	cm/y
1	M	14:9	15:0	101:5	1.3	19.0/28	8.1
2	M	19:1	15:3	121:8	1.0	4.3/6	8.6
3*	F	15.2	13:0	141.6	1.4	8.3/23	4.3
4*	F	15:10	13:6	138.4	1.5	4.5/11	4.9
5	F	17:10	15:0	136.2	1.7	0.3/3	1.2

* Previously treated with hGH. CA - chronological age. BA - bone age. Yrs - years. Mos - months. y - year. cm - centimetre.

Table V. Group 2. Idiopathic multiple pituitary hormone deficiencies: effect of hGH treatment in patients over 15 yrs of age.

No.	Sex	CA	BA	Height	Pre hGH	Growth On hGH	
		Yrs: Mos		cm	cm/y	cm/mos	cm/y
1*	M	15:1	9:0	133.5	2.5	19.1/41	5.6
2*	M	16:1	8:0	121.7	3.1	14.6/25	7.0
3	M	16:6	13:3	137.8	3.4	9.5/19	6.0
4	M	17:5	9:6	136.5	3.7	19.4/30	7.7
5*	F	15:10	12:0	148.2	1.2	10.1/19	6.3
6*	F	17:1	10:0	123.6	1.7	5.6/7	9.6
7	F	18:6	10:0	122.5	0.9	9.2/16	6.9
8	F	19:6	11:0	140.2	6.4	5.9/8	8.8

* Previously treated with hGH. CA - chronological age. BA - bone age. Yrs - years. Mos - months. y - year. cm - centimetre.

Group 2: Eight patients (4 males and 4 females) with multiple pituitary hormone deficiencies (MPHD), the nonorganic, idiopathic type.

Group 3: Four patients (3 males and 1 female) with MPHD due to craniopharyngioma.

All the patients in group 2 and 3 lack GH and gonadotrophins in combination with other pituitary hormone deficiencies.

Results.

The effect of hGH therapy in the three groups of patients is summarized in Tables IV, V and VI. The actual growth "on hGH" administration is expressed as cm/mos (centimetres of linear growth per period of treatment) and calculated to cm/y (linear growth per year). These are compared to "Pre hGH" growth,

Table VI. Multiple pituitary hormone deficiencies (craniopharyngioma): effect of hGH treatment in patients over 15 yrs of age.

No.	Sex	CA	BA	Height	Growth		
					Pre hGH	On hGH	
		Yrs: Mos		cm	cm/y	cm/mos	cm/y
1*	M	17:2	13:0	164.4	2.0	3.6/9	4.8
2*	M	16:6	13:0	154.4	5.2	4.8/9	6.4
3	M	35:1	15:6	158.8	3.5	1.1/4	3.3
4	F	16:6	13:0	153.2	2.8	1.9/7	3.2

* Previously treated with hGH. CA - chronological age. BA - bone age. Yrs - years. Mos - months, y - year. cm - centimetre.

cm/y, where the pre hGH treatment observation period lasted 6 to 12 month.

Table IV presents the response in linear growth to administration of hGH in 5 patients (2 males and 3 females) with IGHD aged 15 yrs or more. The BA of the 2 males was 15 and 15 $^{3}/12$ yrs, of 2 females 13, 13½ yrs and one female with BA 15 yrs. The BA of the first patient was advanced for his CA because of previous treatment with thyroid preparations and androgens, prior to the availability of hGH. Patients nos 3 and 4 had been treated with hGH for three years, prior to this study. Patients 1–4 were at the clinical pubertal stage P_3, patient no.5 had her menarche 1½ years before hGH was instituted and she was a sexually fully mature being at P_5. The growth velocity of all the patients without hGH therapy was less than 2 cm/year. Upon hGH therapy the growth rate increased strikingly in the 2 boys despite a BA of 15 yrs. Patient no.1 grew 19 cm over a period of 28 months, the greatest response occurring in the first year, 11.5 cm. With progressive closure of the epiphyses the growth rate declined. When comparing patient no.2, a 19 yr-old male with a BA of 15 $^{3}/12$ yrs, and pubertal stage 3 with patient no.5, a girl aged 17 $^{10}/12$ yrs with the same BA as the boy, but at pubertal stage 5, it is seen that the boy had a good acceleration of growth and achieved a growth spurt corresponding to his pubertal stage, while the fully mature girl showed almost no growth. In other words the acceleration in growth corresponded to the pubertal stage rather than to the CA or BA.

The problem is somewhat different in patients with multiple pituitary hormone deficiencies (MPHD), namely deficiency of GH, gonadotrophins, thyroid activity and sometimes ACTH.

Table V illustrates the effect of hGH therapy on growth velocity in 8 patients (4 males and 4 females) with MPHD treated at the age of over 15 yrs. Not one of them had pubertal signs. Two males (nos 1 and 2) and 2 females (nos 5 and 6) had been previously treated with hGH. All were treated with 1-thyroxin 0.1 mg/day. Without hGH the 4 males had a yearly growth rate of 2.5–3.7 cm, and upon hGH administration the growth rate was doubled or more. Three of the four females had a very low growth rate, below 2 cm/y, hGH administration increased their growth velocity to 6.3–9.6 cm/y. The oldest female, patient no. 8, who had started thyroxin therapy shortly before, had a growth rate of 6.4 cm/y which increased to 8.8 cm/y upon addition of exogenous hGH. It is again remarkable that despite their advanced CA these patients grew at a rapid rate

Table VII. Multiple pituitary hormone deficiencies: effect of M—S, hGH, and M—S + hGH in patients over 15 yrs of age.

No.	Growth Rate cm/y			
	Basal*	M—S	hGH	FM—S 3 HGH
1	2.5	6.0	7.2	11.6
2	2.5	2.8	4.6	2.7
3	3.7	—	3.9	10.0
4	1.8	—	10.8	6.2
5	3.6	4.8	6.6	16.2
6	2.8	6.6	2.8	9.2
Mean	2.8	5.0	6.0	9.3

* Thyroxin replacement therapy. M—S — Methandrostenolone.

upon institution of hGH therapy. In 5 of the patients (nos 1,2,4,6 and 7) the BA was prepubertal. As has been shown in the patients with IGHD, the response of growth cannot be entirely due to the lag in BA maturation.

Table VI shows 4 patients with MPHD due to craniopharyngioma. Only patient no.3 has not been operated on, and was referred and diagnosed very late. All received replacement therapy with 1-thyroxin 0.1 mg/day. Patient no.3 also received intranasal DDAVP (Minirin® Ferring). None had signs of puberty, but their BA was 13 yrs or more. Upon hGH therapy there was little, if any, acceleration of growth rate.

Conclusions

When comparing the effect of hGH on linear growth velocity in the three groups of patients treated at 15 years or beyond, it is striking that in the first group of patients (Table IV), the boys with IGHD who had advanced pubertal signs reacted to the first course of hGH therapy with a growth velocity typical of puberty. The girls in this group had passed their initial accelerated velocity during previous courses of hGH therapy. This is clearly evident by their taller stature, 141.6, 138.4 cm. In the patients with idiopathic MPHD (Table V) who had no sexual development and their BA corresponding to stage P_1 or P_2, the growth velocity was accelerated by hGH. The patients with MPHD due to craniopharyngioma (Table VI) were of the same age and also prepubertal as were the patients of Group 2, but the growth velocity was only little affected by hGH. In addition to having a different etiology these patients are taller than the patients of Group 2 and their BA is more advanced. However, the BA is less advanced than in the patients with IGHD who did show pubertal signs.

In order to evaluate further the necessity of concomitant sex hormones and hGH action upon the pubertal spurt in patients with MPHD, we administered Methandrostenolone alone (0.02 — 0.05 mg/kg/day) and in combination with hGH to the male patients who had MPHD (Pertzelan et al., 1977).

Table VII summarizes the results of this trial. It is seen that the combined administration of hGH and Methandrostenolone induced a growth velocity greater than with each hormone alone in 4 out of 6 patients.

The proper treatment of patients with hypothalamic pituitary hormone deficiencies with the aim of achieving normal height and puberty is a great challenge to the Paediatric Endocrinologist. Early diagnosis of the hormonal insufficiencies would enable better planning of substitution therapy schemes. The ideal response to such therapy is to delay the epiphyseal closure until normal stature is reached by mimicking the physiological amounts of hormone necessary for the pubertal growth spurt (Aynsley-Green et al., 1976).

REFERENCES

Aynsley-Green, A., Zachmann, M. and Prader, A. (1976). *J. Pediat.* **89**, 992-999.
Blizzard, R.M., Thompson, R.G., Baghdassarian, A., Kowarski, A., Migeon, C.J. and Rodriguez, A. (1974). *In* "The Control of the Onset of Puberty" (M.M. Grumbach, G.D. Grave and F.E. Mayer, eds) pp.342-359. John Wiley & Sons, New York.
Dickerman, Z., Prager-Lewin, R. and Laron, Z. (1976). *Am. J. Dis. Child.* **130**, 634-638.
Greulich, W.W. and Pyle, S.I. (1959). "Radiographic Atlas of Skeletal Development of the Hand and Wrist", 2nd Ed. Stanford University Press, Stanford, California.
Grumbach, M.M., Roth, J.C., Kaplan, S.L. and Kelch, R.P. (1974). *In* "The Control of the Onset of Puberty" (M.M. Grumbach, G.D. Grave and F.E. Mayer, eds) pp.115-166. John Wiley & Sons, New York.
Laron, Z. and Sarel, R. (1970). *Acta andocr. (Copenh.)* **63**, 625-633.
Pertzelan, A., Blum, I., Grunebaum, M. and Laron, Z. (1977). *Clin. Endocrinol.* **6**, 271-276.
Tanner, J.M. (1962). *In* "Growth at Adolescence", pp.28-39. Blackwell Scientific Publications, Oxford.
Tanner, J.M. and Whitehouse, R.H. (1975). *J. clin. Endocr. Metab.* **41**, 788-790.
Tanner, J.M., Whitehouse, R.H., Hughes, P.C.R. and Carter, B.S. (1976). *J. Pediat.* **89**, 1000-1008.
Zachmann, M. and Prader, A. (1970). *J. clin. Endocr. Metab.* **30**, 85-95.

3. HUMAN GROWTH STANDARDS AND AUXOLOGIC VARIANCE

HUMAN GROWTH STANDARDS: CONSTRUCTION AND USE

J.M. Tanner

*Department of Growth and Development, Institute of Child Health
University of London, 30 Guilford Street, London WC1N 1EH, England*

SUMMARY

Growth standards are used as a screening device applied to populations and for assessing the effects of medical or social intervention on the health of particular children. The use of large-scale growth data for comparing one population with another, or with itself at a later time, does not involve the use of standards and should be carefully distinguished from the use of data to create standards. Standards may be either cross-sectional or longitudinal, but the latter are the appropriate ones in all situations except the initial screening. Their construction and use are discussed and the 1976 revision of the longitudinal British standard for height and height velocity are described. Standards are also presented for sitting height and leg length and their comparison, and for height when height of parents is allowed for. The use of such standards in differential diagnosis of clinical cases is illustrated.

I. INTRODUCTION

Auxology is as essential a tool as radiology in the investigation of children's health and disease. First, growth standards are the most powerful screening device we possess for investigating groups of not overtly ill children to see which individuals might benefit from special medical, educational, or social care. Second, growth standards, and especially standards of velocity of growth, may be used to study the response to treatment of a child known to be ill, for example, with growth hormone deficiency, adrenal hyperplasia or kidney disease. The first of these two uses may be called the clinical community medicine or public health *screening use*; the second, the *paediatric use*.

There is a further use of growth data, which does *not* involve the construction of growth *standards*. Growth data can be used — and used most powerfully — as an index of the general health and nutrition of a population or subpopulation. Such data can be compared with simultaneously collected data from other populations or subpopulations or with data collected on the same population or subpopulation after the passage of years or after some remedial social or economic action. The population may be that of a whole country, of which the Cuban study provides the classical model in terms of sampling, training, attendance and anthropometric reliability (Jordan et al., 1975), or of a particular ethnic, social, or occupational group within the country. Thus, children from families whose father (or mother) have different occupations (i.e., from different 'social classes' in the British Registrar-General's terminology) may be compared, as Lindgren (1976) has recently done with such gratifying results in Swedish towns. In such comparisons, conventional statistical methods are used and, though variances and even specific centiles as well as means may be compared, there is no call for the construction of standards, which imply a norm. Some workers, particularly nutritionists, like to have what they refer to as a Reference Population with which data from a variety of surveillance and monitoring programmes may be compared (see Buzina et al., 1977). This is statistically quite unnecessary and tends to lead to intellectual confusion, especially when the reference population is called a Reference Standard, as sometimes occurs. If the intellectual muddle can be avoided, there is no objection to the Reference Population; but we have to be clear about the distinction between these *population* comparisons (followed by *population* treatment) and comparisons of *individual* children with a *standard* (followed by *individual* treatment).

To these clinical standards we now return.

II. STANDARDS FOR HEIGHT, HEIGHT VELOCITY, ETC.

I have already remarked that there are two uses of clinical standards; for screening and for following the effects of treatment. The two uses require somewhat different methodologies and interpretations, and this raises some complications in the construction of standards, which, though clear to Franz Boas, the great pioneer of growth studies, in 1892 (see Tanner, 1959), are still not universally understood by clinicians at the present time.

Studies of growth are classified as cross-sectional or longitudinal. In the former, each child is only measured once and all the children at age six, say, are different from those aged seven. In a longitudinal study, the same child is followed over a period of years; perhaps one only, or perhaps the full span of growth.

Screening surveys may be cross-sectional in nature (though if continuous as in school health work, they may, on the contrary, be longitudinal and strictly comparable to clinical longitudinal studies with treatment classified as "none"). Clinical work is longitudinal. Thus, on the first screening occasion, one needs to compare the height of the child with the heights of children derived from a cross-sectional study. But in clinical work (and one must remember that, after the initial screen, the 'discovered' child at once moves into this category) one needs to compare the height and height velocity of a child, repeatedly seen, with the growth of healthy children of comparable genetic and social background, who in principle should have been repeatedly seen too. In other words, longitudinal-type standards are required.

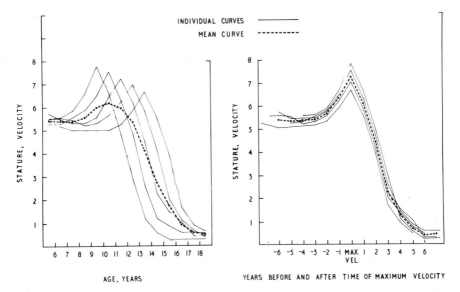

Fig.1 Cross-sectional treatment of velocity curves of five longitudinally followed boys. Left: five curves averaged, producing heavy 'cross-sectional' curve. Right: curves replotted in terms of a developmental age (from Tanner, 1962).

Until about age nine, the cross-sectional and longitudinal standards for height and weight are essentially the same and the complication to which I refer does not arise to a significant extent.

When puberty starts, however, cross-sectional and longitudinal standards part company. As is well known, the curve of height growth accelerates at puberty in the so-called adolescent growth spurt (see, for example, Tanner, 1962). Some children have their growth spurt early, some late, and this distorts cross-sectionally-derived standards in the way implicit in Fig.1. A longitudinally followed child should follow the sharper curve, not the flattened-out and lengthened ones. Thus, a truly average child will not follow the 50th centile at adolescence in the old standards (or indeed in new, albeit cross-sectional standards, such as those of the United States National Center for Health Statistics). He will first fall below and then catch-up the average (that is, if he matures at the average time).

The difficulty about constructing longitudinal standards is that in no longitudinal study are the numbers of children sufficient for the accurate location of the outside centiles, the 97th and 3rd, which are the centiles of most importance in clinical standards. This difficulty can be circumvented by using curves whose *shape* is based on longitudinal data, but whose *amplitude* (in the sense of distance between the centiles) is based on large-scale cross-sectional surveys. In 1966, my colleagues and I introduced a new type of standard based on this tactic (Tanner et al., 1966). When first we published these standards, we left the old cross-sectional centiles in and showed the new longitudinal centiles shaded in grey. But increasing familiarity with the use of the standards and with the misunderstandings engendered in the minds of auxologically-innocent paediatricians (including the author of a textbook on growth and development of children) induced us to reissue the standards in the format shown in Fig.2. Here, the centiles are

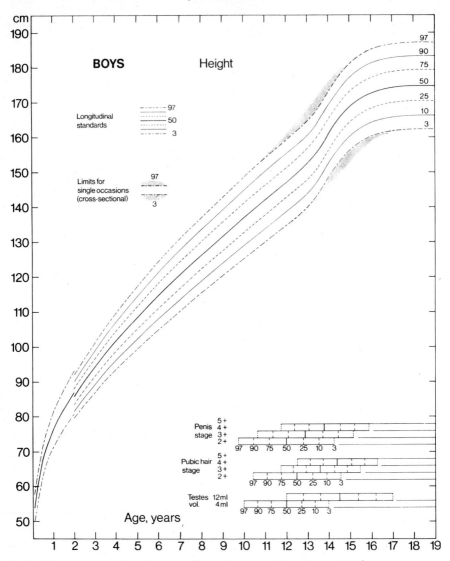

Fig.2 Height standards for clinical use (from Tanner and Whitehouse, 1976).

longitudinal; thus, the average boy who has his growth spurt at the average age does actually follow the 50th centile, in contrast to his situation on the cross-sectional standards. We added shaded areas above the 97th and below the 3rd centiles, however, to accommodate the whole 3rd–97th centile spectrum of cross-sectional data. Thus, in the initial screening, only children outside the shaded area should be considered "perhaps abnormal" (an arbitrary classification, of course, but the one conventionally used for probabilistic reasons). If a child lies within the shaded area, he is (conventionally) normal, but his growth curve on subsequent visits should head into the longitudinal range; if it fails to do so, he is (conventionally) abnormal.

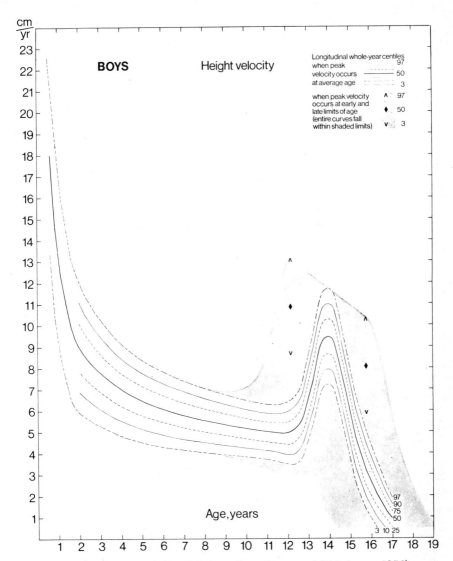

Fig.3 Height velocity standards for clinical use (from Tanner and Whitehouse, 1976).

Standards for the various stages of pubertal development, in terms of "permissible" ages at which testis size 12 ml, or pubic hair stage 3, etc., may be present, are given as well. Note that these standards require the physician to rate, for example, pubic hair development on the conventional 5-point scale and then look up the standard to see if the boy's *age* is within normal limits for this stage of development. Thus, the standards represent ages of *being-in-a-stage*. This is quite different from "standards" of age of *transition* from one stage to another. Such standards could be constructed (via probits or logits in cross-sectional surveys) but would be clinically useless, for the doctor cannot see the moment of first appearance of a stage, only that a child is actually in it. A little reflection

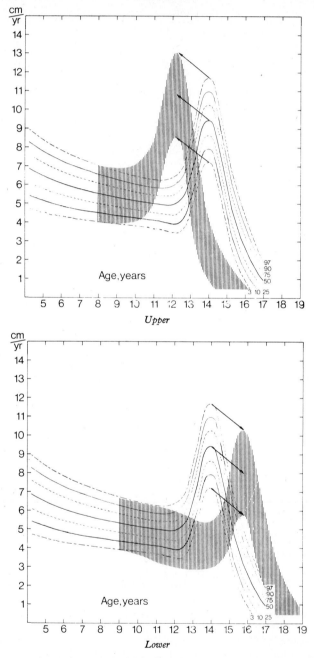

Fig. 4 Construction of longitudinal standards for height velocity. Centiles represent boys having peak velocity at average age of peak. Upper: hatched area represents centiles for boys with peak occurring 2 SD of age before mean age (approximately two years early). Lower: hatched area represents centiles for boys with peak occurring 2 SD after mean age (approximately two years late). (From Tanner and Whitehouse, 1976.)

on the difference between mean ages of transition and their variances, and mean ages of being-in-a-stage and their variances may help clarify in the reader's mind the distinction between population comparisons (of transition ages) and clinical standards (of being-in-a-stage). Population comparisons and clinical standards are more sharply distinguished in relation to puberty stages, because of their discontinuity, than in relation to height.

Figure 3 shows the height velocity standards (Tanner and Whitehouse, 1976). In the velocity standards we give limits for early- and late-maturing children as well as for the average-maturing. The clear centiles represent children all of whom have peak height velocity at the average age for this event. Boys with an earlier peak velocity have a higher peak. The 2 SD limit of "normal" earliness is approximately two years, and the left-hand edge of the shaded area encloses the velocity curve of such a boy. The upper arrow, diamond, and lower arrow represent the 97th, 50th and 3rd centile peak velocities for boys with peaks two years earlier. Similarly the right-hand edge of the shaded area represents boys with peaks two years late. Figure 4 may make clearer how the shaded areas in Fig.3 were constructed. Thus all boys with (conventionally) normal peaks at (conventionally) normal ages will have velocity curves lying within the shaded area. Those outside are abnormal.

In using these velocity charts, it must be remembered that velocities over a whole year are represented, not those over a shorter period, multiplied up. Due to seasonal variations, the shorter-period velocities have higher ranges of normal.

We used centiles rather than standard deviations in these charts because the centile represents the direct probability statement and is the interpretation given by the physician to the patient or his parents ("five per cent of normal children are shorter than your child"). Few paediatricians carry the tables of the normal deviate in their heads and without this facility what exactly *do* they say when a child has a height of -1.65 SD? ("fairly normal," I suppose). Formally, of course, centiles and SD scores are equivalent if, and only if, the distribution of the variate is Gaussian (and indeed the height centiles are best calculated via the SD; see Healy, 1974). If the variate is skewed, like weight, or still more, skinfolds, then SD scores cannot be interpreted at all, and in this sense are erroneous. Centiles then are calculated directly from the distribution, without assuming a distributional form. The use of SD scores as a convenience for saying *how* small is a small child, is different from their use as standards, and entirely justifiable (even in a sense for weight) as a rather arbitrary way of expressing smallness (leading to the statement "he is not now so small in relation to other children as he was when we started treatment"). A full discussion will be found in Tanner (1952).

In weight-for-height (as opposed to weight-for-age) standards, a further problem arises. The distribution of weight of children of a given height range may be different in six-year-olds than in seven-year-olds, say. Thus, the average seven-year-old of height 110 cm may not have the same weight as the average six-year-old of 110 cm. And even if these *average* children do have the same weights, then the *range* of weights of the six- and seven-year-olds may not be the same. This problem, tackled by Correnti (1953) and recently reinvestigated by Buzina *et al.* (1977) is a difficult one. Between the ages of one and nine, age makes a sufficiently small difference to the distributions of weight for given height for its effect to be ignored, allowing a single weight-for-height standard over this age range. But outside that range, changes of body shape with age make

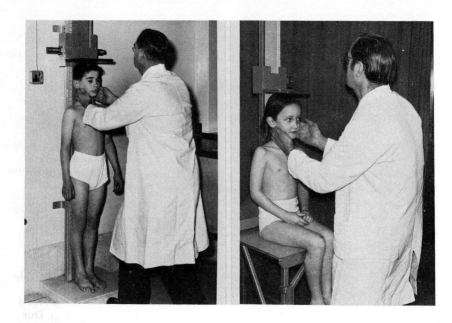

Fig.5 *Fig.6*

Fig.5 Method of measuring stature (from Tanner *et al.*, 1971).

Fig.6 Method of measuring sitting height.

it imperative that weight-for-height charts be constructed within specific age, or better, specific pubertal stage bands.

III. MEASUREMENT TECHNIQUE

No standards of growth are better than the measurements of the children from which they are derived and of the child who is being compared with them. "Garbage in, garbage out" is the motto of the auxologist no less than the biometrician. Measurements of growth, exactly like measurements of growth hormone, begin with modern well-maintained equipment and a trained technician using it. The apparatus for measuring stature and the method of doing it are shown in Fig.5. Similar apparatus is available for measuring the supine lengths of preterm and term neonates and infants up to two years (obtainable from Holtain Limited, Crosswell, Crymmych, Pembrokeshire, Wales). In the figure, the headboard is counter-balanced and the height read from a rack-and-pinion digital counter. More sophisticated apparatus reads on-line to a small computer in which the measurement is checked against previous ones and other measurements, the velocities computed and displayed, and a warning given if sets of limits are passed so that a check may be at once made. In less sophisticated clinics the same checks can be made by a trained recordist. In any clinic there should be one or more technicians who are specially trained as anthropometrists and who measure the child each time he comes, to minimise errors, especially in velocities. The heights

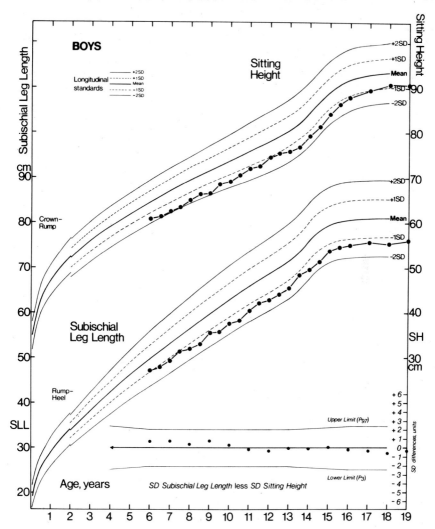

Fig. 7 Standards for sitting height and subischial leg length; a normal child is plotted (Tanner and Whitehouse, *in press*).

of the averagely casual clinic are useless even for accurate clinical purposes, let alone research.

Figure 6 shows the method of measuring sitting height, using the apparatus specially designed for this. Sitting height has long been preferred by anthropologists to the crown-pubis measurement introduced by paediatricians, on the grounds of greater reliability as well as for anatomical reasons. Leg length is best estimated in the clinic as stature less sitting height (subischial leg length, or SLL).

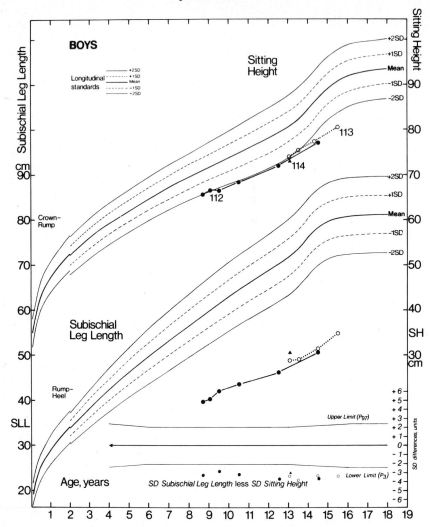

Fig.8 Two cases of hypochondroplasia plotted on the sitting height-subischial leg length standards.

IV. STANDARDS FOR TRUNK AND LIMB LENGTHS

R.H. Whitehouse and I have recently constructed standards for sitting height and subischial leg length, based on our longitudinal material of the last 20 years, comprising about 1000 children. The curves in Fig.7 are longitudinal type standards.

These standards overcome the difficulty inherent in the regression-type standards of sitting height for stature which we used previously (see Tanner, 1973); in case of abnormality, one could not easily determine whether it was that sitting height was abnormally short or the legs abnormally long. The same trouble (as well as all the other troubles of ratio standards) occurs in standards

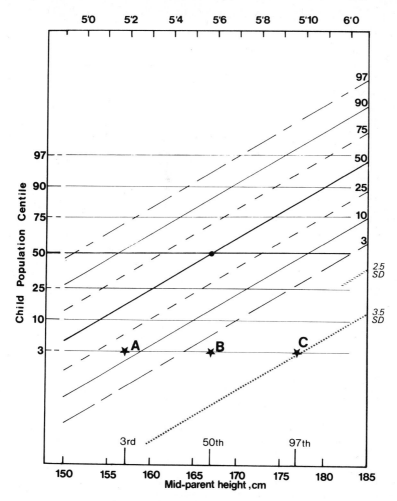

Fig. 9 Standards for child's height allowing for midparent height (from Tanner *et al.*, 1970).

such as upper segment/lower segment. In the new standards one plots the measurements directly. Figure 8 gives an example, where in three cases of hypochondroplasia it can be seen how the sitting height is just within normal limits while the leg length is greatly reduced. The lowest section of the chart displays the limits for normality of leg length for given sitting height. The SD score for leg length is subtracted from that for sitting height and the result plotted at the age concerned. The 3–97th centile limits for this SD difference have been worked out from the covariances of the measurements at each age in normal children. We have found the application of this chart very helpful in our growth disorder clinic in the differential diagnosis of the chondrodystrophies and in the defining of new syndromes.

V. PARENT-ALLOWED-FOR HEIGHT STANDARDS

Better standards (i.e., more powerful ones, in the statistical sense) may be made for subsets of a population if some way of classifying a child into the subset is available. This could be applied in principle to many standards (see Healy, 1974) but in practice has been so far used only in relation to the subsets constituted by parents' heights. Discrete subsets (small, medium, large) are inefficient and undesirable in relation to a continuous variation; what is required is a full regression standard. These have been provided for stature in relation to midparent height from ages two to nine (Tanner et al., 1970). The height of the child is not closely related to midparent height at birth, but the correlation rises rapidly as the child gets on its own growth course after escape from the maternal uterus, and by age two the correlation is high. By arranging suitable scales, a chart can be constructed (Fig.9) which gives the position of the children in a regression system. The child's height *centile* (using the ordinary standards) is entered on the vertical axis, and the midparent height (i.e., the average of parental heights, unadjusted for sex) on the horizontal axis. The point thus plotted is looked up in relation to the regression lines. Thus, child A, a 3rd centile-for-population child with 3rd centile parents, is actually at about the 15th centile when this fact is allowed for. The 3rd centile child B, with 50th centile parents lies at about the 1st centile (test your students' statistical knowledge by asking why) and the 3rd centile child, whose parents are at the 97th centile, is far below the standards and around -3.5 residual standard deviations.

The use of parent-allowed-for standards, especially coupled with a consideration of bone age, may be very helpful in dealing with the diagnosis "small/delay" in growth (see Tanner, 1973). Often such children are within normal limits for their parents, though not for the population. Prediction of adult height (Tanner et al., 1975) is also of use in distinguishing children with delayed growth from those with isolated growth hormone deficiency. The small/delay child predicts within the target range specified by the midparental height centile and its variance, whereas the growth hormone deficient child has an adult prediction well below the parental target range.

These, then, are some of the uses of growth standards in the clinic. Every paediatrician and community medical doctor should be able to interpret charts of growth and growth velocity and should know their limitations and their uses. We are nowhere near this state at present and all human auxologists must be prepared to play their part in a teaching effort of increasing size and importance.

REFERENCES

Buzina, R., Keller, W., Nickaman, M.Z., Tanner, J.M. and Waterlow, J.C. (1977). *Bull. WHO* (in press).
Correnti, V. (1953). *Pediatr. Int.* **3**, 169-192.
Healy, M.J.R. (1974). *Ann. Hum. Biol.* **1**, 41-46.
Jordan, J., Ruben, M., Hernandez, A., Bebelagua, A., Tanner, J.M. and Goldstein, H. (1975). *Ann. Hum. Biol.* **2**, 153-172.
Lindgren, G. (1976). *Ann. Hum. Biol.* **3**, 501-528.
Tanner, J.M. (1952). *Arch. Dis. Child.* **27**, 10-33.
Tanner, J.M. (1959). *In* "The Anthropology of Franz Boas: Essays on the Centennial of his Birth" (W. Goldschmidt, ed) *Am. Anthropol.* **61**, 76-111.
Tanner, J.M. (1962). "Growth at Adolescence", 2nd ed. Blackwell Scientific Publications, Oxford; and C.C. Thomas, Springfield.

Tanner, J.M. (1973). *In* "Textbook of Paediatrics" (J.O. Forfar and G.C. Arneill, eds). Churchill Livingstone, Edinburgh and London.
Tanner, J.M., Goldstein, H. and Whitehouse R.H. (1970). *Arch. Dis. Child.* **45**, 755-762.
Tanner, J.M. and Whitehouse, R.H. (1976). *Arch. Dis. Child.* **51**, 170-179.
Tanner, J.M., Whitehouse, R.H., Hughes, P.C.R. and Vince, F.P. (1971). *Arch. Dis. Child.* **46**, 745-782.
Tanner, J.M., Whitehouse, R.H., Marshall, W.A., Healy, M.J.R. and Goldstein, H. (1975). "Assessment of Skeletal Maturity and Prediction of Adult Height". Academic Press, London.
Tanner, J.M., Whitehouse, R.H. and Takaishi, M. (1966). *Arch. Dis. Child.* **41**, 454-471; 613-635.

THE FITTING OF LONGITUDINAL GROWTH DATA OF MAN

E. Marubini

Laboratory of Health Statistics, First Medical School
University of Naples, Italy

SUMMARY

The purpose of this chapter is to discuss some statistical methods suitable for analyzing longitudinal growth data and their particular merits for studying growth in man. Three different approaches are dealt with and their main features are pointed out; namely: whole description of experimental or survey results by Factor Analysis: inferential properties by fitting polynomial models; clear meanings of the constants with a definite biological relevance by fitting nonlinear models. It is concluded that: (1) further, deep investigations are needed before using the first approach to the analysis of human growth; (2) owing to the fact that the indiscriminate averaging travesties the pattern of growth by oversmoothing the peaks and masking the inherent interindividual variability, and moreover the residual standard deviations are greater than those obtained by fitting nonlinear models, it appears that the use of polynomials to describe entirely human growth is questionable; (3) the third approach seems to be the most fruitful for *describing* the growth curve in man and it is proposed to investigate the possible extension to nonlinear models of the Bayesian approach Fearn (1976) suggested for polynomials.

I. INTRODUCTION

It is easily understood that, in growth studies, the mean of the growth curves of a set of individuals is perfectly suitable for analyzing distributions of various measurements (height, weight, sitting-height, etc.) at different ages, for drawing the average growth curve of the population, for preparing the "distance" standards, and for being able to relate important features of growth to other relevant variables as, for instance, social background, size of family, parental methods, and so on. For such purposes, provided that the average growth curve is not changing with time, it does not matter whether the data are collected by means of cross-sectional or longitudinal researches. However, the crucial requirement is that the sample size is sufficiently large to estimate means with the needed precision (Tanner, 1951). On the other hand, this approach is entirely misleading if the concern is to trace the course of change in the individual subject over time, to identify as many as possible events which may have significance for the individual as determinants of his future course, and to construct "velocity" standards whose important role in clinical works has been widely pointed out (Anderson et al., 1965; Tanner et al., 1966). As a matter of fact, the averaging process not only conceals irregularities of growth rates in individual children, but also can greatly distort common trends such as the "take-off" of adolescence (Tanner, 1962) when these occur at different ages in different individuals. The sudden change at puberty and its different ages of onset can be brought out only by getting information from a cohort on successive occasions over a given number of years and by fitting individual growth curves. Omitting to discuss the ambitious aim of *interpreting* growth by means of mathematical equations (see Marubini, 1977), it may be assessed that there are at least two reasons to follow the last approach: (1) to describe the pattern of growth by conveying information in a relatively reduced number of constants; (2) to keep a reasonable power to the methods of statistical inference; such a power is otherwise reduced if too much redundant information is present (Rao, 1958, 1967; Healy, 1969).

A detailed review of available methods for longitudinal data analysis has been published by Kowalski and Guire (1974). In this chapter, three statistical methods suitable for analyzing data corresponding to the situation they labelled "one (or k)-sample data matrix" will be dealt with. Each of these techniques stresses different aspects concerning the two points above; namely: whole description of experimental or survey results by using Factor Analysis (Esteve and Shifflers, 1974); inferential properties by fitting polynomial models; clear meanings of the constants with a definite biological relevance by fitting nonlinear models.

II. FACTOR ANALYSIS

Although about twenty years ago this multivariate technique was suggested for analyzing one-dimensional processes (Rao, 1958; Tucker, 1958), only recently has it been applied to growth studies (Esteve and Shifflers, 1974, 1976). It yields the reduction of data by fitting a set of orthogonal functions calculated from the data themselves to each longitudinal record. The approach is technically similar to principal component analysis, but the interpretation of results is quite different.

According to Esteve and Sfhifflers (1976) the pertinent model can be written:

$$\mathbf{Y}(t) = \boldsymbol{\mu}(t) + \sum_{k=1}^{d} a_k \mathbf{V}_k(t) + \epsilon(t) \qquad (1 \leq t \leq p) \qquad (1)$$

where $\mathbf{Y}(t)$ is a p-dimensional normal vector whose expected value is $\boldsymbol{\mu}(t)$ and

$V_k(t)$ $(1 \leq k \leq d)$ are unknown functions of time (i.e., the factors). Furthermore, a_k are normal random variables with unit variance and $\epsilon(t)$ is a normal error with dispersion matrix Ψ. Let V be the matrix whose columns are V_k, thus:

$$Y \sim N_p(\mu, \Sigma) \qquad (2)$$

where:

$$\Sigma = VV' + \Psi . \qquad (3)$$

In the two mentioned papers, the authors discuss in detail the theoretical background of their approach and give the results of an analysis carried out to study the weight growth curves of three generations of mice. The curves are obtained by weighing all the animals at the same occasion, namely: at birth (day 0), 4th, 8th, ..., 36th day of age. Such a condition seems to be a drawback to use this method to investigate the growth of man, also in some definite intervals (for instance, from birth to 8–9 years), since it is well known that children often do not come for measurement at exactly the age specified, for instance, on the birthday, but for various reasons they are several days or weeks, early or late. In order to avoid this obstacle, one could utilize a method of individual adjustment as that suggested by Goldstein and Carter (1970). The essence of this method is to fit, by means of the Newtonian divided differences, a second-degree polynomial to pass through measurements attained at three time-points scattered around the target day and to read the corresponding value. However, to evaluate the reliability of such an adjustment device in the different parts of human growth curve and its possible use in the approach the underlying model of which is given by (1), deep investigations are needed.

III. POLYNOMIAL MODELS

Several theoretical, as well as applied, statistical papers are devoted to methods suitable to analyze growth curves by polynomial models; a compact review has been written by Rao (1972) who pointed out also the unsolved problems and the direction in which further research is needed. Furthermore, quite recently: Bowden and Steinhorst (1973) suggested a method to construct tolerance bands which contain a given percentage of the population mean growth curves; Srivastava and McDonald (1974) developed an analysis of longitudinal data when different numbers of measurements are recorded on each individual; Krishnaiah (1975) proposed simultaneous test procedures for the multiple comparisons of polynomial coefficients of several growth curves when the residuals are autocorrelated. The fitting of polynomials has the property that the mean curve (discussed in the Introduction) is equivalent to the *mean constant curve* (obtained by fitting a given function to each subject's measurements, estimating the constants for each of them, and finally averaging the calculated constants) (Merrel, 1931). As far as human growth is concerned this may be a critical point in practice, since, as it has been previously said, indiscriminate averaging travesties the pattern of growth by oversmoothing the peaks and masking the inherent interindividual variability the estimate of which is often the main goal of the study.

In order to investigate the reliability of this approach from the point of view of the goodness of fitting, a third-degree polynomial was fitted to height growth data during adolescence (Marubini, 1977) and the results were compared with those obtained by fitting the logistic function (see next section). The overall residual standard deviations (which are due to seasonal and diurnal variation and to measurement error) attained by fitting the polynomial model were about three times greater for boys, and twice for girls, than those attained by fitting logistic. Even worse were the results of the runs test.

Table I. Results of fitting of the same set of longitudinal data by means of third-degree polynomial and logistic model.

Function fitted		3rd degree polynomial		Logistic	
Sex	No. of subjects	Residual SD (cm)	Percentage of significant runs tests $a = 0.025$	Residual SD (cm)	Percentage of significant runs tests $a = 0.025$
M	40	min. 0.30 pooled 1.08 max. 1.65	37.5	min. 0.19 pooled 0.39 max. 0.81	0.00
F	32	min. 0.23 pooled 0.82 max. 1.31	56.3	min. 0.18 pooled 0.38 max. 0.64	6.25

These findings are in contradiction with Goldstein's assessments (1971) and therefore it appears that the use of polynomials to describe human growth is questionable.

IV. NONLINEAR MODELS

Apart from the above-mentioned property concerning definite relations of constants to biological conceptions, the nonlinear models most commonly used to fit human longitudinal records possess other very convenient features: (1) relatively simple form; (2) good agreement between the observed values and those predicted by the function.

Table II reports some models widely adopted.

A detailed discussion of the properties and the biological meaning of the constants of these models can be found in the papers by Jenss and Bayley (1937), Deming (1957), Deming and Washburn (1963), Marubini *et al.* (1971), Bock *et al.* (1973) and Bock and Thissen (1976).

Owing to the complex form of the human growth process as a whole, some authors have thought it fruitful to break up the curve into three (Count, 1943; Laird, 1967) or two (Deming, 1957; Israelsohn, 1960) cycles. These, according to Courtis (1937), are periods of specific maturation during which the elements and forces acting upon the growth process are nearly stable.

The first four functions of Table II have been used to fit cycles of growth curve and, with regard to those used in the adolescent period, it is easily seen that they satisfy some basic requirements, namely: lower and upper asymptote, a

Table II. Mathematical functions widely used for fitting growth data. t = age in decimals. f = values reached at maturity. P = lower asymptote, i.e., value reached before the adolescent spurt. $a, a_1, k, b, b_1, b_2, b_3, c_1, c_2, c_3, p$ = constant to be estimated.

	Function	Anthropometric variable	Growth period	Authors
1.	$f(t) = c + dt - \exp(a - bt)$	length; weight supine length; weight supine length; skull circumference; weight	birth–6 yr birth–8 yr birth–48 wk	Jenss and Bayley (1937) Deming and Washburn (1963) Manwani and Argawal (1973)
2.	$f(t) = at^b$	skeletal nose height	5–12 yr	Meredith (1958)
3.	$f(f) = P + \dfrac{K}{1 + \exp[-b(t-c)]}$ (logistic function)	height height height; sitting height leg length; biacromial and biliac diameter	12 yr–adulthood adolescence adolescence adolescence	Count (1943) Marubini et al. (1971) Marubini et al. (1972) Tanner et al. (1976)
4.	$f(t) = P + K \exp[e^{-b(t-c)}]$ (Gompertz function)	height weight height	adolescence adolescence adolescence	Deming (1957) Laird (1967) Shohoji and Pasternack (1973) Pasternack and Shohoji (1976)
5.	$f(t) = a_1 \{1 + \exp[-b_1(t-c_1)]\}^{-1} +$ $+ (f - a_1)\{1 + \exp[-b_2(t-c_2)]\}^{-1}$	recumbent length	1 yr–adulthood	Bock et al. (1973)
6.	$f(t) = a_1 \cdot (q\{1 + \exp[-b_1(t-c_1)]\}^{-1} +$ $+ p\{1 + \exp[-b_2(t-c_2)]\}^{-1}) + (f - a_1) \cdot$ $\cdot \{1 + \exp[-b_3(t-c_3)]\}^{-1}$	height	1 yr–adulthood	Bock and Thissen (1976)

Table III. Mean differences between PV values (cm/yr) obtained by using estimated take-off age and those given by altering take-off to earlier or later ages.

Model	Starting points before and after estimated take-off age	MALES				FEMALES			
		Height	Sitting height	Leg length	Biacromial diameter	Height	Sitting height	Leg length	Biacromial diameter
Gompertz	12 months before	−0.48*** (16)	−0.33** (14)	−0.30** (16)	−0.08 (14)	−0.20 (12)	−0.05 (14)	−0.03 (12)	−0.04* (12)
	6 months before	−0.23* (19)	−0.12** (16)	−0.17** (19)	−0.04 (18)	−0.05 (12)	0.02 (13)	0.01 (12)	−0.02 (13)
	6 months after	0.40* (21)	0.04 (19)	0.09 (21)	0.02 (20)	−0.02 (14)	−0.08 (12)	−0.08 (14)	0.03 (13)
	12 months after	0.58** (21)	0.18 (19)	0.11 (20)	0.11* (17)	−0.09 (14)	−0.21 (9)	−0.003 (12)	0.03 (13)
Logistic	12 months before	−0.30*** (16)	−0.21** (14)	−0.21** (16)	−0.03 (14)	−0.09* (12)	−0.03 (14)	0.01 (12)	−0.04 (12)
	6 months before	−0.16*** (19)	−0.08** (16)	−0.11** (19)	−0.02* (18)	−0.02 (12)	0.01 (13)	0.01 (12)	−0.01 (13)
	6 months after	0.15*** (21)	0.02 (19)	0.15 (21)	0.02 (19)	0.08 (14)	0.19 (12)	−0.01 (14)	0.07 (14)
	12 months after	0.74*** (21)	0.32* (19)	0.08 (20)	0.05 (14)	0.27 (14)	−0.12 (9)	−0.09 (11)	0.09 (13)

* $0.01 < p < 0.05$; ** $p < 0.01$; df shown by numbers in parenthesis.

Table IV. Mean differences between age of PV values (yr) obtained by using estimated take-off age and those given by altering take-off to earlier or later age.

Model	Starting points before and after estimated take-off age	MALES				FEMALES			
		Height	Sitting height	Leg length	Biacromial diameter	Height	Sitting height	Leg length	Biacromial diameter
Gompertz	12 months before	−0.36** (16)	−0.12* (14)	−0.38** (16)	−0.11** (14)	−0.33** (12)	−0.19* (14)	−0.42** (12)	−0.36** (12)
	6 months before	−0.19* (19)	−0.04 (16)	−0.20** (19)	−0.02 (18)	−0.15** (12)	−0.15 (13)	−0.21** (12)	−0.22** (13)
	6 months after	0.02 (21)	0.03 (19)	0.13** (21)	−0.01 (20)	0.14** (14)	−0.06 (12)	0.24** (14)	−0.04 (13)
	12 months after	−0.01 (21)	−0.01* (19)	0.40** (20)	−0.18 (17)	0.26** (14)	0.26** (9)	0.61** (12)	0.10 (13)
Logistic	12 months before	−0.16** (16)	0.01 (14)	−0.27** (16)	0.07 (14)	−0.19** (12)	−0.01 (14)	−0.31** (12)	−0.23** (12)
	6 months before	−0.06** (19)	0.06 (16)	−0.15** (19)	0.09 (18)	−0.07** (12)	−0.04 (13)	−0.16** (12)	−0.14** (13)
	6 months after	−0.04 (21)	−0.07 (19)	0.02 (21)	−0.16** (19)	0.01 (14)	−0.22* (12)	0.16 (14)	−0.20** (14)
	12 months after	−0.21* (21)	−0.30* (19)	0.27** (20)	−0.31*** (14)	0.00 (14)	0.04 (9)	0.63** (11)	−0.14 (13)

* $0.01 < p < 0.05$; ** $p < 0.01$; df shown by numbers in parenthesis.

single point of inflection, and a different time-origin for each individual. The approach involving the fitting of different functions in different cycles carries the risk of wrong division; for instance, in fitting the adolescence curves, a difficult task concerns the age estimation of the onset of puberty regardless of the device used (Frish and Revelle, 1971; Resele and Marubini, 1972). The extent to which the logistic, as well as the Gompertz curve are altered in their fits by placing the take-off point systematically earlier or later than a "real" estimate has been investigated by Marubini et al. (1972) with regards to height, sitting-height, leg-length, and biacromial diameter. In this way, a set of measured values were included or excluded as one would if he made a mistake in assessing the take-off point at the beginning of a curve-fitting exercise.

Table III shows the effect on peak velocity (PV) and Table IV the analogous effect on age at PV.

It can be seen that errors of less than one year in choosing the "take-off" point do not have a relevant effect on the estimates of both the maximum velocity and the age at which this maximum arises. Furthermore, in the discussion, the authors concluded that logistic, on the whole, fitted the data better than the Gompertz. It was slightly more stable under errors in lower asymptote and it produced fewer biased fits as judged by the runs test. Subsequently, by fitting the logistic function to the individual longitudinal records of the Harpenden Growth Study, Tanner et al. (1976) have studied the difference in pattern of growth of height, sitting-height, leg-length, and biacromial and biiliac diameter during adolescence in both boys and girls. Furthermore, the relationship between the estimates of the parameters and the development of pubertal characters has been deeply investigated. As far as the height is concerned, Figs 1 and 2 give the curves plotted from the mean values of the constants (i.e., the *mean constant curves*) respectively for distance and velocity. The take-off points are marked with arrows. Before the take-off point, and for 0.5 year after it, the empirical mean curves (from row data) are given by a dotted line. The empirical mean distance and velocity curves come to coincide with the fitted *mean-constant* curves nearly 0.5 years after take-off. Thus, the logistic has a lower asymptote (where the velocity is null), whereas the spurt takes place from the basis of a preexisting growth curve (of velocity about 4 cm/yr). A number of authors (Count, 1943; Laird, 1967; Bock et al., 1973) have endeavoured to graft the curve of the adolescent spurt onto a preexisting growth curve considered to continue, though with decreasing force, throughout the growing period. To this end, Bock et al. (1973) have used a two-logistic model (see: no.5 of Table II), one riding on the back of the other, to fit supine length measurements from one year to maturity in subjects of the Fels Research Institute. The first curve represents preadolescent growth, and the two are added together after the second begins. It is now clear that, using this model, the risk of wrong division of the curve in different cycles is avoided and the complete pattern of growth is reduced to a simple functional form with five constants to be estimated, since K is assumed to be known.

Unfortunately, this model fared badly on the runs test and recently, Bock and Thissen (1960) stated:

> "the two-component model can provide unbiased comparisons between groups only when the longitudinal measures are made with roughly the same time spacing, a requirement that is not easily met when comparing data from different growth studies.

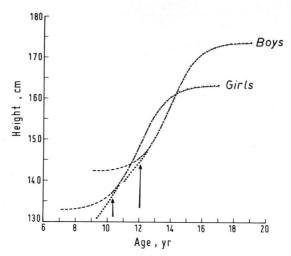

Fig. 1 Mean-constant height curves, girls and boys, resulting from logistic fit. The arrows indicate take-off points, girls at 10–3, boys 12–0 years. Dotted lines are empirical means prior to take-off + 0.5 yr (from Tanner *et al.*, 1976).

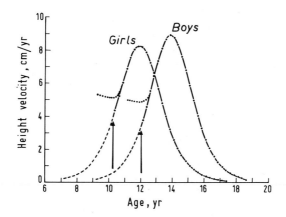

Fig. 2 Mean-constant height velocity curves, girls and boys, resulting from logistic fit. The arrows indicate take-off points. Dotted lines are empirical means prior to take-off + 0.5 yr (from Tanner *et al.*, 1976).

"This lack of robustness of the two-component logistic model has two sources: 1) the model is not capable of fitting exactly all systematic trends in individual growth ... 2) the least squares method used by Bock *et al.* (1973) to estimate the parameters of the model, does not allow for the patterned correlation structure of the residuals and, therefore, does not weigh the observations correctly."

Moreover, it has been shown (Marubini, 1977) that the velocity curves derived from the model no. 5 (Table II), drawn by resorting to the constants given by Bock *et al.* (1973), does not compare favorably with the typical one reported by Tanner *et al.* (1966), particularly in early childhood.

Owing to the inability of the two-component model to represent accurately the growth, particularly from birth to five years of age, Bock and Thissen (1976) fitted a three-logistic function with nine constants. After allowing for the three additional constants, they found either the average residual standard deviations significantly smaller than those of the two logistic model or a decrease of the heteroscedasticity of the residuals and a reduction of their serial correlation. Furthermore, as can be seen by a glance at Table V, the three-component model allows the investigators to give estimates of age at PV for both boys and girls more in line with those attained in other samples and by other approaches, than the two-component one.

Table V. Age at adolescence PHV from various studies (from Bock and Thissen, 1976).

Authors	Males	Females
Deming (1957) (Gompertz-CRC)	13.4	11.4
Marubini *et al.* (1972)		
(Gompertz-Harpenden)	14.0	11.7
(Logistic-Harpenden)	14.2	11.9
Tanner *et al.* (1976) (Logistic-Harpenden)	14.4	11.5
Bock *et al.* (1973)		
(Double logistic-Fels)	13.0	11.0
(Triple logsitic-BGS)	13.9	11.7

By considering the feature of the correlation matrix of estimators, Bock and Thissen (1976) concluded:

"It is possible that some of the parameters, especially those of the first component, are so highly correlated to be redundant. If so, the model could be simplified by expressing two or more parameters as functions of a single parameter with a consequent simplification of the model."

V. CONCLUDING REMARKS

Recently, Fearn (1975), developing the theory by Lindley and Smith (1972), suggested a Bayesian approach to the analysis of growth curves by polynomial models which involves two stages. In the first one, after assuming that the set of n_i data measured on the i-th individual are independently and normally distributed about his own curve (with design matrices X_i and dispersion matrices $\sigma_i^2 I$), separate polynomials are fitted to each individual's longitudinal record and the constants vectors (β_i) are estimated. In the second stage, after supposing that the β_i are independently and normally distributed with a common mean vector β and covariance matrix C, the estimate of vector β is derived according to Smith (1973). Such an estimate is a weighted mean of the estimates of β_i and each weight takes into account both the covariance matrices (which, in turn, involve either X_i or $\sigma_i^2 I$) of the estimated β_i and the covariance matrix C. A crucial point of this approach is the prior distribution either of the unknown variances σ_i^2 or the covariance matrix C whose role is essential in defining the efficiency of the estimate of β. It is well known that, after fitting *nonlinear models*, the *mean constant curve* is drawn by means of the vector which is obtained by computing the simple arithmetic mean of the components of the estimated β_i; it seems, therefore, that investigations which allow extension of Fearn's approach to *nonlinear models* could be very fruitful in order to permit use of the Bayesian approach to describe the growth patterns of man in the most efficient way available.

REFERENCES

Anderson, M., Hwang, S. and Green, W.T. (1965). *J. Bone Jt. Surg.* **8**, 1554-1564.
Bock, R.D. and Thissen, D. (1976). *Proc. 9th Intern. Biometric Conference* **1**, 431-442.
Bock, R.D., Wainer, H., Petersen, A., Thissen, D., Murray, J. and Roche, A. (1973). *Hum. Biol.* **45**, 63-80.
Bowden, D.C. and Steinhorst, R.K. (1973). *Biometrics* **29**, 361-372.
Count, E.W. (1943). *Hum. Biol.* **15**, 1-32.
Courtis, S.A. (1937). *Growth* **1**, 155-167.
Deming, J. (1957). *Hum. Biol.* **29**, 83-122.
Deming, J. and Washburn, A.M. (1963). *Hum. Biol.* **9**, 556-563.
Esteve, J. and Shifflers, E. (1974). *Ann. Zool. Ecol. Anim.* **6**, 449-461.
Esteve, J. and Shifflers, E. (1976). *Proc. 9th Intern. Biometric Conference* **1**, 463-480.
Fearn, T. (1975). *Biometrika* **62**, 89-100.
Frish, R.E. and Revelle, R. (1971). *Hum. Biol.* **43**, 140-157.
Goldstein, H. (1971). *Proc. 13th Intern. Congress of Pediatrics*, 39-42.
Goldstein, H. and Carter, B. (1970). "Compte Rendu de la X Réunion des Équipes Chargées des Études sur la Croissance et le Développement de l'Enfant Normal", C.I.E., Davos.
Healy, M.J.R. (1969). *Biometrics* **25**, 411-413.
Israelsohn, W.J. (1960). *In* "Human Growth" (J.M. Tanner, ed), pp.21-42. Pergamon Press, Oxford—London—New York—Paris.
Jenss, R.M. and Bayley, N. (1937). *Hum. Biol.* **9**, 556-563.
Kowalski, C.J. and Guire, K.E. (1974). *Growth* **38**, 131-169.
Krishnaiah, P.R. (1975). *Proc. of the 40th Session of International Statistical Institute*, 488-492.
Laird, A.K. (1967). *Growth* **31**, 345-355.
Lindley, D.V. and Smith, A.F.M. (1972). *J.R. Statist. Soc.* **B34**, 1-41.
Manwani, A.H. and Agarwal, K.N. (1973). *Hum. Biol.* **45**, 341-349.
Marubini, E. (1977). *In* "Human Growth. A Comprehensive Treatise" (F. Falkner and J.M. Tanner, eds) *(in press)*.
Marubini, E., Resele, L.F. and Barghini, G. (1971). *Hum. Biol.* **43**, 237-252.
Marubini, E., Resele, L.F., Tanner, J.M. and Whitehouse, R.H. (1972). *Hum. Biol.* **44**, 511-524.
Meredith, H.V. (1958). *Child Dev.* **29**, 19-34.
Merrell, M. (1931). *Hum. Biol.* **3**, 37-69.
Pasternack, B. and Shohoji, T. (1976). *In* "Essays in Probability and Statistics" (S. Ikeda *et al.*, eds) pp.559-577. Shinko Tsusho Co. Ltd., Tokyo.
Rao, C.R. (1958). *Biometrics* **14**, 1-17.
Rao, C.R. (1967). *Proc. 5th Berkely Symp.* **1**, 355-372.
Rao, C.R. (1972). *Biometrics* **28**, 3-22.
Resele, L.F. amd Marubini, E. (1972). *A.b.d.c.e.* **7**, 187-200.
Shohoji, T. and Pasternack, B. (1973). *Health Phys.* **25**, 17-27.
Smith, A.F.M. (1973). *J.R. Statist. Soc.* **B35**, 67-75.
Srivastava, J.N. and McDonald, L.L. (1974). *Sankhya Ser.* **A36**, 251-260.
Tanner, J.M. (1951). *Hum. Biol.* **23**, 93-159.
Tanner, J.M. (1962). "Growth at Adolescence". Blackwell Scientific Publications, Oxford.
Tanner, J.M., Whitehouse, R.H. and Takaishi, M. (1966). *Arch. Dis. Child.* **41**, 454-471.
Tanner, J.M., Whitehouse, R.H., Marubini, E. and Resele, L.F. (1976). *Ann. Hum. Biol.* **3**, 109-126.
Tucker, L.R. (1958). *Psychometrika* **23**, 19-23.

UNITED STATES GROWTH CHARTS

A.F. Roche and P.V.V. Hamill

The Fels Research Institute, Yellow Springs, Ohio 45387
and
National Center for Health Statistics, Rockville, Maryland, USA

SUMMARY

Recently, the National Center for Health Statistics in Bethesda, Maryland, has published a set of growth charts based on data obtained from a national survey and from the longitudinal study at the Fels Research Institute. These carefully collected data have been treated statistically by fitting spine functions to construct mathematically smoothed percentiles. It is expected that the resultant charts will be applicable to the United States population for a relatively long period because there is evidence that secular increases have probably stopped in the country as a whole. Large quantities of these charts have been printed for free distribution and their worldwide use has been recommended by the World Health Organization and the Food and Agriculture Organization.

I. INTRODUCTION

New growth charts considered suitable for the total United States population, and indeed for other populations also, have been published recently by The National Center for Health Statistics, Washington (1976). These charts, that are applicable from birth to 18 years, give reference data for weight, recumbent length or stature, weight by recumbent length or stature, and head circumference. They are intended for use in clinical circumstances so that growth parameters can be assessed by comparison with the known status of a large, well-defined population.

For many years, the growth charts used most commonly in the United States have been those based on data from Iowa and Boston (Stuart and

Meredith, 1947; Meredith, 1948a,b). Despite their general usefulness, there has been concern about applicability of these charts to present-day children because of the secular trends that have occurred in the United States and other developed countries (Tanner, 1962; Meredith, 1963). Several government committees in the United States recommended the preparation of new charts, based on current data from the National Center for Health Statistics (NCHS) and other appropriate data. The new charts were prepared by groups from the National Center for Health Statistics and the Center for Disease Control with the assistance of Professor R. Reed of Harvard and under the leadership of Dr. P.V.V. Hamill of the National Center for Health Statistics. As a result, 14 charts were produced, each containing seven smoothed curves for the 5th, 10th, 25th, 50th, 75th, 90th, and 95th percentiles.

The basic scaling is metric; pounds and inches are provided as subordinate units. The total age range was divided into two parts (birth to 36 months; 2 to 18 years). The data for the earlier period were obtained from the Fels Research Institute, Yellow Springs, Ohio, where approximately 800 children have been studied serially from birth to ages that range from 3 to 48 years. These data have minor sampling biases because the children were mainly middle-middle to lower-middle class socioeconomically, and were living in villages or small cities in Southwest Ohio; these biases are considered insufficient to disqualify these data which were judged more appropriate than other data available. This judgment is supported by the finding that, at overlapping ages, the differences between Fels data and those from the National Center for Health Statistics are minor.

The charts for the period 2 to 18 years are constructed with data from three surveys made by the National Center for Health Statistics between 1962 and 1974. The data from these three separate surveys could be combined due to the similarity of measurement and sampling procedures. Each survey included a stratified probability sample with nearly complete coverage of those selected. The recorded data from each individual were weighed to obtain national estimates.

II. MEASUREMENTS

A. Fels Data

Nude body weight was measured using a regularly calibrated beam balance. Recumbent length was measured by two examiners working as a team. The participant was stretched fully on a specially constructed measuring table with his or her head touching the fixed headboard. An attempt was made to flatten any lumbar lordosis. Keeping the knees extended, the examiner brought the footboard firmly against the soles of the feet so that these were vertical. Head circumference was measured with a steel tape about 2.5 cm superior to the glabella and over the maximum diameter posteriorly. The tape was kept horizontal and drawn snugly. Although the data were serial, each set of observations was made independently and on each visit every child was measured twice by two anthropometrists working independently. The only exception was in the measurement of recumbent length. For this, the anthropometrists worked cooperatively, but exchanged roles to repeat the measurement. The interobserver differences are very small.

B. *National Center for Health Statistics Data*

Weight was measured on a scale that printed the weight directly on a permanent record. The NCHS weights include light standardized clothing weighing approximately 0.05 kg at 2 years progressing to 0.09 kg from 3 to 5 years and 0.11–0.03 kg from 6 to 18 years. Stature was measured with the child's stockinged feet together and both back and heels against an upright bar. A horizontal bar was brought down snugly on the examinee's head; a camera in the same plane recorded the identification number next to the pointer on the scale, giving a precise reading. This eliminated parallax and reduced observer and recording error.

III. CURVE SMOOTHING

Curve smoothing is desirable when such charts are being prepared because the smoothed curve may represent better the true data devoid of random errors of measurement. A mathematical technique was used so that the smoothing process would be quantifiable and reproducible. The method chosen was least squares cubic spine polynomial smoothing of the observed percentiles (de Boor and Rice, 1968). In this mathematical procedure, polynomials, fitted to a few successive points at a time, are connected at selected points known as knots. At each knot, the pair of consecutive polynomials have identical values, slopes and accelerations. There was concern that the method might not develop a coherent relationship between different percentile lines. However, when the same fixed knots were used for all percentiles, there was good parallelism between the smoothed percentile lines. Many trials were made using both fixed and variable knots. The optimal number and placement of knots was determined by knowledge of the properties of the data and pragmatic tests. The ultimate choices were made by comparing smoothed curves with the observed data.

IV. GROWTH CHARTS

A. *0 to 36 Months*

The charts of recumbent length, body weight and head circumference by age do not require explanation. Weight by length is a relatively new type of chart resulting, in part, from the work of van Wieringen (1972). This presentation assumes approximate age independence. To prepare this part of the chart, all data from birth to 48 months were pooled and rearranged in length intervals of 2 cm. When the chart is applied, any child aged less than 48 months can be assessed by finding the appropriate recumbent length and weight on the sex-appropriate graph. His or her percentile for weight by length is obtained by comparison with all children of the same sex irrespective of age.

B. *2 to 18 Years*

These charts are similar in format but head circumference is omitted and weight by stature is based on a restricted group (Fig.1). It has been assumed weight by stature is age-independent until marked changes in body size and proportions occur during pubescence. In the cross-sectional NCHS data, the pubescent growth spurt could not be identified in individuals nor were there data

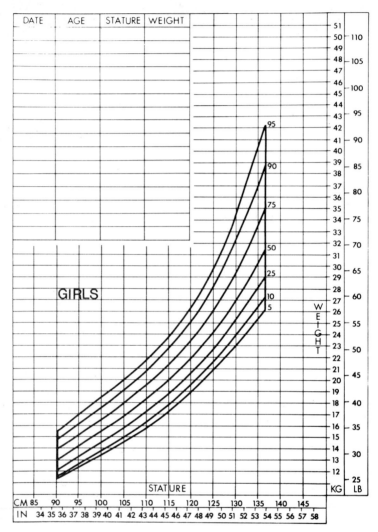

Fig. 1 The National Center for Health Statistics percentiles of weight for stature in pre-pubescent girls.

relating to secondary sex characters. A few girls have early pubescent changes in stature or weight by 8 years, but their effect on the total data base would be very small. In choosing a maximum age range over which weight-by-length charts could be used independent of age, judgment was based on the knowledge that early maturers tend to be tall (Shuttleworth, 1938). It was considered the age range could be extended with little risk of including pubescent children if the tallest childen were excluded. Consequently, girls with statures above the 95th percentile at 8 years (137 cm) were excluded, but the remainder of the girls were included until the chronological age of 10 years. Similarly, boys were excluded whose statures were greater than the 95th percentile at 9.5 years

Table I. Mean stature and weight differences (in cm and kg) between HANES I and HES II or HES III, by sex: seven selected percentiles.

Surveys compared	Percentile						
	5th	10th	25th	50th	75th	90th	95th
	Mean stature difference, male						
HANES I – HES II	0.31	0.66	−0.34	−0.16	0.14	0.05	0.15
HANES I – HES III	0.76	0.06	0.47	0.24	0.37	−0.02	0.33
	Mean stature difference, female						
HANES I – HES II	0.54	−0.05	−0.22	0.18	0.12	0.31	0.08
HANES I – HES III	0.79	0.45	−0.03	−0.51	−0.42	0.14	−0.16
	Mean weight difference, male						
HANES I – HES II	−0.13	0.10	0.02	0.17	0.51	0.37	1.09
HANES I – HES III	0.39	0.89	0.60	0.34	0.78	−0.86	0.43
	Mean weight difference, female						
HANES I – HES II	0.30	0.19	0.02	−0.13	−0.29	−0.31	−0.98
HANES I – HES III	−0.05	0.62	0.35	−0.19	0.24	1.47	4.51

HANES I = Health and Nutrition Examination Survey, 1971-1972. HES II = Health Examination Survey, 1962-1965. HES III = Health Examination Survey, 1977-1970.

(145 cm) or whose ages exceeded 11.5 years. These graphs of weight by stature should not be used for children with signs of pubescence or for those taller than, or older than, the limits noted above.

V. LIMITATIONS OF THE CHARTS

These charts are of very high quality but not perfect. The data from birth to 36 months, from the Fels Research Institute, have a slightly reduced variance in body weight and in weight by length than in overlapping NCHS samples, because of limitations in the Fels sampling design. The NCHS data for stature from 2 to 3 years are imperfect, because a mixture of recumbent length and stature measurements were made between these ages. The median value is about 1.5 cm higher at 2 years than it would have been if all the measurements had been of stature.

VI. SECULAR TREND

The data on which the charts are based indicate the secular trend in growth has stopped, or almost stopped, in the United States. This is important biologically; in addition, it has practical implications with regard to the expected period during which the charts will be useful.

Bakwin and McLaughlin (1964) and Damon (1968) were the first to suggest this trend was slowing or had, in fact, stopped in privileged groups in the United States. Their findings, together with reports from the Child Research

Council in Denver (Maresh, 1972) and the Fels Research Institute (Roche et al., 1975) show secular trends have ceased in the upper and middle socioeconomic groups of the United States population. The present findings that differences are small and inconsistent between the three NCHS surveys spread over a period of 12 years and, each including a probability sample for the total United States population, show the secular trend has stopped for the total population except for slight changes at lower percentiles (Table I.).

VII. APPLICABILITY

These new growth charts are applicable to all United States children because of the data base from which they were derived. Also, they are considered applicable to children in other countries on the assumption that, with the exception of some small unusual groups, children of all countries would grow similarly in stature, weight, and head circumference if provided with the same environment. There is considerable evidence this is true for preschool children (Habicht et al., 1974). These charts have been recommended by WHO and FAO for worldwide use. Within the United States, they are being distributed free by Ross Laboratories.

Admittedly, if these growth charts were used to assess underprivileged children, in general, these children will be bound to be below the median values shown on the charts. It is important, however, that this be recognized and the causes of their deficit corrected. The alternative is to compare such children with local reference data or their siblings. In this case, their growth is likely to be judged adequate. It is not.

REFERENCES

Bakwin, H. and McLaughlin, S.D. (1964). *Lancet* 2, 1195-1196.
Damon, A. (1968). *Am. J. Phys. Anthropol.* 29, 45-50.
de Boor, C. and Rice, J.R. (1968). "Least Squares Cubic Spine Approximation I-Fixed Knots". Ind. Technical Report No.20, Computer Science Dept., Purdue University, W. Lafayette.
Habicht, J.-P., Yarbrough, C., Martorell, R., Malina, R.M. and Klein, R.E. (1974). *Lancet* 1, 611-615.
Maresh, M.M. (1972). *Am. J. Phys. Anthropol.* 36, 103-109.
Meredith, H. (1948a). "Physical Growth Record for Boys". Bureau of Health Education, American Medical Association, Chicago.
Meredith, H. (1948b). "Physical Growth Record for Girls". Bureau of Health Education, American Medical Association, Chicago.
Meredith, H.V. (1963). *In* "Advances in Child Development and Behavior" (L. Lipsitt and C. Spiker, eds) Vol.I, pp.69-114. Academic Press, New York.
Roche, A.F., Wainer, H. and Thissen, D. (1975). "Skeletal Maturity. The Knee Joint as a Biological Indicator!". Plenum Publishing Co., New York.
Shuttleworth, F.K. (1938). Monogr. Soc. Res. Child Dev. 3 (3).
Stuart, H. and Meredith, H. (1947). *Yearb. Phys. Anthropol.* 46, 175-196.
Tanner, J.M. (1962). "Growth and Adolescence", 2nd ed. Blackwell Scientific Publications, Oxford.
van Wieringen, J.C. (1972). "Seculaire Groeiverschuiving: Lengte en Gewicht Surveys 1964-1966 in Netherlands in Historisch Perspectief". Netherlands Institute voor Praeventive Geneeskunde TNO, Leiden.

VARIABILITY BETWEEN POPULATIONS IN GROWTH DURING CHILDHOOD: COMPARATIVE AUXOLOGY OF CHILDHOOD

P.B. Eveleth

University Museum, University of Pennsylvania
and
W.M. Krogman Center for Research in Child Growth and Development
Children's Hospital of Philadelphia, Philadelphia, Pennsylvania 19104, USA

SUMMARY

Comparative auxology is the study of variability between populations during the growth period. Differences between populations may exist in body size, body shape, body composition, maturity and rate of growth. In body size, European children, well-off Africans, and those of European and African ancestry, tend to be larger than those living in Asia or of Asiatic ancestry. In the upper socioeconomic groups and urban populations, one finds that children are larger, on average, than in the lower socioeconomic or rural groups. Characteristic differences between populations in body proportions are seen in childhood, such as Africans having longer legs and longer arms to sitting height than European or Asiatics. Marked differences are also seen between populations in rate of maturation as evidenced by skeletal ossification.

I. INTRODUCTION

I would like to introduce a new phrase for the topic described as 'auxologic variance in human populations': that is, comparative auxology. Comparative auxology would be analogous to comparative anatomy except that the comparisons are made between populations rather than species. It would consist of analysis resulting from comparisons between two or more populations, either geographic or temporal, to demonstrate variability between populations. My purpose is to look at trends in the differences and similarities between human populations during the growth period.

Comparisons between populations are based on mean values. Thus, they are necessarily influenced by sample selection, and necessitate the use of standardized methods of measurement. The latter was one of the aims of the International Biological Programme. International cooperation is needed in growth studies, not nationalism. Differences between populations need to be real ones, and not those introduced by differing measuring techniques. The only way we can be confident of the results of our comparisons will be by following a standard manual: that is, the IBP volume, "Human Biology, A Guide to Field Methods" (Weiner and Lourie, 1969) or the World Health Organization manual on anthropometry (WHO/NUTR/70129). Both present the same standard methods. It is expected that one of the results of international congresses such as the present one will be standardization of methods used by auxologists throughout the world.

Differences between young adult populations are seen in body size, body shape, and body composition. They result from differences in their gene pools, in their environments, and in the interaction between the two. During childhood, relative differences between populations in size, shape, and composition may, or may not, be in the same direction as those in adult populations. Furthermore, additional differences between child populations in level of maturity and rate of growth must be considered. From the differences between populations during childhood we cannot necessarily predict differences in size or shape between these same adult populations. A population that is taller, for example, than another during childhood may be so either because its members will end up taller as adults, or because they are simply more advanced towards maturity at a given age. Sometimes, however, it is the case that the same size or shape characteristics that differentiate adult populations are clearly visible in childhood.

II. BODY SIZE

Childhood is a period of gradual deceleration in growth in body size following the very high velocity of the perinatal growth period. Annual height increments continue to decrease gradually until puberty. Weight increments, after initial deceleration, again increase around 2 to 4 years of age. Childhood is that period of height deceleration between infancy and adolescence, and extends approximately from 1 to 9 or 10 years of age. Since this period encompasses the age most commonly used to monitor the health and nutrition of a population by using growth analysis, it is quite rich in information from many populations around the world. The discussion here is based on data presented in the appendix tables in Eveleth and Tanner (1976) and the original sources are also fully referenced there.

A. Height and Weight

Differences between populations in length and weight means are considerable even in 1-year-old children. The greatest means are found in European populations, and those of European descent as Americans and Australians, in the well-off populations of Ibadan and Lagos (Nigeria), Shiraz (Iran), Beirut (Lebanon), La Plata (Argentina), Formosa, and in black Americans in San Francisco and Washington (Table I).

Populations with the lowest 1-year-old length and weight means are New Guineans, rural Africans (Keneba, Lower Shire, Pangani Basin), Indians, rural

Table I. One-year-old height and weight means of boys.[a]

Country	Height (cm)	Weight (kg)
Netherlands	77.0	10.5
Sweden (Stockholm)	76.5	10.2
Finland	76.6	10.4
Poland (Warsaw)	77.2	10.7
United Kingdom	76.3	10.2
USA (San Francisco)	75.8	10.3
Australia (Sydney)	77.5	10.9
Nigeria (Lagos)	76.4	10.2
Nigeria (Ibadan)	77.6	10.1
Iran	75.6	10.2
Lebanon (Beirut)	75.8	10.5
Argentina (La Plata)	75.1	10.3
Formosa	75.0	9.0
USA (San Francisco, blacks)	75.5	10.4
USA (Washington, blacks)	76.1	9.6
Brazil (São Paulo, slum)[b]	70.0	8.2
Peru (Quechua Indians)	69.7[d]	8.6[d]
Guatemala (Ladinos)[c]	70.0	8.0
Indonesia (Jakarta, poor)	71.0	8.2
Philippines (rural)	70.8	8.0
India (Maharastrians)	68.0	6.5
New Guinea (Lumi)	68.1	7.7
Tanzania (Pangani Basin)	68.7	8.2

[a] Source references in Eveleth and Tanner (1976). [b] Oliveira et al. (1973). [c] INCAP (1969).
[d] Both sexes.

Filipinos, Gautemala Ladinos, Quechua Indians, Colombians, and the poor of Jakarta and São Paulo (Oliveira et al., 1973) (Table I). Those mentioned are the ones from which I have data: it does not mean they are the only ones.

I have selected a few large geographical areas and have subtracted one of the lowest mean lengths I found in each area from one of the highest (Fig.1). In Africa, for example, 9.9 cm and 1.9 kg separate the average Tanzanian boy from the Pangani Basin from the average well-off boy in Ibadan. In India, Maharastrian boys average 7.9 cm and 2.9 kg below the middle-income boys of north India, and in the Far East 4.0 cm and 1.9 kg mark the early advantage of Taiwanese over the Indonesian poor in Jakarta. In Europe, by contrast, the greatest differences in length means is only 0.9 cm, and in weight means 0.9 kg, and that is by comparing Norwegian boys in Bergen and Spanish boys of lower socioeconomic groups.

Children in Asia, according to data reported so far, are similar to Europeans in size only during the first months of life, and in many samples are smaller at birth. They fall progressively behind as they get older, which might appear offhand to be the result of infants growing up in poor environments. Well-off children in Japan, Formosa and Hongkong do fare somewhat better, but they are still below American means in height at 3 or 4 years, and, after 1 year of age, are lower

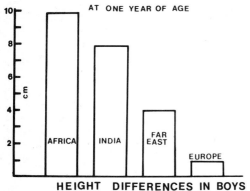

Fig.1 Differences in mean lengths in boys at one year of age of some selected populations throughout the world.

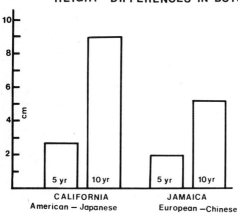

Fig.2 Height differences at 5 and 10 years of age showing greater height in boys of European ancestry compared to those of Asiatic ancestry. (Based on data from Greulich, 1957; Tuddenham and Snyder, 1954; Ashcroft and Lovell, 1964).

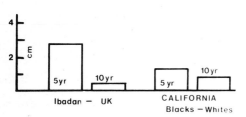

Fig.3 Height differences at 5 and 10 years of age showing greater height in boys of African ancestry compared to those of European ancestry. (Based on data from Wingerd *et al.*, 1973; Janes, 1974 and unpublished; Tanner *et al.*, 1966).

in weight. (Formosans are below at 5 years for height and 7 years for weight.) It is possible that the effects of a poor environment are felt from one generation to the next, so that mothers who have experienced poor growth themselves are more likely to have small infants (Ounsted and Ounsted, 1973; Johnstone and Inglis, 1974). More evidence is needed from well-nourished populations, however, to show whether the smaller size amongst Asiatic children reflects actual genetic differences. We must note that smaller size is also observed amongst Asiatics living outside Asia: Chinese in Jamaica, Japanese in California, Japanese in Brazil,

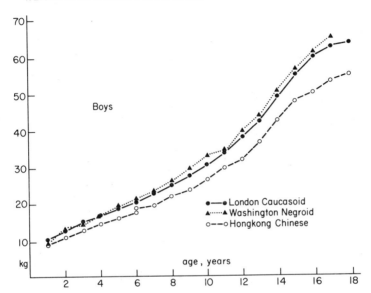

Fig.4 Differences in weight in three major population groups. American blacks in Washington, D.C., are somewhat heavier than British children after 5 years of age, while middle-class Chinese in Hongkong are lighter in weight. (Redrawn from Eveleth and Tanner, 1976. Based on data from Tanner *et al.*, 1966; Chang, 1969; Verghese *et al.*, 1969.)

and Orientals in San Francisco. All of these are descendants of immigrants and have been experiencing an improvement in their environment. Figure 2 shows the deficit in height at 5 and 10 years of age where I have subtracted mean height of Californians of Japanese ancestry in San Francisco (Greulich, 1957) from that of Californians of European ancestry in Berkeley (Tuddenham and Snyder, 1954): well-off Jamaicans of Chinese ancestry are compared in a similar manner with Jamaicans of European ancestry (Ashcroft and Lovell, 1964). Differences that were considerable at 5 years of age were even greater at 10 years.

In contrast, it appears that Negro children of African ancestry are likely to be taller than children of European ancestry and possibly heavier as well when they grow up in a good environment. Numerous investigators across the United States agree in their conclusions that American blacks are taller than American whites living in the same localities (Barr *et al.*, 1972; Garn and Clark, 1976; Garn *et al.*, 1973; Hamill *et al.*, 1970; Krogman, 1970; Rauh *et al.*, 1967; Robson *et al.*, 1975). Height differences between Ibadan, Nigerian boys and English boys at 5 and 10 years of age, and between black and white California boys (Wingerd *et al.*, 1973) in Fig.3, show the greater Negro heights. In contrast to the comparison between boys of European and Asiatic descent, the differences at 10 years are smaller than those at 5 years.

Naturally, it is not possible to show comparison between a large number of populations in a single figure; therefore, three samples have been selected to rep-

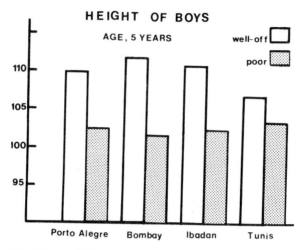

Fig.5 Mean height of 5-year-old boys contrasting growth in well-off and poor socioeconomic groups in some developing countries. (Based on data from Janes, 1974 and unpublished; H. Bourtouline-Young, unpublished; Costa *et al.*, 1970; Udani, 1963).

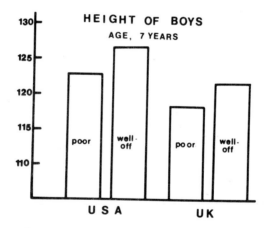

Fig.6 Height means of boys in the United States and United Kingdom contrasting growth in well-off and poor socioeconomic groups. (Based on data from Goldstein, 1971 and Hamill *et al.*, 1972).

resent major ethnic groups in order to show weight differences from 1 through 18 years of age (Fig.4). American blacks are somewhat heavier than English children after 5 years of age, while middle-class Chinese in Hongkong are always lower in weight.

B. *Socioeconomic Differences*

It is assumed that families in the high and middle socioeconomic groups provide their offspring with better nutrition, child care, and take better advantage of health and medical services than families at the lower levels. But not all children in well-off samples have similar height and weight means. For example, middle to upper socioeconomic groups from Hongkong, Tokyo, Bangkok, and Tunis, have means clearly below the European populations. In any single country, children from families belonging to the high or middle socioeconomic groups are, on average, larger in body size than those in the lower socioeconomic groups. Differences between socioeconomic groups in height and weight are found both in developing countries and in highly industrialized ones. Well-off preschool children in Shiraz, Baghdad, Beirut, Tunis, Bombay, Delhi, and other areas of India, Ibadan, Lagos, Addis Ababa, Porto Alegre (Brazil), Santiago, and Bogota, are larger than their lower-class peers. Figure 5 shows height of 5-year-old boys in four of these cities. Many of the well-off groups are very similar to Europeans in weight and height (Eveleth and Tanner, 1976).

Industrialized countries show similar class discrepancies. Seven-year-old children in the British National Child Development Study were on average 3.3 cm taller than children from unskilled working class families (Goldstein, 1971). In the United States, both the Health Examination Survey and the Ten-State Nutrition Survey showed that, as family income and/or education increased, so did the body size of the children (Garn and Clark, 1975; Hamill *et al.*, 1972) (Fig.6).

In contrast to this, however, are reports of a higher incidence of obesity amongst the lower socioeconomic groups in Britain and the United States (Stunkard *et al.*, 1972; Whitelaw, 1971), though not in developing countries. This puzzling observation may be related to type of diet. In Kent, English children of the lower social class had lower total nutrient intake than higher-class children, but consumed amounts of carbohydrate and added sugar equal to those of the higher class resulting in a poorer quality lower-class diet (Cook *et al.*, 1973).

C. Urban-rural Differences

Differences between population within countries are also seen between the urban and rural areas. Children in European cities are generally larger than those in the country, and this may actually reflect an economic differential. In the United States, when the samples were standardized for income level, the rural-urban height and weight differences disappeared (Hamill *et al.*, 1972).

There are data to show rural-urban differences from several European countries, some of which are shown in Fig.7. Finnish boys in the city were taller than those in the country. In Rumania and Greece even greater height differences were shown.

Differences also appear in children of migrants from the country to the city. In Poland, according to Panek and Piasecki (1971), children who grew up in the industrial new town of Nowa Huta averaged 4.3 cm taller and 1.5 kg heavier at ages 4 to 14 years than children who were raised in the farm villages from whence the migrants originated (Fig.8). These results suggest better conditions in the new town, but we cannot rule out some selection of migrants as well (Tanner and Eveleth, 1976).

In developing countries the cities are often newer, and since they are continually fed by migrants from the traditional rural zones, they grow rapidly. On the edge of the cities rise the slums of the urban poor. Here the question is whether such urban slums are better or worse than the villages of the impover-

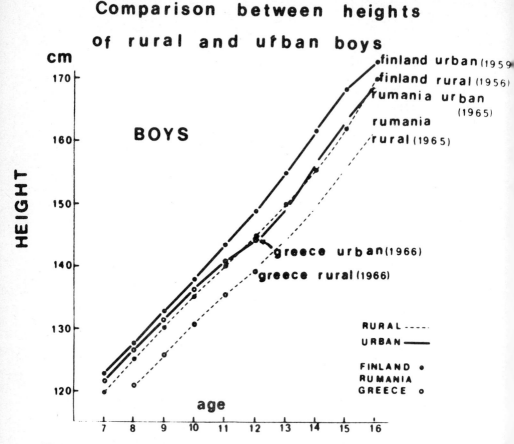

Fig. 7 Height growth in boys using samples from urban and rural regions. Urban boys are taller at all ages than those in the rural areas. The Greek sample goes from 6 to 12 years of age; the Rumanian one, from 12 to 16 years. The Finnish urban sample is from Helsinki. (Based on data from Backström-Jarvinen, 1964; Valaoras and Laros, 1969; Cristeçu, 1969. Redrawn from Eveleth and Tanner, 1976.)

ished countryside that the migrants have left. In Nigeria, Costa Rica, India, and South Africa (Bantu), urban-slum children had heights and weights that were not significantly greater or less than those of rural children (Tanner and Eveleth, 1976). Better-off urban children in Jamaica and Costa Rica, though, were considerably taller and heavier than rural children as one would expect (Eveleth and Tanner, 1976).

III. BODY SHAPE

In characteristic body shape, differences between populations are as well marked as differences in body size. During growth, shape changes as one dimension grows at a different rate from another. Variability between populations in shape may frequently be observed during childhood, but not always.

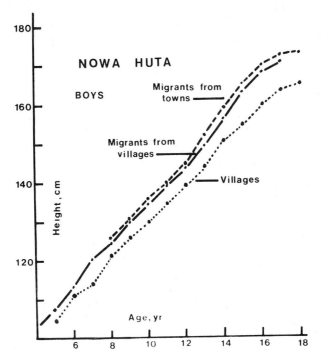

Fig. 8 Height growth in boys born of migrants from farm villages to a new industrial town, Nowa Huta, Poland, compared to those born of migrants from towns and those who remained in the villages. Boys whose families migrated were taller than those who stayed in the villages. (Based on data from Panek and Piasecki, 1971. Redrawn from Eveleth and Tanner, 1976.)

I have analyzed shape differences and shape changes by computing linear regressions of one body proportion on another using the means for each age group in a population. Fuller analysis of shape will be found elsewhere (Eveleth, in preparation).

A. *Sitting Height to Subischial Leg Length*

In sitting height to leg length the characteristic body proportions of West Germans, Japanese, and American Negroes, are clearly established during childhood. In Fig. 9 we see that, at 1 year of age, American Negroes have decidedly longer legs to sitting height than Europeans. At 5 years, Japanese (and Chinese as well) have relatively shorter legs, but they do not maintain throughout childhood the same relationship with the other populations. This is seen by the difference in the slopes of the regression lines. Whereas their relative leg length to trunk length is only somewhat less than it is in Germans at 5 years of age, it becomes increasingly smaller with age as the Japanese develop characteristic short legs.

B. *Other Lengths*

American Negroes not only have longer legs to sitting height, but they also have longer arms to sitting height, and this we can see in children (Eveleth and

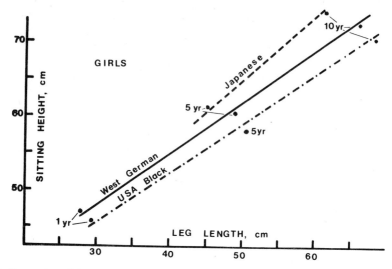

Fig.9 Regression of sitting height on subischial leg length for Japanese, West German, and American black girls from 1 to 10 years. Blacks consistently have longer legs to sitting height. (Based on data from Tokyo Dept. Maternal and Child Health, 1970; Spranger *et al.*, 1968; Verghese *et al.*, 1969.)

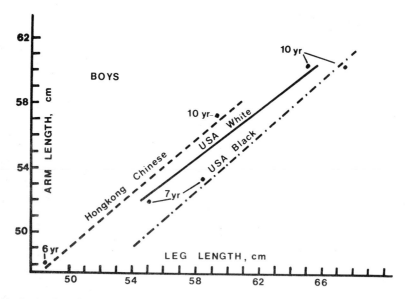

Fig.10 Regression of arm length on subischial leg length for Hongkong Chinese, American whites, and blacks from 6 to 10 years. Chinese have relatively longer arms to legs than other groups. (Based on data from Krogman, 1970 and Chang, 1969.)

Tanner, 1976). On the other hand, Philadelphians of European ancestry and Chinese in Hongkong are rather comparable to each other during childhood in this proportion. In arm length relative to leg length, however, Chinese have longer

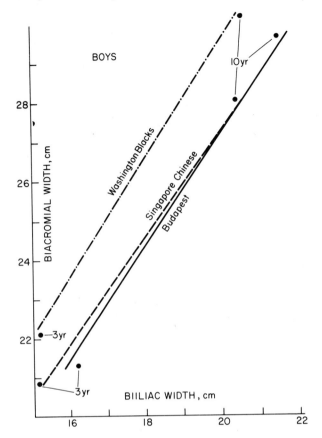

Fig.11 Regression of biacromial width on biiliac width in American Negro, Singapore Chinese and Budapest boys. Negroes have narrower hips to shoulders at ages 3 to 10 years. (Based on data from Eiben *et al.*, 1971; Wong *et al.*, 1972, and unpublished; Verghese *et al.*, 1969.)

arms to legs than either Africans or Europeans, as seen in Fig.10. While in overall size Asiatics at any one age have shorter limbs, their relative proportions even in childhood distinguish them from Europeans or Africans.

C. *Shoulder Width to Hip Width*

In the relationship of shoulder width to hip width, Chinese children do not differ a great deal from European children, as represented here by Budapest (Fig.11). Children of African ancestry in Africa and the USA, however, do have narrower hips to shoulders than Chinese or Hungarians, even as early as 3 years of age.

IV. SKELETAL MATURATION

Variability between populations in level of maturity may be estimated by skeletal maturation. Methods for rating skeletal maturation are based on specified populations. The Greulich-Pyle standards, based on a well-off American sample,

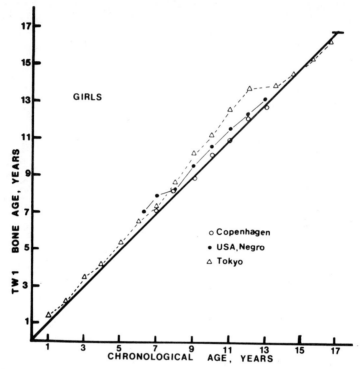

Fig.12 TW I bone age plotted against chronological age for Tokyo, Philadelphia Negro, and Copenhagen girls. Negro and Japanese girls are advanced over British standards during adolescence. (Based on data from Malina, 1970; Andersen, 1968; Ashizawa, 1970.)

are about nine months advanced over the British children of the Tanner-Whitehouse system (Tanner *et al.*, 1975, p.18). With this in mind, we can compare other populations rated by one or the other of these two methods, or by the appearance of ossification centers which does not involve a population-specific standard.

There is some evidence pointing to earlier maturation of very young children of African ancestry, but in Africa and Jamaica this advancement is not continued throughout childhood (Levine, 1972; Marshall *et al.*, 1970; Masse and Hunt, 1963; Michaut *et al.*, 1972). In the United States Negroes, even those of low-income families were advanced in the first appearance of ossification centers, that is, up to about 7 years of age (Garn *et al.*, 1972; Garn and Clark, 1975). From 9 to 13 years at least, black girls in Philadelphia were more mature than Philadelphia white girls (Malina, 1970). Shown in Fig.12, are the Philadelphia black girls against the British standard.

Chinese and some Japanese populations tend to be delayed in skeletal maturation during childhood (Chang *et al.*, 1967; Greulich, 1957; Kimura, 1972a,b; Kondo and Eto, 1972; Low *et al.*, 1964) although the Tokyo populations were very comparable to the British standard (Kimura, 1972b; Ashizawa, 1970) (Fig.12). By 8 years, some of these populations were beginning to mature faster than either the British or American standards. By 13 years, all were well in advance of the standards.

Children living at high altitudes appear to be retarded in skeletal maturation. Populations such as the Quechua Indians in Peru (Frisancho, 1969), the Sherpas and Tibetans (Pawson, 1974) suffer from hypoxia, cold weather, possible poor nutrition, and are the slowest maturing populations recorded.

Throughout this discussion I have not made a strong effort to distinguish between cause and effect. I have not indicated whether genetic or environmental factors may be held responsible for the various effects I have described. Although the causal theme underlies the discussion of population differences, it was not the purpose of the present chapter to attempt to sort out these complexities.

REFERENCES

Andersen, E. (1968). "Skeletal Maturation of Danish Schoolchildren in Relation to Height, Sexual Development and Social Conditions". Universitatsforlaget, Aarhus.
Ashcroft, M.T. and Lovell, H.A. (1964). *Trop. Geogr. Med.* **4**, 346-353.
Ashizawa, K. (1970). *Bull. Mem. Soc. Anthropol. Paris* **6**, 265-280.
Backström-Jarvinen, L. (1964). *Ann. Paediatr.* (Suppl.) **23**, 116 pp.
Barr, G.D., Allen, C.M. and Shinefield, H.R. (1972). *Am. J. Dis. Child.* **124**, 866-874.
Chang, K.S.F. (1969). "Growth and Development of Chinese Children and Youth in Hongkong". University of Hongkong.
Chang, K.S.F., Chan, S.T., Low, W.D. and Ng, C.K. (1967). *Far East Med. J.* **3**, 289-293.
Cook, J., Altman, D.G., Moore, D.M.C., Topp, S.G., Holland, W.W. and Elliott, A. (1973). *Br. J. Prev. Soc. Med.* **27**, 91-99.
Costa, R.M.B., Duarte, L.J.V. and Romeu, A.C. (1970). "Levantamento do Estado Nutricional de Crianças en Idade Preescolar". Servicio de Nutricão Escolar, Porto Alegre.
Cristesçu, M. (1969). "Aspecte ale Cresteril si Dezvoltării Adolescentilor din Republica Socialista România." Editura Academiei Republicii Socialistă România, Bucharest.
Eiben, O., Hegediis, G., Bánhegyi, M., Kis, K., Monda, M. and Tasnády, I. (1971). "Growth and Development of Budapest Kindergarten and Schoolchildren". (In Magyar, English summary). Tempo, Budapest.
Eveleth, P.B. and Tanner, J.M. (1976). "Worldwide Variation in Human Growth". Cambridge University Press, Cambridge.
Frisancho, A.R. (1969). *Hum. Biol.* **41**, 365-379.
Garn, S.M. and Clark, D.C. (1975). *Pediatrics* **56**, 306-319.
Garn, S.M. and Clark, D.C. (1976). *Am. J. Public Health* **66**, 262-267.
Garn, S.M., Clark, D.C. and Trowbridge, F.L. (1973). *Am. J. Dis. Child.* **126**, 164-166.
Garn, S.M., Sandusky, S.T., Nagy, J.M. and McCann, M.B. (1972). *J. Pediatr.* **80**, 965-969.
Goldstein, H. (1971). *Hum. Biol.* **43**, 92-111.
Greulich, W.W. (1957). *Am. J. Phys. Anthropol.* **15**, 489-515.
Hamill, P.V.V., Johnston, F.E. and Grams, W. (1970). "Height and Weight of Children: United States". Vital Health and Statistics Series 11, **no.104**, US Govt. Printing Office, Washington, D.C.
Hamill, P.V.V., Johnston, F.E. and Lemeshow, S. (1972). "Height and Weight of Children: Socioeconomic Status". Dept. Health, Education and Welfare Publ. No. (HSM) 73-1601, Vital Health and Statistics Series 11, **no.119**, US Govt. Printing Office, Washington, D.C.
INCAP (1969). "Evalucaión Nutricional de la Población de Centro America y Panamá". Guatemala.
Janes, M.D. (1974). *Trop. Geogr. Med.* **26**, 389-398.
Johnstone, F. and Inglis, L. (1974). *Br. Med. J.* **3**, 659-661.
Kimura, K. (1972a). *Acta Anat. Nippon* **47**, 358-372.
Kimura, K. (1972b). *J. Anthropol. Soc. Nippon* **80**, 319-336.
Kimura, K. (1976). *Ann. Hum. Biol.* **3**, 149-156.
Kondo, S. and Eto, M. (1972). *In: Proceedings of Meeting for Review of the US-Japan Cooperative Research on Human Adaptabilities".* Japan Society for the Promotion of Science and National Science Foundation, Kyoto.

Krogman, W.M. (1970). *Monogr. Soc. Res. Child Dev.* 35, 1-80.
Levine, E. (1972). *Hum. Biol.* 44, 399-412.
Low, W.D., Chan, S.T., Chang, K.S.F. and Lee, M.M.C. (1964). *Child Dev.* 35, 1313-1336.
Malina, R.M. (1970). *Hum. Biol.* 42, 377-390.
Marshall, W.A., Ashcroft, M.T. and Bryan, G. (1970). *Hum. Biol.* 42, 419-435.
Massé, G. and Hunt, E.E. (1963). *Hum. Biol.* 35, 3-25.
Michaut, E., Niang, J. and Dan, V. (1972). *Ann. Radiol.* 15, 767-779.
Oliveira, Y., Mourão, F.A. and Marcondes, E. (1973). *J. Pediatria* 38, 256-264.
Ounsted, M. and Ounsted, C. (1973). "On Fetal Growth Rate". W. Heinemann Medical Books, London.
Panek, S. and Piasecki, E. (1971). *Mater. Pr. Antropol.* 80, 1-249.
Pawson, I.G. (1974). "The Growth and Development of High Altitude Children with Special Emphasis on Populations of Tibetan Origin in Nepal". Ph.D. Thesis, Pennsylvania State University.
Rauh, J.L., Schumsky, D.A. and Witt, M.T. (1967). *Child Dev.* 38, 515-530.
Robson, J.R.K., Larkin, F.A., Bursick, M.P.H. and Perri, K.P. (1975). *Pediatrics* 56, 1014-1020.
Spranger, J.A., Ochsenfarth, H.P., Kock, H.P. and Henke, J. (1968). *Z. Kinderheilkd.* 103, 1-12.
Stunkard, A.J., d'Aquili, E., Fox, S. and Filion, R.D.L. (1972). *J. Am. Med. Ass.* 221, 579-584.
Tanner, J.M. and Eveleth, P.B. (1976). *In* "Man in Urban Environments" (G.A. Harrison and J.B. Gibson, eds). Oxford University Press, London.
Tanner, J.M., Whitehouse, R.H. and Takaishi, M. (1966). *Arch. Dis. Child.* 41, 454-471; 613-635.
Tanner, J.M., Whitehouse, R.H., Marshall, W.A., Healy, M.J.R. and Goldstein, H. (1975). "Assessment of Skeletal Maturity and Prediction of Adult Height (TW2 Method)". Academic Press, London and New York.
Tokyo Department of Maternal and Child Health (1970). "Physical Status of Japanese Children in 1970". Institute of Public Health, Tokyo.
Tuddenham, R.D. and Snyder, M.M. (1954). "Physical Growth of California Boys and Girls from Birth to Eighteen Years". University of California Press, Berkeley and Los Angeles.
Udani, P.M. (1963). *Indian J. Child Health* 12, 593-611.
Valaros, V. and Laros, K. (1969). *IATRIKI* 5, 266-276.
Verghese, K.P., Scott, R.B., Teixeira, G. and Ferguson, A.D. (1969). *Pediatrics* 44, 243-247.
Weiner, J.S. and Lourie, J.A. (1969). "Human Biology: A Guide to Field Methods". Blackwell Scientific Publications, Oxford.
Whitelaw, A.G.L. (1971). *Hum. Biol.* 43, 414-420.
WHO/NUTR/70.129 (1970). "Nutritional Status of Populations: A Manual of Anthropometric Appraisal of Trends". Geneva.
Wingerd, J., Solomon, I.L. and Schoen, E.J. (1973). *Pediatrics* 52, 555-560.
Wong Hock Boon, Tye Cho Yoke and Quek Kai Miew (1972). *J. Sing. Pediatr. Soc.* 14, 68-89.

SOME CHARACTERISTICS OF THE POSTWAR SECULAR GROWTH IN THE NETHERLANDS

J.C. van Wieringen

University Children's Hospital, Nieuwe Gracht 137, Utrecht, Netherlands

SUMMARY

Statistics on heights of conscripts provide very suitable biometrical research data with regard to the study of secular growth changes. Details of the postwar trend in the Netherlands, where the whole height-spectrum was shifting upwards while the changes were more rapid than ever before, indicate that a genetic threshold has not been reached. This is found in a population where young adults, having a median height over 180.0 cm, belong to the tallest on earth.

We may assume that, in history, there have always been undulating changes in the growth pattern of populations, i.e., an alternate succession of positive and negative shifts. We do not know very much about the "wavelength" of these undulations, but the impression is that it can be very short.

The growth pattern of a particular population is a suitable indicator of the health status of that population. Experience all over the world reveals that an increase of the mean (or median) height at a given age during the growth period (called a positive secular growth shift) coincides with a decline in morbidity and mortality. It is highly probable that it is better nutrition (with respect to quantity and quality) that leads to an earlier skeletal and sexual maturation, greater growth velocity, and taller adult stature. While nutrition is also important in the general improvement of the resistance to diseases, the considerable reduction in number, seriousness and duration of diseases in childhood itself stands for a decline in growth-inihibiting factors. The positive secular trend in the last century shows that, at this moment, in the Netherlands, there is living quite another population — with respect to its biological appearance — when compared with our ancestors of three to four generations ago. A hundred years ago, the generally existing inhibition of growth indicated the "dimensions of poverty", whereas at the moment, the still continuing positive secular trend indicates for the Netherlands the dimensions of wealth, welfare and health.

In those countries where industrialization transformed society, we see the ascending part of a wave that is characterized by its very long duration, and therefore is known as "age trend".

The prolonged duration of the positive trend in a number of countries inevitably evokes questions like: (1) Is the end in sight? (2) Is the distribution of height still diminishing? (3) Accepting better nutrition and decrease in morbidity as the directly acting causal factors, what environmental conditions are concerned with maintaining this trend in a modern postwar society?

As the statistics on heights of conscripts are very accurate, representative for the whole nation and — in the Netherlands — available for almost every draft since the middle of the 19th century, this material, although restricted according to age and sex, provides the best biometrical research data with regard to these questions.

Table I. First drafts in which median heights exceeded 170.0, 177.0 and 180.0 cm respectively. Dutch conscripts according to province.

Province	> 170.0 cm	> 177.0 cm	> 180 cm
Groningen	∼1905	1958	1969
Friesland	∼1903	1953	1969
Drenthe	1920	1962	1972
Overijsel	1917	1962	1972
Gelderland	1917	1964	1975
Utrecht	∼1909	1962	1974
N. Holland	∼1909	1957	1972
Z. Holland	1915	1961	1974
Zeeland	1924	1966	1976
N. Brabant	1933	1969	—
Limburg	1936	1972	—
Netherlands	1917	1964	1975

A century ago, median height of 19-year-old males was 165.0 cm. The median height of the 1975 draft, as measured in 1973, when the boys were 18 years of age, exceeded for the first time 180.0 cm (Table I). Now 9 of the 11 provinces have a median height over 180.0 cm. Between 1860 and 1900, the distribution of the heights diminished rapidly, because the percentages of the shortest statures diminished, while those over 180.0 cm hardly increased. Between 1900 and World War II, the distribution diminished slightly. For the postwar drafts, the shifts of specified centile-values were calculated per 5-year intervals (Table II). Up to the 1960 draft, the shift of the values of the lower centiles remained a little more rapid than those of the higher centiles, but after 1960, the higher centile-values seem to change at least at the same speed. In the last decade, the ratio of the statures over 200 cm increased from 1 to 3 per 1000 conscripts.

Table II. Increments in mm of height-centiles. Dutch conscripts, 1950-1975 drafts.

Centiles	5-year intervals				
	1950-55	1955-60	1960-65	1965-70	1970-75
P 3	17	11	16	9	13
P 10	16	9	14	12	12
P 50	12	7	14	13	14
P 90	13	6	13	13	17
P 97	11	6	16	12	20

As the trend keeps on at a higher speed than ever before, and still the tallest statures participate in the process, it will be clear that the genetically-determined maximum of height as yet has not been attained. Therefore, if it is reported for particular countries or populations, respectively, that the prolonged positive trend appears to have come to an end, at this moment one is inclined to ascribe this to a levelling off of the socioeconomic and sociohygienic living conditions, rather than a genetic threshold.

In the postwar period, a very regular annual increase of 2-3 mm took place all over the height spectrum. This induces the question as to what conglomerate of factors, in a modern society, can be found as an explanation for the fact that still every year-group of 18-year-olds again is taller than the preceding one.

The still existing influence of some of the following conditions is found to be proven in the literature, while the influence of other factors appears to be evident, or merely reasonable.

(1) It is proven that the general nutrition pattern in the Netherlands gradually improved in the postwar decades.

(2) Leaving out of consideration qualitative aspects of artificial formula and breastfeeding, it is true that by promoting breastfeeding too rigidly, as happened in the first years after World War II, a lot of infants did not get their caloric requirements in the first months of life.

(3) Reduction of family size plays a role, as can be concluded from studies in which the heights of children according to the number of siblings were compared. As vital statistics reveal, in the Netherlands nuptial fertility is still decreasing.

(4) A contributive factor may be that the prenatal start of every year group occurs in mothers of age groups that are taller and show lower morbidity and mortality rates than preceding age groups.

(5) Epidemics and endemics of most communicable diseases have been eradicated, either by mass vaccination or by other nation-wide measures.

(6) Improvement of medical care lessened the seriousness and duration of diseases in childhood.

(7) Modern developments in sanitation, nutrition technology, housing conditions, etc., probably still play a role in health and nutrition.

(8) The general care for children in families with medical and social troubles is better ensured than before by the extension of social facilities.

(9) In 1930, 20% of 16-year-old boys and 15% of their female counterparts received secondary education. In 1970, these percentages were 70 for boys and 54 for girls. The considerable increase of the number of adolescents attending

secondary schools may be beneficial to the general care of the subsequent generation.

The conclusion is, that, at the moment, there are still a lot of factors in favour of a continuation of the positive secular trend, while the very rapid shift of the whole height-spectrum during the last decade indicates that a genetic maximum of median height has not been reached in the Dutch population. On the other hand, the recent economic crisis may have a negative influence, especially on the adolescent growth spurt, an effect that may be expected to become visible in the heights of the 18-year-old conscripts of the drafts to be measured in the next years.

RACIAL VARIATIONS IN BODY HEIGHT AND WEIGHT

A.M. Budy and M.P. Mi

University of Hawaii, 1960 East West Road, Honolulu, Hawaii 96822, USA

SUMMARY

Data on individual body height and weight for the total resident population in Hawaii during 1942-1943 were analyzed. These represented seven major racial groups, namely: Caucasian, Hawaiian, Chinese, Filipino, Japanese, Puerto Rican and Korean. The Part-Hawaiian and Others were the two additional groups included. The distribution of body height over a wide range of age was adequately depicted by a fifth-order polynomial regression equation for males and females separately within each racial group. Similar equations were also derived for body weight. These equations were found to be heterogeneous statistically among all racial groups. Observed mean values of body height and weight were tabulated by sex and racial group and by year of age from one to 20 years.

I. INTRODUCTION

Information on the distribution of body height and weight by age is useful for investigation in many disciplines. The routine assessment of body growth of a child is usually based on the accepted standards for the population. Since it is recognized that each racial group has a characteristic pattern of body growth and development, it becomes necessary to make interpolation in assessment for individuals with other racial classification that has not yet been studied extensively. Body measurements are also universally used in the assessment of nutritional status.

The standards of body growth and development can be derived from one of two basic approaches. The first is a prospective approach characterized by the longitudinal study of a selected sample of individuals who are measured initially at a given age and followed for subsequent measurements at preset age intervals. A growth curve is then constructed based on the repeated observations of the same group of individuals. The second approach is a survey type, in which individuals of varying ages are selected randomly from the population for measurement at the time of sampling. Each individual represents a single observation and forms a distribution with others measured at the same age. The mean measurement estimated from such a distribution may be taken as an age-specific population average. These averages are based on different numbers of observations if sampling is made in proportion to the age structure of the population.

It is now generally accepted that body height and weight in man are determined by heredity and environmental factors (Rao et al., 1975; Rashad et al., 1975). It emphasizes that data based on small samples, not randomly selected, may not be representative of the population. The present report using the second approach, describes data on body height and weight of a total population, of a large number of individuals of various racial groups residing in Hawaii during 1942-43.

II. MATERIALS AND METHODS

Data on individual body height and weight for the resident population in Hawaii during 1942-43 were available from the Household Registration file which was established by the Office of Civil Defense. A total of 289,250 individuals residing on the island of Oahu were used. The detailed description of the file and other public records which were used for the development of a population data base for biomedical research was presented elsewhere (Mi, 1967; Mi et al., 1976). Seven major racial groups were represented in Hawaii, namely: Caucasian, Hawaiian, Chinese, Filipino, Japanese, Puerto Rican and Korean. Two additional groups were identified: Part-Hawaiian and Others. Part-Hawaiians were estimated as 50% Hawaiian, roughly 25% Caucasian and 25% Oriental, mostly Chinese, based on evidence from birth certificate as well as gene frequency calculations for the ABO and Rh blood groups (Morton et al., 1967), and from the analysis of dermatoglyphic data (Rashad and Mi, 1971). Because the remaining racial groups and interracial crosses were small in number, they were pooled into a single group of Others.

Body weight was reported in pounds and height in feet and inches in the original document. For each individual, sex and the age at which information was recorded was also available. The distribution of age and body height or weight was analyzed for males and females separately within each racial group. Two statistical procedures were employed. One dealt with data classified by age in 5-year intervals for analysis by the least-squares method for unequal subclass numbers (Harvey, 1960). The statistical model underlying the analysis is given as follows:

$$Y_{ij} = \mu + a_i + e_{ij}$$

where Y_{ij} = the jth observation in the ith age class; μ = the overall mean when equal frequencies existed among all age classes; a_i = the effect of the ith age class expressed as deviation from μ; and e_{ij} = random error associated with each individual measurement and assumed to be independent and normally distributed. The other procedure was the fitting of a polynomial regression equation by regarding the successive powers of the actual age as separate variables. The model is:

$$Y_i = b_0 + b_1 X_i + b_2 X_i^2 + \ldots + b_k X_i^k + e_i$$

where Y_i = the value of body height or weight of the ith individual at age X_i; b_0 = intercept; b_k = partial regression coefficient of Y on the kth variable, X_i^k; and e_i = random error. The highest degree of polynomial was set to five. Both procedures were designed to cope with the nonlinearity of the relationship between age and body height or weight as expected over a wide range in age.

III. RESULTS AND DISCUSSION

A total of 275,674 individuals were used in the final analysis after the removal of records with incomplete information. Table I gives the sample size by sex for each racial group. Based on data classified by age in 5-year inter-

Table I. Number of individuals by race and sex.

Racial Group	Male	Female
Caucasian	40,173	32,350
Hawaiian	4,528	4,136
Chinese	13,862	10,929
Filipino	19,280	5,290
Japanese	47,412	43,540
Korean	2,768	2,283
Puerto Rican	2,759	1,955
Part-Hawaiian	18,012	18,927
Others	3,974	3,496
TOTAL	152,768	122,906

vals, the least-squares estimate of the overall mean of body height and weight and their standard deviations by sex for all racial groups were obtained. Since these samples represented all individuals of varying age from less than one to over 90 years residing in Oahu during 1942 to 1943 and the age structure of each sex differed among racial groups as shown in Table II, it was not appropriate to consider the mean body height or weight estimated in this model as racial parameters for comparison. However, a general observation was that Caucasians and Hawaiians were taller and heavier than the Oriental group at a given age class. Based on the model in which individual data were grouped by age in 5-year intervals, the differences among age classes were found to account for 84 to 90% of the variation in body height, and 75 to 88% in body weight in various racial groups for male subjects. The corresponding figures for body height and weight in female subjects were 82 to 88% and 64 to 82%, respectively. It is apparent that this model is of limited value in estimating body height or weight from age for the population because subjects with varying age but falling in the same 5-year interval would have the same expectation in body height or weight.

In linear simple regression of body height or weight on age gave a relatively poor fit, but the higher-order polynomials improved the fitting significantly. The

Table II. Mean age in years ($\bar{x}$) and standard deviation (s).

Racial Groups	Males		Females	
	$\bar{x}$	s	$\bar{x}$	s
Caucasian	31.12	15.83	30.79	16.93
Hawaiian	31.79	18.66	31.64	18.81
Chinese	31.29	20.10	25.68	16.56
Filipino	32.06	12.71	20.14	15.26
Japanese	26.26	17.94	24.66	16.60
Puerto Rican	28.20	17.69	24.09	16.92
Korean	33.78	22.12	26.10	16.76
Part-Hawaiian	18.36	14.65	19.53	15.42
Others	17.33	14.74	15.39	12.92

fifth-order polynomial was found to explain nearly 90% of variation in body height and 80% in body weight. Because it was based on actual age, the fifth-order polynomial was considered to be far better for describing the distribution of body height or weight of a specified population than the approach in which individuals were grouped in the 5-year intervals. The polynomial regression model, which was based on fewer degrees of freedom, but accounted for the same or a higher proportion of variation, was statistically more desirable. The fifth-order polynomial models for various racial groups were tested statistically for homogeneity within sex. Three test criteria used were: equality of variances, equality of coefficients, and equality of intercepts. The results showed that these racial groups were significantly different from each other in the distribution of body height and weight over a wide range of age from infancy to old age.

In Tables III and IV are the observed mean values of body height and weight for males by race and by year of age from one to 20 years. A similar display of body measurements for females is shown in Tables V and VI. The sample size at each age was comparable between males and females within each race but varied greatly among racial groups. The number of individual observations of one sex, at a given age, ranged from 344 to 888 for Caucasian, 42 to 95 for Hawaiian, 115 to 347 for Chinese, 105 to 242 for Filipino, 418 to 1553 for Japanese, 26 to 75 for Puerto Rican, 15 to 92 for Korean, 442 to 1030 for Part-Hawaiian, and 71 to 280 for Others.

Published data on human somatic measurements have been compiled comparing populations in different parts of the world (Meredith, 1968; 1969a,b; 1970; 1971). Most of the studies were made between 1950 and 1965. The population-based comparison of many racial groups residing contemporarily in the same geographic area, as presented in this chapter, was intended to provide additional background information for the study of human variation.

ACKNOWLEDGEMENT

This work was supported in part by US PHS Grants CA 15655, CP 53511 and HD-04275.

Table III. Mean body height (in cm) for males.[a]

Age (yr)	CAU	HAW	CHI	FIL	JAP	PRC	KOR	PHW	OTH
1	79.26 (8.78)	74.43 (9.21)	77.45 (7.73)	74.05 (7.14)	76.54 (7.74)	77.67 (6.39)	82.72 (10.13)	76.79 (10.05)	76.78 (8.80)
2	89.77 (8.40)	88.36 (12.20)	87.51 (7.37)	85.18 (9.77)	87.19 (8.24)	84.02 (8.08)	85.09 (10.42)	86.38 (9.58)	86.21 (9.21)
3	98.19 (8.42)	94.27 (11.67)	96.75 (7.88)	91.42 (10.77)	94.55 (7.46)	91.36 (10.92)	97.03 (6.31)	95.21 (10.52)	94.96 (7.83)
4	105.34 (8.46)	101.20 (10.50)	102.16 (7.20)	98.31 (11.41)	100.73 (7.13)	99.97 (7.85)	102.07 (13.82)	102.68 (9.21)	101.66 (9.88)
5	111.92 (8.36)	108.49 (10.57)	111.16 (6.54)	104.73 (7.74)	108.15 (5.86)	108.28 (3.99)	110.42 (8.65)	110.48 (8.42)	109.20 (8.22)
6	117.70 (7.84)	116.65 (7.81)	113.26 (6.80)	110.67 (7.54)	112.25 (6.54)	112.85 (7.20)	114.24 (6.46)	114.87 (8.80)	114.16 (7.57)
7	123.24 (9.69)	120.30 (10.75)	120.47 (7.66)	114.63 (9.24)	118.19 (7.29)	118.36 (8.01)	120.36 (8.50)	122.10 (9.23)	119.98 (8.76)
8	128.62 (9.68)	125.55 (12.15)	125.67 (7.77)	119.64 (10.18)	123.14 (7.79)	120.62 (10.26)	126.63 (8.04)	126.47 (9.83)	124.94 (9.03)
9	132.81 (9.60)	130.76 (10.47)	129.90 (7.22)	125.83 (8.68)	127.94 (7.26)	130.27 (7.36)	129.59 (6.55)	132.10 (10.14)	128.31 (8.60)
10	138.42 (9.59)	135.09 (12.04)	135.02 (7.77)	129.09 (9.24)	132.49 (7.23)	133.57 (8.53)	135.67 (8.65)	136.38 (10.06)	135.21 (9.27)
11	142.53 (9.87)	140.90 (8.92)	138.50 (7.26)	134.72 (9.99)	136.94 (7.62)	139.05 (11.34)	139.32 (7.95)	141.91 (9.14)	138.60 (9.77)
12	148.58 (10.11)	147.78 (8.90)	144.04 (8.57)	139.22 (9.61)	142.51 (8.25)	142.54 (11.82)	147.74 (9.93)	147.18 (9.91)	142.39 (9.67)
13	155.34 (10.69)	153.60 (11.27)	151.45 (9.13)	146.69 (10.34)	150.03 (9.01)	146.92 (10.20)	153.15 (9.98)	154.60 (10.85)	150.46 (10.05)
14	161.57 (10.63)	160.34 (12.16)	157.35 (8.81)	153.08 (10.36)	157.01 (7.97)	160.28 (8.83)	159.38 (7.61)	161.25 (9.30)	160.55 (9.87)
15	167.80 (9.45)	166.62 (7.85)	163.72 (7.72)	158.08 (7.27)	161.47 (6.77)	161.22 (10.19)	163.00 (7.57)	166.42 (8.54)	162.21 (10.74)
16	171.58 (9.51)	170.69 (5.81)	166.09 (6.39)	162.44 (6.56)	163.90 (6.00)	165.04 (8.29)	167.37 (7.03)	169.98 (8.07)	167.07 (7.92)
17	173.44 (7.97)	169.63 (6.47)	167.89 (6.27)	163.25 (7.13)	165.32 (5.64)	166.65 (7.59)	169.44 (6.67)	172.02 (6.92)	169.14 (7.55)
18	174.80 (7.54)	170.51 (6.57)	168.31 (5.54)	164.74 (7.58)	165.74 (5.63)	170.05 (7.24)	169.25 (5.75)	172.82 (6.88)	169.24 (7.10)
19	175.90 (7.47)	173.13 (6.48)	169.14 (5.81)	164.79 (5.05)	165.85 (5.72)	170.76 (7.17)	170.84 (5.72)	173.13 (7.42)	171.69 (8.13)
20	176.35 (7.35)	171.76 (6.81)	168.94 (6.20)	163.99 (5.53)	165.79 (5.81)	171.04 (7.30)	169.33 (5.05)	173.01 (6.70)	172.56 (8.10)

[a] Standard deviations in parentheses. [b] Racial groups: CAU — Caucasian; HAW — Hawaiian; CHI — Chinese; FIL — Filipino; JAP — Japanese; PRC — Puerto Rican; KOR — Korean; PHW — Part-Hawaiian; OTH — Others.

Table IV. Mean body weight (in kg) for males.[a]

Age (yr)	CAU	HAW	CHI	FIL	JAP	PRC	KOR	PHW	OTH
1	12.60 (2.02)	11.16 (1.92)	11.33 (1.80)	10.72 (2.30)	11.09 (1.74)	11.05 (2.71)	11.52 (1.18)	11.98 (2.51)	11.61 (2.05)
2	14.50 (2.12)	13.76 (2.75)	13.33 (1.88)	12.54 (2.18)	13.32 (2.78)	13.19 (2.03)	14.67 (2.25)	13.96 (2.60)	13.70 (2.39)
3	16.63 (2.61)	15.79 (2.91)	15.11 (2.17)	14.14 (2.52)	14.92 (2.15)	14.91 (2.78)	15.48 (2.34)	15.81 (2.78)	15.49 (2.73)
4	18.21 (2.81)	20.02 (3.45)	16.59 (2.68)	15.76 (4.36)	16.41 (2.20)	16.81 (2.79)	17.58 (2.78)	17.44 (3.29)	18.05 (11.18)
5	20.48 (3.02)	19.61 (4.30)	19.08 (2.68)	17.22 (2.94)	18.29 (2.62)	18.57 (2.60)	19.16 (1.99)	19.61 (3.33)	19.18 (2.68)
6	22.18 (3.98)	22.65 (3.87)	19.72 (2.88)	19.13 (3.56)	19.75 (2.91)	19.81 (3.12)	20.92 (3.94)	21.28 (3.54)	20.59 (3.56)
7	24.86 (4.69)	24.14 (4.57)	22.39 (4.20)	20.91 (3.65)	21.84 (3.18)	22.00 (4.41)	22.26 (3.29)	24.28 (4.92)	23.47 (4.15)
8	28.12 (5.18)	28.01 (6.80)	24.81 (4.68)	22.73 (3.78)	24.14 (3.67)	24.40 (3.80)	25.03 (3.46)	27.00 (5.43)	25.66 (4.91)
9	30.67 (6.26)	31.12 (5.57)	26.85 (4.94)	25.56 (4.79)	26.65 (4.18)	28.25 (5.36)	27.06 (4.77)	29.96 (6.25)	28.16 (5.67)
10	34.06 (6.63)	34.40 (7.18)	30.48 (6.71)	27.93 (5.37)	29.32 (4.69)	30.00 (6.41)	30.52 (5.04)	33.25 (6.90)	31.22 (6.64)
11	36.14 (6.99)	37.04 (8.17)	32.71 (6.44)	30.92 (6.48)	32.07 (5.43)	34.71 (7.31)	33.23 (3.89)	36.77 (7.29)	34.47 (7.29)
12	40.73 (8.83)	44.60 (10.31)	35.57 (7.20)	34.01 (7.24)	35.61 (6.74)	37.53 (7.06)	40.36 (8.74)	40.62 (8.11)	37.12 (7.39)
13	45.76 (9.83)	49.01 (10.09)	40.70 (8.22)	39.47 (6.63)	40.68 (7.81)	44.65 (22.91)	44.12 (7.26)	46.85 (9.61)	41.28 (7.59)
14	51.37 (10.59)	53.17 (10.09)	45.48 (8.94)	45.41 (7.55)	46.38 (7.62)	49.62 (9.58)	49.39 (7.30)	51.61 (9.37)	49.42 (9.74)
15	57.36 (10.39)	60.26 (9.26)	51.44 (8.65)	49.50 (6.95)	51.13 (7.19)	54.45 (8.51)	52.59 (7.49)	58.19 (9.06)	52.52 (9.19)
16	61.84 (10.70)	64.69 (8.63)	54.52 (7.77)	53.89 (6.57)	54.11 (6.36)	58.10 (7.39)	57.87 (7.67)	62.52 (9.29)	57.85 (7.36)
17	64.12 (8.61)	67.97 (11.74)	56.71 (7.82)	55.38 (7.53)	55.97 (6.24)	59.83 (6.36)	59.53 (7.56)	65.95 (9.32)	61.12 (8.24)
18	66.86 (8.81)	69.02 (9.67)	58.39 (7.21)	57.80 (5.60)	57.35 (7.75)	64.36 (8.29)	61.21 (8.78)	67.17 (9.14)	63.86 (8.73)
19	69.20 (9.41)	70.09 (8.20)	59.49 (8.26)	59.20 (7.41)	57.84 (6.25)	66.54 (8.58)	62.51 (7.31)	68.78 (12.11)	65.43 (9.18)
20	70.45 (9.03)	70.36 (10.58)	59.76 (7.92)	58.94 (9.39)	58.21 (6.48)	67.81 (10.19)	63.84 (7.57)	69.40 (9.84)	66.81 (9.33)

[a] Standard deviations in parentheses. [b] See footnote to Table III.

Table V. Mean body height (in cm) for females.[a]

Age (yr)	CAU	HAW	CHI	FIL	JAP	PRC	KOR	PHW	OTH
1	77.44 (9.36)	73.73 (11.29)	77.14 (7.18)	73.41 (9.40)	75.10 (8.45)	74.07 (9.05)	77.63 (7.48)	75.56 (9.50)	76.63 (8.23)
2	88.89 (7.44)	83.77 (10.67)	85.67 (10.03)	83.01 (9.46)	85.62 (8.25)	79.91 (12.30)	86.00 (9.18)	86.06 (10.20)	85.28 (10.62)
3	96.80 (8.40)	93.67 (10.64)	95.54 (7.55)	90.55 (9.96)	93.51 (7.51)	93.13 (8.12)	97.59 (5.24)	94.33 (10.01)	94.20 (9.17)
4	103.22 (8.54)	100.62 (11.92)	101.51 (7.33)	98.43 (8.19)	100.03 (7.58)	98.79 (10.44)	103.66 (6.03)	101.82 (8.45)	101.06 (7.77)
5	112.22 (9.28)	108.17 (9.30)	109.42 (6.57)	103.98 (8.96)	107.39 (6.62)	105.99 (7.36)	108.54 (8.12)	109.84 (9.02)	108.71 (7.20)
6	116.65 (9.13)	114.08 (8.67)	113.86 (7.38)	109.43 (7.32)	111.30 (7.14)	112.55 (6.98)	112.63 (8.46)	114.26 (8.26)	112.32 (10.38)
7	122.78 (8.79)	121.30 (9.49)	118.25 (7.71)	114.47 (8.78)	116.78 (7.48)	119.05 (8.76)	118.82 (8.89)	120.54 (8.85)	118.00 (10.40)
8	128.83 (9.42)	128.95 (9.47)	124.73 (8.39)	119.64 (9.03)	122.40 (7.49)	122.13 (10.34)	126.25 (10.51)	127.22 (9.16)	125.11 (10.70)
9	132.32 (10.60)	132.53 (10.65)	129.72 (9.20)	125.24 (10.78)	127.88 (7.36)	128.29 (11.81)	131.14 (6.82)	131.25 (8.86)	130.23 (8.57)
10	139.37 (8.94)	138.16 (9.67)	134.94 (8.36)	131.53 (9.86)	133.51 (8.02)	136.26 (9.24)	135.19 (7.47)	137.17 (10.25)	135.53 (10.58)
11	145.06 (9.86)	144.20 (12.68)	141.34 (7.78)	137.23 (10.51)	139.54 (7.73)	142.60 (8.83)	142.12 (7.69)	143.98 (9.00)	140.64 (10.25)
12	151.81 (9.80)	151.01 (8.69)	147.82 (7.79)	143.55 (9.39)	145.59 (7.61)	146.36 (10.22)	147.03 (9.91)	150.52 (9.20)	148.76 (8.64)
13	155.98 (8.45)	156.08 (9.44)	152.23 (6.23)	148.20 (8.51)	149.92 (6.87)	152.30 (9.76)	152.25 (7.96)	155.68 (7.96)	152.33 (8.92)
14	159.50 (7.16)	158.64 (9.57)	154.74 (6.28)	150.15 (7.55)	152.77 (6.19)	153.54 (6.50)	156.90 (5.28)	157.98 (7.45)	155.48 (8.24)
15	160.99 (6.35)	159.80 (6.79)	156.12 (5.46)	150.97 (6.64)	154.16 (5.73)	155.52 (7.27)	157.16 (4.64)	160.19 (6.61)	157.68 (6.01)
16	162.37 (6.47)	160.98 (7.71)	156.98 (5.06)	151.82 (6.54)	154.19 (5.73)	156.14 (8.03)	157.74 (5.80)	160.82 (6.28)	158.43 (6.78)
17	161.80 (6.91)	161.24 (5.87)	157.25 (5.04)	152.98 (6.73)	154.43 (5.35)	157.22 (8.15)	158.31 (5.88)	161.30 (6.25)	158.79 (6.98)
18	161.00 (6.42)	161.98 (6.48)	157.56 (5.77)	152.58 (6.74)	154.68 (5.35)	157.27 (5.66)	158.70 (4.72)	160.97 (6.51)	157.68 (6.96)
19	161.74 (6.42)	160.42 (6.24)	157.08 (5.02)	150.90 (7.62)	154.43 (5.45)	156.21 (7.10)	158.36 (5.50)	161.27 (6.51)	160.29 (7.10)
20	162.42 (7.14)	160.23 (9.55)	157.45 (5.60)	152.12 (6.94)	154.18 (5.58)	156.63 (6.58)	158.35 (5.00)	160.94 (5.81)	159.25 (7.03)

[a] Standard deviations in parentheses. [b] See footnote in Table III.

Table VI. Mean body weight (in kg) for females.[a]

Age (yr)	CAU	HAW	CHI	FIL	JAP	PRC	KOR	PHW	OTH
					Racial Group[b]				
1	11.80 (1.91)	10.70 (2.21)	10.91 (1.58)	9.96 (1.66)	10.53 (1.76)	11.36 (1.91)	11.59 (1.29)	11.27 (2.23)	11.27 (2.20)
2	13.96 (2.20)	12.98 (2.25)	13.03 (2.03)	12.38 (1.96)	12.92 (3.81)	12.46 (2.07)	13.05 (1.59)	13.40 (2.37)	13.34 (2.31)
3	15.86 (2.41)	14.67 (2.72)	14.80 (2.39)	13.50 (2.11)	14.54 (5.21)	14.68 (2.32)	15.76 (2.14)	15.19 (2.77)	14.75 (2.71)
4	17.62 (2.88)	16.69 (3.69)	16.04 (2.48)	15.98 (10.43)	15.80 (2.22)	15.95 (2.74)	16.93 (2.28)	16.96 (3.12)	16.65 (3.53)
5	19.86 (3.12)	19.49 (4.78)	17.56 (2.69)	16.64 (2.86)	17.88 (2.60)	16.86 (2.66)	18.64 (3.64)	19.24 (4.21)	18.46 (2.78)
6	21.84 (3.97)	21.34 (3.68)	19.52 (3.44)	18.12 (2.72)	19.17 (2.72)	18.96 (3.18)	20.13 (4.01)	20.86 (3.75)	19.78 (3.31)
7	24.49 (4.86)	24.43 (4.47)	21.37 (3.30)	20.22 (3.48)	21.31 (3.30)	21.77 (4.38)	22.67 (4.21)	23.85 (4.85)	22.29 (4.54)
8	27.53 (5.12)	27.08 (5.31)	23.72 (4.01)	23.46 (4.96)	23.77 (3.71)	24.31 (4.57)	24.56 (3.78)	27.06 (5.57)	24.87 (5.25)
9	30.05 (6.35)	31.40 (6.69)	26.59 (5.15)	25.57 (5.07)	26.38 (5.17)	30.04 (21.49)	28.75 (5.96)	29.53 (6.26)	27.81 (5.84)
10	34.27 (7.29)	35.68 (8.93)	28.91 (5.52)	28.97 (6.98)	29.33 (5.70)	31.44 (5.90)	29.63 (4.52)	33.55 (7.28)	31.65 (7.85)
11	38.11 (8.24)	40.70 (8.44)	32.71 (6.29)	32.60 (7.18)	33.14 (6.24)	36.88 (9.43)	34.68 (6.82)	38.09 (8.30)	35.54 (7.99)
12	43.33 (9.20)	45.81 (9.67)	37.85 (7.34)	37.15 (7.14)	37.56 (6.74)	42.58 (10.16)	40.06 (8.73)	43.68 (9.25)	42.25 (9.55)
13	47.10 (8.92)	52.83 (10.39)	40.91 (6.42)	43.04 (9.00)	41.10 (6.91)	46.27 (9.14)	42.86 (6.38)	48.32 (8.27)	46.02 (8.58)
14	50.61 (7.90)	53.16 (8.82)	43.81 (6.29)	45.55 (7.22)	44.26 (5.98)	47.65 (5.62)	46.70 (6.54)	51.60 (9.24)	49.16 (9.49)
15	52.07 (7.64)	57.00 (7.80)	45.65 (6.42)	45.78 (6.40)	46.03 (6.22)	50.61 (7.67)	48.51 (6.16)	53.70 (7.92)	50.28 (7.84)
16	53.98 (7.50)	58.34 (8.70)	46.79 (6.82)	47.91 (7.14)	47.02 (6.02)	53.08 (7.64)	49.30 (5.46)	55.08 (8.31)	52.46 (8.46)
17	54.15 (7.60)	61.93 (8.78)	46.40 (5.62)	49.39 (6.90)	47.79 (5.83)	53.54 (6.37)	50.75 (5.97)	57.03 (9.28)	53.07 (9.00)
18	53.76 (8.22)	61.13 (7.90)	47.73 (6.09)	49.10 (9.26)	48.09 (5.87)	54.32 (10.16)	52.36 (6.52)	57.08 (9.55)	53.07 (8.13)
19	54.76 (7.55)	61.27 (11.20)	47.48 (5.73)	49.82 (7.92)	48.40 (5.74)	53.16 (8.05)	51.45 (5.70)	57.34 (10.62)	55.76 (11.06)
20	54.99 (10.53)	62.28 (12.55)	48.34 (6.37)	51.20 (7.71)	48.23 (7.72)	55.64 (9.21)	51.16 (7.94)	57.56 (10.01)	56.24 (22.43)

[a] Standard deviations in parentheses. [b] See footnote to Table III.

REFERENCES

Harvey, W.R. (1960). "Least-Squares Analysis of Data with Unequal Subclass Numbers", U.S.D.A. ARS 20-8.
Meredith, H.V. (1968). *Child Dev.* **39**, 335-377.
Meredith, H.V. (1969a). "Body Size of Contemporary Groups of Eight-year-old Children Studied in Different Parts of the World". Monographs of the Society for Research in Child Development, **Vol.34**, No.1. University of Chicago Press.
Meredith, H.V. (1969b). "Body Size of Contemporary Youth in Different Parts of the World". Monographs of the Society for Research in Child Development, **Vol.34**, No.7. University of Chicago Press.
Meredith, H.V. (1970). *Child Dev.* **41**, 551-600.
Meredith, H.V. (1971). *Am. J. Phys. Anthropol.* **34**, 89-132.
Mi, M.P. (1967). *In: Proceedings of the Third International Congress of Human Genetics* (J.F. Crow and J.V. Neal, eds) pp.489-496. John Hopkins Press, Baltimore.
Mi, M.P., Onizuka, S. and Okinaga, N.S.G. (1976). *In: Proceedings of the Fifth Hawaii International Conference of Systems Sciences* (A. Lew, ed) pp.404-406. Western Periodicals Co.
Morton, N.E., Chung, C.S. and Mi, M.P. (1967). "Genetics of Interracial Crosses in Hawaii". S. Karger, Basel.
Rao, D.C., MacLean, C.J., Morton, N.E. and Yee, S. (1975). *Am. J. Hum. Genet.* **27**, 509-520.
Rashad, M.N. and Mi, M.P. (1971). Abstract presented 4th Int. Congr. Human Genetics.
Rashad, M.N., Yamashita, T. and Mi, M.P. (1975). *Proc. Am. J. Hum. Genet.* **27**, 74A.

GROWTH OF CHILDREN IN ARCTIC CONDITIONS AND IN NORTHERN EUROPEAN COUNTRIES

Genetic, Nutritional and Socioeconomic Aspects

H.K. Åkerblom

Department of Pediatrics, University of Oulu, 90220 Oulu 22, Finland

SUMMARY

A review is presented on growth studies done by several research groups among Eskimo and Lapp children. A striking increase in birth weights has occurred both among Greenland and Canadian Eskimos in the last few decades. Secular growth has occurred among Lapps as well. The investigators attribute the changes to improvements in nutrition and living conditions. The adult height of these populations is still, however, below that of, e.g., Scandinavians. The secular trend in growth continues also in Scandinavian and Finnish children. The author finally reports results of his study group on school-aged children living in a northern Finnish province.

In recent years, several research groups have published growth studies, particularly of the cross-sectional type, on children living in arctic conditions. The central question has been the extent and timing of the secular trend. In the following, some of these studies are reviewed, and own data are presented on the growth of children in a northern, although not arctic country, Finland.

I. ESKIMO CHILDREN

A. *Greenland Eskimos*

Greenland Eskimos have for a long time been the target of interest for Danish investigators. The largest series is comprised in the study of Littauer *et al.*,

(1976), who measured heights and weights of 4,249 subjects in the years 1962-64. The mean birth weights and lengths were similar to values found for Danish children. However, later in life Greenland children were shorter than Norwegian children of the same age and sex, and the height difference increased throughout the growth period to about 12 cm for both sexes by the end of adolescence. The mean height of adult Greenlanders was 163.9 cm for males and 153.2 cm for females, both considerably below Danish standards. Littauer et al. compared their data with those of an old study from Greenland, Hansen's of the years 1885-88, and concluded that secular growth in this period had been approximately 4-5 cm for Greenland males and 2 cm for females. As there was no difference in height to be seen in the two studies for the age groups over 35 years, the authors considered it likely that secular growth among Greenlanders is of relatively recent origin, likely beginning after 1940, probably because of a change of nutrition and welfare. Kromann and Degnbol (1976) recorded the birth weights and lengths of 1,007 infants in Greenland during the years 1955-75, and found the readings significantly higher than those of a previous study from Greenland, that of Fog-Poulsen (1956) (Table I). Kromann and Degnbol's data were, as a whole, similar to Danish birth weights from the year 1970.

Table I. Birth weight (in g) of Greenland and Danish children (Kromann and Degnbol, 1976).

	Greenland children			Danish children (1970)
	Fog-Poulsen (1955)		Kromann and Degnbol (1955-75)	
Boys	3340		3398	3445
Girls	3161	$p < 0.001$	3324	3329
Total	3251		3361	3388

B. Canadian Eskimos

Similarly, in Canadian Eskimos a striking increase in birth weights and height measurements in children has been observed in recent years. The mean birth weight of full-term Eskimo babies born in Coppermine, a western central arctic town in the North-West Territories, was, in 1958-61, 3.22 kg (range 2.39 – 4.32 kg), whereas, in 1965-67, it was 3.50 kg (range 2.39 – 5.00 kg), $p < 0.01$ (Schaefer, 1970). According to Schaefer, the growth acceleration seen to varying degrees in different Eskimo groups from various places in the Canadian Arctic appeared most closely to parallel the increase in the *per capita* annual sugar consumption, which had more than quadrupled in one decade in some of the trading areas of the Canadian central and eastern Arctic. The observation of the association between prenatal growth acceleration, that is, secular changes of birth weight and percentage of "large babies", of distinct population groups, and, on the other hand, their sugar consumption, was originally made by Ziegler (1966). The increase in birth weight is, according to Ziegler's theory, explained by the enhanced hormonogenic effect of sugar and other quickly absorbed "refined" carbohydrates, an effect also potentiated by the proteins (Ziegler, 1976).

II. LAPP CHILDREN

In the north of Europe, children belonging to another interesting ethnic group, the Lapps, have been subject to extensive growth studies, some of them conducted within the International Biological Programme/Human Adaptability research plan. The Lapps live in the northernmost areas of Norway, Sweden and Finland, and some of them in northwestern Soviet Union. The Lapps can be divided into a few subpopulations according to the place of living, language, etc.. One of these subpopulations is the Skolt Lapp group, living in northern Finland.

A. Skolt Lapps

The Skolts are one of the last groups in a developed country, still keeping genetic isolation both from the other Lapp groups and the Finns. This genetic isolation and the poor socioeconomic conditions of the Skolt Lapps were the main reasons for the Scandinavian IBP/HA studies on their population biology during the years 1967-71. In this Lapp population, it has been possible to determine the relationship between secular changes and environmental factors alone. In most other populations, however, the environmental changes are so closely interrelated with increasing heterozygosity, that it is not possible to give a final answer about the main cause of the secular trend (Skrobak-Kaczynski and Lewin, 1976). Studies among the Skolts have shown that the growth rate of the Skolt children is less than that of the Finnish children from the first years of life. Their height at the age of 14 to 15 years is about 8-9% below that of Finnish children. In one of these studies, done on Skolt Lapp children, Forsius (1973) showed them to have a slower longitudinal growth, and Skolt boys, in addition, a weaker motor proficiency, than Finnish children living as far north. Low stature was also more noticeable among the Skolt boys than the girls. Forsius's findings support the theory that boys less successfully than the girls withstand the rigours of an unfavourable environment (Greulich, 1957; Graffer and Corbier, 1966). Changes in body dimensions of the Skolt Lapps seems to be closely related to the alterations of their economic situation (Skrobak-Kaczynski et al., 1974).

B. Other Lapps

Skrobak-Kaczynski and Lewin (1976) studied the secular trends in two groups of Lapps in northern Finland, the already-mentioned Skolt Lapps and the so-called Inari Lapps. The Inari Lapps are another subpopulation of Lapps living in northern Finland, with somewhat better socioeconomic conditions and less isolation than the Skolts. The stature of both Skolt and Inari Lapps showed a considerable increase from those born in the years 1885-1905 to those born in 1945-50, the increase amounting to 10.6 cm ($p < 0.01$) in Skolt males and 5.0 cm ($p < 0.01$) in Inari Lapp males. Mellbin (1962) studied all 450 Lapp children, aged 7-14 years who, during the years 1957 to 1960, attended nomad boarding schools in the country of Norrbotten in northern Sweden. The body height was a great deal less in Lapp children than in Swedish children. A clear increase in height was observed in children born 1950-53 as compared with those born 1942-46, the difference being most pronounced in the far north. Mellbin attributed the differences in growth and development especially to environmental changes, which had taken place more rapidly in the northernmost areas. Mellbin maintained that the environment exerts its influence during the time before the beginning of school.

Strong secular trends are thus observable in the Lapps, too. For instance, Lapp children of school age have at present the same stature Norwegian children had 50 years ago (Skrobak-Kaczynski, 1976). According to Skrobak-Kaczynski and Lewin (1976), the possibility cannot be excluded that, over a shorter or longer period of time, the difference in stature between Lapps and Scnadinavians will be completely eliminated. The older opinion, considering the body characteristics of Lapps as genetically fixed facial marks, has thus lost support. This older opinion held that any shortness of stature in Scandinavians was explained by Lappish admixture, and any increase of stature in Lapps by penetration of Scandinavian genes (Skrobak-Kaczynski and Lewin, 1976).

To summarize, the above investigators consider that the rate of changes in growth has been too rapid to have a genetic origin, and the alterations in growth are generally attributed to improvement in, above all, nutrition and also in other socioeconomic conditions. Hereby, it is of interest to note that both Eskimos and Lapps have traditionally had a periodically high intake of protein, and their main nutritional change during the last decades has been an increase in carbohydrates and total calories.

III. SCANDINAVIAN AND FINNISH CHILDREN

Several recent studies show that the secular trend continues also in children of these populations, and a few of these studies are mentioned in the following as examples.

A. Scandinavians

Karlberg et al. (1976) made a prospective, longitudinal study on the somatic development of 212 randomly selected Swedish children, born between 1955 and 1958. A secular trend was established for physical growth, pubertal development and ossification. In the comparison of the growth data with those of older Swedish studies, the increase of height was greater than that of weight. According to Karlberg et al., there were small differences in body weight between Nordic children, whereas their values were increasingly higher with age in comparison with British standards. The authors further noted that there are pronounced similarities in the timing and pattern of pubertal development between contemporary European investigations (Karlberg et al., 1976).

B. Finns

The Healthy Child Research Group initiated in 1953 a growth study on 2,800 Finnish children comprising pediatric, anthropometric, odontological, and psychological aspects (Hallman et al., 1971). The study included both a cross-sectional and (in a smaller sample) a longitudinal part. A distinct secular growth was observed in height and weight at all ages. For instance, the mean height of 7-year-old boys increased by 1 cm in 2 decades and 3 cm in 4 decades (Takkunen, 1962). The secular trend for menarche has been in Finnish girls 6 months per decade (Kantero and Widholm, 1971). Preliminary results from a later study of growth in 1970 shows that the secular trend continues (Kantero and Ojajärvi, 1976).

The growth charts presently in use in Finland are based on results of the cross-sectional part of the Healthy Child Research Programme.

In a country like Finland, long in the direction south-north, and the climate being harsh in the north, the living conditions have always been poorer in the north than in the south. At the time (in the years 1955-57) of establishing the height and weight curves presently in use in Finland, children in the northern part of the country were somewhat shorter as compared to children from southern Finland, the differences between the height means being significant in boys upwards from the 11th year and in girls in the 7th-9th year (Kantero and Ojajärvi, 1976). However, as the differences were not greater than about one standard deviation, the same growth charts were considered valid for the whole country.

Study on the growth of children living in a northern Finnish province

In the clinical and public health work in the north of Finland the question often arises as to whether children in the north of Finland are shorter than the average Finnish values would predict. In connection with a nutrition survey, conducted in 1975, we had the opportunity to do a cross-sectional study on the growth of over 600 school-aged children, living in the northern province of Oulu (latitudes 64° to 66°). The children were from three communities representing both urban and rural living conditions. The percentage distribution of height, in boys aged 7-15 years, is given in Table II, and that of weight in Table III, as compared to presently-used Finnish growth charts. For height, one can state that the

Table II. Height distribution in boys of a northern Finnish province as compared to Finnish growth charts (Åkerblom et al., 1977).

Age (yr)	N	< 16th percentile	Height (% values) 16–84th percentile	> 84th percentile
7 – 9	118	14.4	63.6	22.0
10 – 11	94	20.2	66.0	13.8
12 – 15	107	11.2	62.6	26.2
7 – 15	319	15.1	63.9	21.0

Table III. Weight distribution in boys of a northern Finnish province as compared to Finnish growth charts (Åkerblom et al., 1977).

Age (yr)	N	< 16th percentile	Weight (% values) 16–84th percentile	> 84th percentile
7 – 9	119	11.8	58.0	30.2
10 – 11	95	10.5	63.2	26.3
12 – 15	107	7.5	57.9	34.6
7 – 15	321	10.0	59.5	30.5

Table IV. Height distribution in girls of a northern Finnish province as compared to Finnish growth charts (Åkerblom et al., 1977).

Age (yr)	N	< 16th percentile	Height (% values) 16—84th percentile	> 84th percentile
7 – 9	120	18.3	66.7	15.0
10 – 11	80	8.7	67.5	23.8
12 – 15	85	11.8	68.2	20.0
7 – 15	285	13.7	67.4	18.9

Table V. Weight distribution in girls of a northern Finnish province as compared to Finnish growth charts (Åkerblom et al., 1977).

Age (yr)	N	< 16th percentile	Weight (% values) 16—84th percentile	> 84th percentile
7 – 9	120	13.3	66.7	20.0
10 – 11	81	9.9	66.7	23.4
12 – 15	85	4.7	78.8	16.5
7 – 15	286	9.8	70.3	19.9

north Finnish boys were, on the whole, not below the average distribution. About 30% of all boys had a weight located above the 84th percentile of the chart, whereas the corresponding percentage for height was 21. The percentage distribution, of height, in girls aged 7—15 years is given in Table IV, and that of weight in Table V. As in boys, the percentage of height, and particularly of weight, readings over the 84th percentile was somewhat higher than expected. The prevalence of obesity was 4.1% among boys and 1.8% among girls.

Our cross-sectional study thus showed that, at least now, height and weight values of children from northern Finland are located in a rather normal way on Finnish growth charts.

IV. FUTURE VIEWS

Several interesting questions about the factors affecting the growth of children in northern and arctic conditions remain to be investigated; for instance, the possible effect of great alterations in light and darkness on growth velocity and pubertal development. With regard to some less developed populations, living in the Arcitc, we can also foresee tasks for future research on growth: How to identify the biologically and socially "sensitive" ages and conditions, through which the greatest improvement of growth and development can be obtained with least effort (Thomson, 1970).

REFERENCES

Åkerblom, H.K., Hasunen, K., Akerblom, L., Siltala, E-L., Makëlä, J. and Wasz-Höckert, O. (1977). *In preparation.*
Fog-Poulsen, M. (1956). *Ugeskr. Laeg.* **118**, 529.
Forsius, H. (1973). *Acta Paediatr. Scand. Suppl.* 239.
Graffer, M. and Corbier, J. (1966). *Courrier* **16**, 1-25.
Greulich, W.W. (1957). *Am. J. Phys. Anthropol.* **15**, 489-515.
Hallman, N., Bäckström, L., Kantero, R-L. and Tiisala, R. (1971). *Acta Paediatr. Scand. Suppl.* 220.
Kantero, R-L. and Ojajärvi, P. (1976). *Nordic Council Arct. Med. Res. Rep.* No.14, 77-83.
Kantero, R-L. and Widholm, O. (1971). *Acta Obstet. Gynecol. Scand. Suppl.* 14.
Karlberg, P., Taranger, J., Engström, I., Lichtenstein, H. and Svennberg-Redegren, I. (1976). *Acta Paediatr. Scand. Suppl.* 258.
Kromann, N. and Degnbol, B. (1976). *Nordic Council Arct. Med. Res. Rep.* No.16, 3-15.
Littauer, J., Sagild, U., Jespersen, C.S. and Andersen, S. (1976). *Nordic Council Arct. Med. Res. Rep.* No.14, 46-61.
Mellbin, T. (1962). *Acta Paediatr. (Uppsala) Suppl.* 131.
Schaefer, O. (1970). *Can. Med. Assoc. J.* **103**, 1059-1068.
Skrobak-Kaczynski, J. (1976). *Nordic Council Arct. Med. Res. Rep.* No.14, 62-69.
Skrobak-Kaczynski, J. and Lewin, T. (1976). *In* "Circumpolar Health", Proceedings of the III International Symposium, Yellowknife, N.W.T. (R.J. Shephard and S. Itoh, eds), pp.239-247. University of Toronto Press, Toronto.
Skrobak-Kaczynski, J., Lewin, T. and Karlberg, J. (1974). *Nordic Council Arct. Med. Res. Rep.* No.8, 17-46.
Takkunen, R-L. (1962). *Ann. Paediatr. Fenn. Suppl.* 19.
Thomson, A.M. (1970). *Am. J. Dis. Child.* **120**, 398-403.
Ziegler, E. (1966). *Helv. Paediatr. Acta Suppl.* 15, 1.
Ziegler, E. (1976). *Helv. Paediatr. Acta* **31**, 365-373.

BIOLOGICAL RESPONSE TO SOCIAL CHANGE: ACCELERATION OF GROWTH IN CZECHOSLOVAKIA

M. Prokopec and V. Lipková

Institute of Hygiene and Epidemiology, Prague
and
Research Institute of Hygiene, Bratislava, Czechoslovakia

SUMMARY

Secular growth trends are examined in Czech and Slovak conscripts, whose mean height increased by 7.9 cm from 1921 to 1971. Geographical distribution and variability of stature is shown on diagrams: mean stature in Czechoslovakia increases from east to west. Growth acceleration in Czech and Slovak children is analysed on the basis of three state-wide growth studies conducted in 1951, 1961, and 1971. Substantially increased growth rates in Slovak children (who were smaller in 1951 and in 1971 remained slightly smaller than the Czech children) are discussed in view of the rapid industrial development and deep social changes that have taken place in Slovakia during the past 25 years. The value of various growth factors, particularly those applicable to the Czech and Slovak growth studies, is analysed.

A remarkable phenomenon may be traced in most of the industrialized populations in the past 100 years and recently also in some developing nations: the children have been getting taller and growing to maturity more rapidly. There are many factors responsible for this secular growth trend and for the acceleration of growth, which means that certain height and weight levels are now attained at an earlier age than in the past. This developmental trend seems to be correlated with improved nutrition, control of infectious diseases, improved and more widespread health care, and improved hygienic conditions (Prokopec, 1974b; Eveleth and Tanner, 1976). Some of the growth acceleration factors, divided into

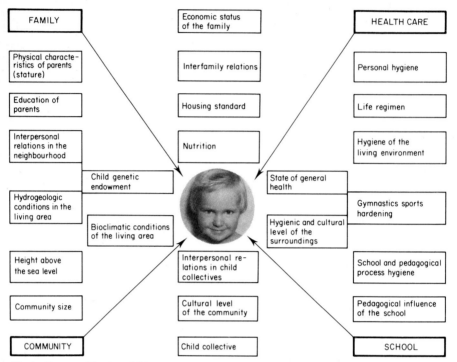

Fig.1 Factors affecting child growth.

four main subgroups or streams, are shown in Fig.1.

The first of them is *family* (upper left corner), which includes child's genetical endowment, the level of nutrition and living standard, economic and educational levels, physical characteristics of the parents, family relations, and interpersonal relations in the surroundings.

The second one is *community* (lower left corner), involving hydrogeologic and bioclimatic conditions of the area including its height above sea level and its size and cultural level.

The third is *health care* (upper right corner) which covers personal hygiene, daily regimen of life style, and the sanitary level of the living environment. There is included also the personal health status of the child.

The fourth factor is *school* (lower right corner) and associated factors, such as pedagogic system of education influences, collective establishment for children, interpersonal relations in these establishments, hygiene of the pedagogic process, physical training and sport, hygienic and cultural level of the school environment.

Some of these factors have been involved in various research programmes and their effects on child growth and development have already been experimentally studied or tested (however, there are still others, which remain to be systematically analysed).

A good information about the body height of the male population, the secular trend towards a higher stature, and the geographic distribution of the mean

Biological Response to Social Change

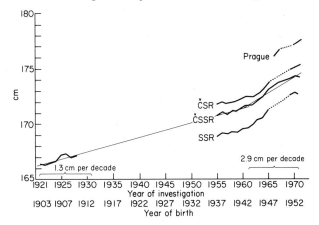

Fig.2 Height increase in Czech and Slovak 18-year-old conscripts during the past 50 years.

body height in the country, is provided by measurements on army conscripts (the data were kindly supplied by the Ministry of National Defense) (Fig.2). The mean height of 18-year-old males in Czechoslovakia has risen, during the past 50 years, by 7.9 cm. The gain was approximately 1.3 cm per decade from 1921 to 1960. Since 1960, this trend toward a higher mean stature has increased to 2.9 cm per decade. If, on the basis of these trends, we were to predict the mean height increments in young adult males for the next 20 years, we should probably chooose a half-way figure, i.e., (1.3 + 2.9 = 4.2/2 = 2.1) 2.1 cm per decade. This judgement is supported by the fact that 18-year-old males from Prague have attained a height of about 178 cm, which, on the basis of this prediction, might be expected in young men after the 20-year period. It furthermore follows that body height in the Czech Socialist Republic (CSR) is greater — exceeding the countrywide mean by 0.5 to 1.0 cm — than in the Slovak Socialist Republic (SSR), where it is about 2 cm below the countrywide mean (Prokopec, 1973b, 1974a).

Persons above 20 are on the average shorter with every year of age, and the overall age-related trend of stature-decrease in adults corresponds well with the trend of a stature-increase in the past decades, but in the opposite sense (Prokopec, 1974b). For every 10 years after the age of 20, the mean height of the contemporary male population decreases by about 1.25 cm (Fig.3).

The gross situation in the mean heights of the Czech and Slovak conscripts is as follows: Prague 178 cm, Czech districts 175, West and Middle Slovakia 173, East Slovakia 172.(Fig.4).

There is quite a variability in the mean stature values even in Prague itself. Less than 5% of short-statured young men below 168 cm live in the northwest districts of Prague, whereas more than 10% live in the southern and northeastern districts that belong to industrial centres. A similar picture of stature diversity in various Prague districts is to be seen in Fig.5, showing the incidence of tall young men (over 180 cm).

The bodily growth of children and youths has been systematically measured in Czechoslovakia since 1951 within the framework of countrywide anthropometrical studies of representative samples taken at 10-year intervals. Each study

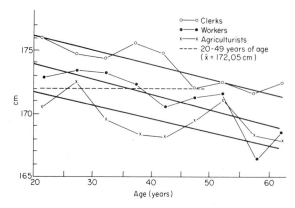

Fig.3 Height decrease in adults of different occupations with age.

involved over 200,000 Czechoslovak children selected randomly from medical centres and schools and measured by pediatricians, child nurses, or teachers of biology and physical training, according to precise guidelines and after personal instruction of those concerned. The data were always worked out separately for the CSR and the SSR (Fetter *et al.*, 1954; Prokopec, 1964).

Height of boys and girls aged 6 to 18 years is shown in Fig.6. In addition to the 1951, 1961 and 1971 curves for children from 6 to 18, there are also included the curves for children from 7 to 14, constructed from a study conducted in 1895 by J. Matiegka in cooperation with schoolteachers, as well as the results obtained in Prague in 1971 on some 5,000 children from 6 to 18. The boy-height curves for 1951, 1961 and 1971, and Prague 1971, lie above each other at more or less equal intervals, with the exception of the 15 to 18-year segment, where the Prague data display considerable variation because of a small number of representatives in this age category. Nevertheless, the distance between the 1951 and 1961 curves is greater than that between 1961 and 1971 curves up to the age of 15 years (Prokopec *et al.*, 1973a).

Growth has accelerated by about 2 to 3 months of age since 1961 and by about 8 months since 1951. As compared with 1895, the growth acceleration at the age of 7 to 14 years amounts to 2 to 2.5 years; in other words, the difference mean heights in children of equal age amounts to 10 to 15 cm.

The girl curves obtained in individual studies lie, as in the case of boys, above each other at regular, 1 to 2 cm, intervals, except for the curve for Prague girls above 15 and the 1961 curve for 18-year-olds. In 1961, the girls did not display an increase in height after the age of 16, whereas in 1971 the height curve continued to grow up to the age of 18. In 1971, equal height values were attained by girls ¼ of a year earlier than in 1961, 1/3 to 1 year earlier than in 1951 and 2 to 2½ years earlier than in 1895.

The results demonstrate that the mean height at every age has increased as compared with 10 years ago, and the considerably higher mean values for Prague boys and girls indicate that the possibilities of mean acceleration and increase have not yet been exhausted in Czechoslovakia, even for the remaining greater part of the youth.

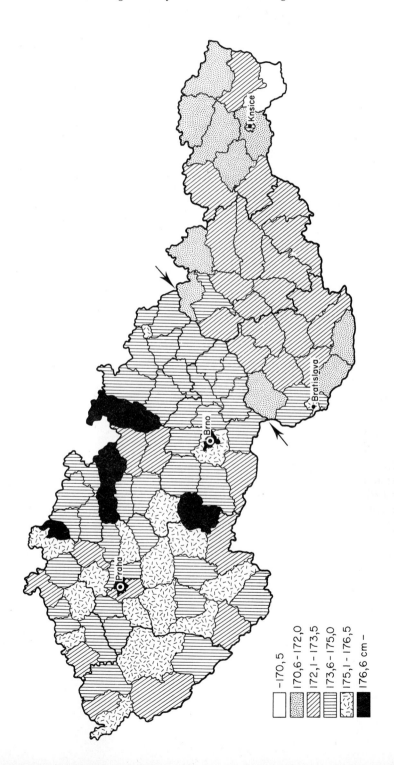

Fig. 4 District mean heights in 18-year-old conscripts in Czechoslovakia in 1971.

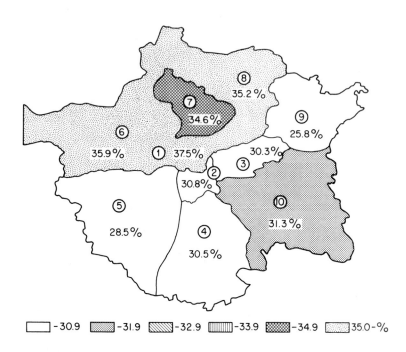

Fig.5 Percentage of Prague conscripts with stature over 180 cm in various parts of the city.

Weight of boys and girls aged 6 to 18 years is shown in Fig.7. Up to 15 years, the 1971 curve for boys passes at equal distance between the 1961 curve and the curve for Prague boys. The difference amounts to approximately 0.5 kg and 2.0 kg at the younger and older ages, respectively. The 1951 curve runs lowest at approximately the same distance up to 9 years and then declines toward still lower values. From the age of 16, the 1961 weight curve for boys forms a plateau and the curve for the Prague sample has an irregular course due to a small number of children measured. As compared with 1951, the mean weight of boys aged 6 to 18 years has increased by 1 to 4 kg in 1971. As compared with 1895, the differences at the age of 7 and 14 amount to about 4 and 13 kg, respectively, which means that an equal mean weight is reached today by Czech boys 1.6 to 2.3 years earlier than in 1895.

The position of the weight curves for girls from the individual studies is similar to that for boys, with the only difference that, at the age of 18, the curves for 1951, 1961 and 1971 converge at a value of 59 kg. The weight of 18-year-old girls has practically not changed since 1951, probably because of rational dieting and subsequent deliberate weight reduction. The only change is that the same weight has been attained at a lower age. Prague girls are heaviest in the age groups between 6 and 16. As compared with 1895, Czech girls in 1971 reached equal weight values 2 to 2.3 of a year earlier. At an equal age, this

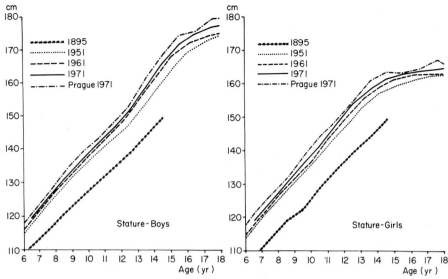

Fig.6 Growth acceleration in Czech boys and girls.

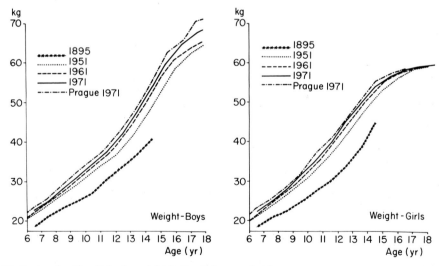

Fig.7 Acceleration in body weight in Czech boys and girls.

gives a mean difference of 4 to 11 kg.

On the whole, the height and weight curves for boys and girls, with the exception of Prague children, display a smooth course without any necessity of additional smoothing out, which speaks for the good work of all the persons collaborating in the data collection, as well as for accurate data processing, and confirms the laws of growth which the curves express by their characteristic course.

Table I. Body height in boys and girls from 3 to 18 years of age in 1951 and 1971 in Czechoslovakia.

Age (yr)	Boys				Girls			
	1951	1971	Difference cm	%*	1951	1971	Difference cm	%*
CSR								
3	95.3	96.5	1.2	1.26	94.0	95.6	1.6	1.70
4	101.9	103.8	1.9	1.86	101.1	103.1	2.0	1.98
5	108.4	111.0	2.6	2.40	107.5	110.3	2.8	2.60
6	114.3	116.7	2.4	2.10	113.6	116.4	2.8	2.46
7	120.4	122.7	2.3	1.91	119.9	122.1	2.2	1.83
8	126.0	128.2	2.2	1.75	125.2	127.7	2.5	2.00
9	131.4	133.8	2.4	1.82	130.5	133.1	2.6	2.00
10	136.1	138.6	2.5	1.84	135.6	138.4	2.8	2.06
11	140.7	143.7	3.0	2.13	141.4	144.8	3.4	2.40
12	144.7	148.4	3.7	2.55	146.6	150.8	4.2	2.86
13	150.1	154.6	4.5	3.00	152.3	156.0	3.7	2.43
14	156.7	161.6	4.9	3.13	156.4	159.9	3.5	2.24
15	163.0	168.2	5.2	3.07	159.0	161.9	2.9	1.82
16	168.4	172.4	4.0	2.37	160.7	162.9	2.2	1.37
17	171.6	174.9	3.3	1.92	161.7	163.5	1.8	1.11
18	173.4	176.4	3.0	1.73	162.1	163.8	1.7	1.05
SSR								
3	93.8	95.8	2.0	2.13	92.7	94.7	2.0	2.16
4	100.0	103.1	3.1	3.10	99.4	102.5	3.1	3.12
5	106.4	110.1	3.7	3.48	105.9	109.5	3.6	3.40
6	111.9	115.9	4.0	3.57	111.1	114.9	3.8	3.42
7	117.7	121.4	3.7	3.14	116.9	120.8	3.9	3.34
8	123.2	126.9	3.7	3.00	122.3	126.5	4.2	3.43
9	128.2	131.9	3.7	2.89	127.1	131.6	4.5	3.54
10	132.4	137.0	4.6	3.47	132.3	137.0	4.8	3.55
11	137.8	141.3	3.5	2.54	138.3	142.9	4.6	3.33
12	141.8	146.4	4.6	3.24	143.4	149.1	5.7	3.97
13	146.9	152.8	5.9	4.02	148.7	154.9	6.2	4.17
14	152.5	159.4	6.9	4.53	153.1	158.6	5.5	3.59
15	158.9	166.4	7.5	4.72	156.4	160.6	4.2	2.68
16	165.9	171.0	5.8	3.07	158.5	161.7	3.2	2.02
17	168.5	173.9	5.4	3.20	159.8	162.2	2.4	1.50
18	170.4	175.2	4.8	2.82	160.1	162.4	2.3	1.44

* The 1951 value is taken as 100%.

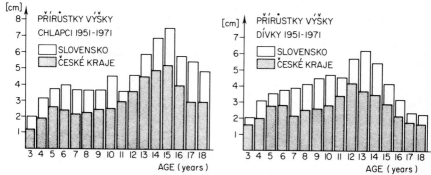

Fig.8 Increments of the mean body height in Czech and Slovak boys and girls between 1951 and 1971.

All that has been said until now concerned children from the CSR. Now we are approaching a very interesting point — which is the different rate of growth in Slovak and Czech children during the time period between 1951 and 1971 (Table I). The Slovak children were on the average smaller in 1951 than the Czech ones. As a matter of fact, they remained smaller also in 1971, but the difference between the means diminished, with a tendency to fuse sometime in the not-very-distant future. Figure 8 shows differences between the 1951 and 1971 height means in both groups, separately for boys and for girls. The 20-year "increments" of the Czech boys are indicated by dark columns and those of the Slovak boys by white columns. Greater differences (increments) are in the Slovak boys in each age group from 3 to 18 years. Note that the greatest "increments" in both groups are detectable in the puberty period and then also at the age between 5 and 6. A similar picture is to be seen in girls. Slovak girls grew also much faster than the Czech girls between 1951 and 1971 (Prokopec *et al.*, 1977).

Should we explain the rapid growth of the Slovak children during the 20 years from 1951 to 1971 by the enormous industrial development which took place in Slovakia in the same period? European countries serve sometimes in some respects as a laboratory for the developing countries. This might be one such example. The change from 1951 to 1971 can even be expressed in statistical terms (gross national product, energy output, mean wages, number of employees, food consumption, etc.). The Slovak example speaks of a hidden strength and vigour in the nation. Industrial achievements are accompanied by improvements in health care, education, and socioeconomic factors. Children are taking lunch at school. There is no longer need for children to work at home, which was still a common practice in rural districts, say 20 years ago. Buses bring children to school and back, even in the most remote villages. The fact that the Slovak children were, still in 1951, much below their theoretically optimal growth level, that they were, so to say, "unsaturated" in growth, was very important. This found them sensitive to the slightest improvements and made them react quickly once the growth-depressive factors had been put aside.

Certainly, many other factors may be involved. It is interesting to note that the number of children per family diminished considerably in both Czech and Slovak Socialist Republics during the critical period. In 1951, for every 100

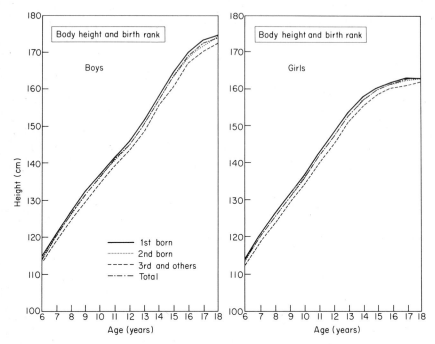

Fig.9 Body height of boys and girls from 6 to 18 years of age according to birth rank.

births 16 children in the CSR and 26 in the SSR were fourth or over in birth rank. These numbers dropped by 1971 to 4% in the CSR and to 14% in the SSR. This fact alone may contribute to a higher stature in children. This will become more apparent from Fig.9, showing our previous findings from the 1951 study, in which the highest curve represents the mean heights of first-born children; those of second-born children coincided well with the general mean, but those of third-born children and others were on the average significantly smaller in all age groups from 7 to 18 years of age (Prokopec, 1969).

Other growth-influencing factors were also derived from our state-wide study. In Fig.10, mean heights of Czech children from 1951 are presented, subdivided by their midparent heights in each age group from 6 to 18. Regression lines show how the mean heights of children correlate with their midparent heights. Small parents produce predominantly small, and tall parents predominantly tall children. If we omit the differences between boys and girls, we may arrive at the following rule to understand the parental influence on a child's height: for each cm by which the midparent height differs in either direction from the adult mean height (men and women together), the child is expected to be taller or smaller in comparison with the average child of corresponding age; between 6 and 9 by 0.4 cm, between 10 and 15 by 0.5 cm, and between 16 and 18 by 0.6 cm.(Prokopec, 1972, 1973c).

Similarly to other European countries, in the past 80 years, the age of menarche has been decreasing in Czech girls from about 16 to 13 years (12.8

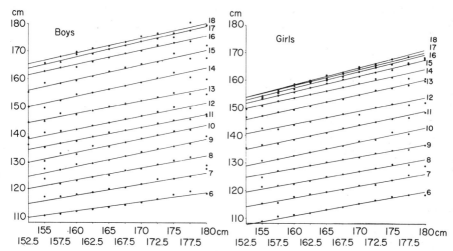

Fig.10 Midparent height and mean height of sons and daughters from 6 to 18 years of age (CSR, 1951).

in Prague girls in 1962; Prokopec, 1967, 1974b). In contrast to Norway and Britain, this trend has not yet been reported to have stopped in Czechoslovakia. A revision is to be done. The fact that children under 12 are often allowed to watch TV programmes for adults may participate in an earlier onset of mental and sexual maturation.

The development of axillary and pubic hairiness in both sexes, or mammae in girls and mammillae and genitalia in boys, was scored according to 5 stages (0 to 4) in several thousands of Czech and Slovak boys and girls in 1962. The mean height of boys with infantile stage of pubic hairiness, as seen in Fig.11, is far below the general mean. Boys of stage 2 between 10 and 13 years of age are well above the gross mean height (or weight), but are quite below the average if this stage of pubescent development is present at the age of 15 or even 17. At the age of 14, there are still boys with an infantile stage of pubescent development; on the other hand, there are also boys of stage 4. This means that, in the same school class, quite immature and fully mature boys sit side by side, competing in their mental and physical capacities and abilities, in their school progress, and in physical education and endurance. This poses quite a problem for their parents and teachers (Prokopec, 1967).

REFERENCES

Eveleth, P.B. and Tanner, J.M. (1976). "Worldwide Variation in Human Growth". Cambridge Univ. Press, Cambridge.
Fetter *et al.* (1954). *In* "Handbook of Pediatricians in Child Health Centres" (in Czech). Stát. Zdrav. Nakl., Prague.
Prokopec, M. (1964). *Indian Pediatr.* 1, 100-111.
Prokopec, M. (1967). "Development of Youths in the Adolescence Period — Report for the Ministry of Health" (in Czech) pp.287, Prague.
Prokopec, M. (1969). *Anthropologie* 7, 27-32.
Prokopec, M. (1972). *Čs. Pediatr.* 27, 432-435 (in Czech with English summary).

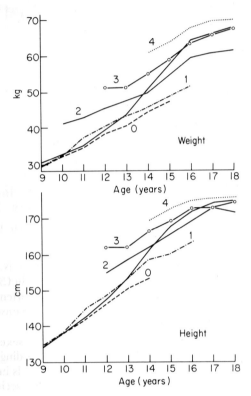

Fig.11 Body weight and height of Czech boys between 9 and 18 years of age according to stage of pubic hair development (0 – 4).

Prokopec, M., Suchy, J. and Titlbachova, S. (1973a). *Čs. Pediatr.* **28**, 341-346 (in Czech with English summary).
Prokopec, M. (1973b). *Demosta* **6**, 50-53.
Prokopec, M. (1973c). *In* "Compte Rendu de la XIe Réunion des Équipes Chargées des Études sur la Croissance et le Développement de l'Enfance Normale – Centre Internationale de l'Enfance", pp.197-199. Paris.
Prokopec, M. (1974a). *Demografie* **16**, 117-123 (in Czech).
Prokopec, M. (1974b). *Czech Central Council of the ČSTV, Prague* **3**, 19-46 (in Czech).
Prokopec, M., Lipková, V. and Zlamalová, H. (1977). *Demografie* **19** *(in press)*.

CHANGES IN BODY MEASUREMENTS AND PROPORTION OF CHILDREN, BASED ON KÖRMEND GROWTH STUDY

O.G. Eiben

Department of Anthropology, Eötvös Loránd University, Budapest, Hungary

SUMMARY

Two cross-sectional growth studies have been carried out at Körmend (West Hungary) in 1958 and 1968. On the basis of very detailed anthropometric investigations of boys and girls (1656 subjects studied in 1958, 1720 in 1968) aged 3—18 years, it was found that boys and girls in 1968 were taller, heavier, and rather more linear than in 1958. The proportional changes occurring during growth were analyzed by the help of the unisex phantom constructed by Ross and Wilson (1974), thereby registering the changes in the children's physique. This phantom is strongly recommended for the analysis of proportional changes.

I. INTRODUCTION

Nowadays, and more especially in 1977, in the Children's International Year, we turn our attention towards the growth problems of children. Children grow up only once: bearing this in mind, we grown-ups should do everything in order to provide optimal conditions for the children's biological-psychological process and thereby the development of children's personality. This is what stimulates the human biologist to learn in even more detail the growth process, body development, features of maturation, alongside with the proportional changes taking place during growth.

The phenomenon is principally analyzed by longitudinal growth studies; still cross-sectional growth analyses would also yield important information, especially when cross-sectional studies are carried out at the very same locality: the so-called follow-up investigations. The genetic endowments of a group of children representing a certain population, which significantly determine growth, do in fact very slowly change; on the other hand, the environmental factors, which similarly influence this process, change relatively quickly. Accordingly, a follow-up measurement some 10 years later may register significant changes (Eiben, 1969a).

The present contribution is aimed to answer the following questions:

(1) What changes have taken place, demonstrable in body dimensions, in the youth of Körmend between 1958 and 1968; and

(2) What proportional changes accompanied the change of growth process, and how these may be identified.

II. MATERIAL AND METHOD

I carried out cross-sectional growth studies at Körmend in 1958 (K-58) and in 1968 (K-68). Körmend is the chief town of a district in West Hungary with 7,500 inhabitants in 1958, and 9,000 in 1968. Between 1958 and 1968, the lapse of 10 years brought significant changes in the town, i.e., it became industrialized with definite signs of urbanization (Eiben, 1961, 1963, 1969a, 1969b).

The K-58 investigation involved 1,656 boys and girls between the age of 3 and 18 years. The K-68 investigation involved 1,720 children similarly between the age of 3 and 18 years (Tables I and II).

During both investigations, I proceeded according to a very detailed anthropometric programme. The present chapter is designed to discuss body weight, the most important length and width measurements, along with chest circumference. The K-68 investigation was made within the IBP/HA. In the course of mathematical-statistical elaboration, the usual parameters were calculated, by which the results of the two investigations, as a first step, were comparable. The results are shown in Tables I and II.

The changes in body proportions taking place during growth, in the present case the differences between the results of K-58 and K-68 and the proportional differences apparent between the boys and girls, have been analyzed by the unisex phantom constructed by Ross and Wilson (1974). This adult unisex phantom is a metaphorical model derived from both sexes. We can describe it by over eighty length, breadth, girth and skinfold values and standard deviations based on male-female size-adjusted averages appropriate for a unisex model with an arbitrary 170.18 cm height, 64.58 kg weight, and 18.78% body fat. The phantom is used by expressing obtained measurements by application of the following general formula:

$$z = \frac{1}{s}\left\{\ell \frac{(170.18)^d}{h} - p\right\}$$

where
- z is a proportionality standard score;
- s is a prescribed standard deviation for a given item for the phantom;
- ℓ is any obtained measurement on a subject for a given item;
- 170.18 is the phantom height constant in cm;
- h is the subject's obtained height in cm;
- d is a dimensional constant based on geometrical considerations;
- $d = 1$ for all lengths, breadths, girths and skinfold thickness;
- $d = 2$ for areas and static strength measures which are related to the area of cross-section of muscle tissue; and

Table I. Body measurements of Körmend boys in 1958 and 1968.

Age (years)	Weight (cm)					Height (cm)				Length of the upper extremity (cm)				Length of the lower extremity (cm)				Biacromial width (cm)				Bicristal width (cm)				Chest circumference (cm)			
	K–58			K–68		K–58		K–68		K–58		K–68		K–58		K–68		K–58		K–68		K–58		K–68		K–58		K–68	
	N	x̄	s	N	x̄	s	x̄	s	x̄	s	x̄	s	x̄	s	x̄	s	x̄	s	x̄	s	x̄	s	x̄	s	x̄	s	x̄	s	x̄	s
3	13	14.5	1.6	12	14.8	1.9	94.5	4.4	98.0	3.7	38.1	1.8	39.0	1.8	45.7	2.6	48.0	1.8	22.5	1.1	22.2	0.6	17.5	0.8	16.2	0.9	52.1	2.4	52.5	2.1
4	24	16.8	2.3	22	15.6	1.8	100.1	4.4	101.7	4.1	40.6	3.5	41.2	3.6	49.9	3.9	51.0	2.9	23.7	1.8	23.0	1.3	18.7	1.3	17.1	1.1	54.6	2.8	53.5	2.1
5	38	18.2	2.2	35	18.0	2.5	107.7	5.1	109.4	5.3	44.5	4.4	45.9	3.0	54.6	3.4	56.4	4.1	24.8	1.2	24.5	1.3	19.5	1.0	18.0	1.0	56.0	2.4	55.9	2.9
6	49	19.3	2.3	41	20.4	3.2	112.4	4.8	114.9	5.4	47.9	3.9	48.0	3.2	57.9	3.0	60.6	3.9	25.4	1.3	24.9	1.6	19.8	0.9	18.7	1.0	56.6	2.9	56.7	2.8
7	79	19.7	2.6	53	21.7	2.3	116.9	4.8	120.6	4.9	50.2	3.1	51.1	3.3	60.8	3.7	64.1	3.8	26.5	1.3	26.1	1.3	20.5	1.2	19.2	1.1	57.9	2.6	57.7	2.6
8	71	23.5	3.8	53	24.3	3.6	123.4	5.4	126.1	5.3	53.0	3.1	54.1	3.3	64.9	3.9	68.1	3.7	27.6	1.4	27.1	1.9	21.5	1.5	20.0	1.2	60.1	3.5	59.9	3.5
9	61	25.1	3.5	67	27.1	5.6	128.6	5.8	131.2	6.3	56.1	2.9	56.6	3.4	68.8	4.3	71.5	4.2	29.1	1.4	28.3	2.0	22.5	1.5	20.9	1.3	62.2	3.0	62.5	6.1
10	65	29.1	4.7	51	30.4	4.8	134.2	6.6	137.3	7.1	59.3	3.5	59.5	4.1	72.2	4.3	74.6	4.6	30.1	1.6	29.5	1.7	23.7	1.9	21.8	2.0	64.7	3.9	64.0	5.8
11	67	30.0	4.1	60	32.7	5.8	135.6	6.6	141.2	7.5	59.9	3.3	61.7	4.6	73.8	4.3	78.4	5.3	30.4	1.7	30.3	2.0	23.7	1.6	22.6	2.0	65.7	3.7	66.0	4.5
12	41	33.7	5.9	57	35.4	5.6	143.8	7.5	145.3	6.7	63.3	3.7	63.4	4.4	78.2	4.5	80.3	4.6	31.7	2.1	31.2	1.9	24.8	1.7	23.1	1.6	69.0	4.4	67.8	5.1
13	59	38.0	5.4	84	39.6	7.3	148.8	6.8	152.1	7.3	66.1	3.5	67.3	4.7	82.0	4.8	84.4	4.8	33.1	1.7	32.7	2.1	26.3	1.9	24.1	1.6	72.3	4.2	71.4	5.9
14	66	41.3	9.0	85	44.5	8.1	153.0	8.9	156.8	8.4	67.6	4.7	69.8	4.9	83.6	5.2	86.3	5.2	34.1	2.5	34.2	2.4	27.2	2.5	24.9	1.8	74.6	6.1	75.0	5.9
15	50	50.2	8.9	140	51.6	9.5	161.6	8.3	164.0	8.5	72.0	3.5	73.3	4.7	88.4	4.3	90.1	5.0	36.3	2.2	36.1	2.6	29.3	2.0	26.2	2.1	80.3	5.8	78.9	7.1
16	66	54.4	6.6	109	56.6	9.9	164.8	5.8	167.7	7.1	73.7	3.6	75.2	4.0	89.6	3.8	91.4	4.4	37.7	1.9	36.8	2.2	29.6	1.7	27.2	2.0	83.1	4.5	82.9	6.5
17	53	57.2	8.5	89	60.2	6.6	166.5	7.1	171.1	6.5	73.8	3.8	77.3	3.5	89.9	4.8	92.7	3.9	37.9	2.1	38.1	2.1	30.7	2.0	27.8	1.5	84.8	5.7	85.4	5.4
18	44	61.8	8.2	25	59.9	4.7	169.9	7.4	171.1	7.4	76.2	3.8	76.4	4.2	91.6	4.3	91.8	4.3	38.9	1.9	38.6	2.0	31.6	2.2	28.2	1.7	88.3	4.4	86.0	3.4

Table II. Body measurements of Körmend girls in 1958 and 1968.

Age (years)	Weight (kg)						Height (cm)					Length of the upper extremity (cm)					Length of the lower extremity (cm)					Biacromical width (cm)					Bicristal width (cm)					Chest circumference (cm)				
	K−58			K−68			K−58		K−68			K−58		K−68			K−58		K−68			K−58		K−68			K−58		K−68			K−58		K−68		
	N	x̄	s	N	x̄	s	x̄	s	x̄	s		x̄	s	x̄	s		x̄	s	x̄	s		x̄	s	x̄	s		x̄	s	x̄	s		x̄	s	x̄	s	
3	18	14.3	2.1	17	13.4	1.8	95.0	7.1	94.2	3.7		38.2	3.6	38.1	2.6		45.8	4.6	46.6	2.7		23.3	1.8	21.8	1.2		18.3	1.2	16.1	1.0		53.2	2.8	51.5	2.9
4	33	16.1	1.8	33	16.1	2.3	100.0	4.4	103.1	4.3		40.6	3.0	42.4	3.1		50.8	3.1	53.3	3.5		23.8	1.3	23.3	1.2		18.9	1.1	17.5	0.8		53.8	2.6	53.2	2.9
5	22	17.6	2.3	20	17.3	2.4	105.3	3.4	109.3	5.3		43.8	2.2	44.9	3.1		53.3	2.4	56.9	4.5		24.0	1.2	23.8	1.2		19.8	1.3	17.7	0.8		55.6	2.2	53.5	2.9
6	41	20.5	3.4	26	21.1	3.9	113.6	4.5	115.7	4.3		47.9	2.5	49.5	2.6		59.6	3.2	61.4	3.0		25.9	1.2	25.4	1.0		20.5	1.3	18.7	1.4		57.6	3.1	56.5	4.6
7	103	20.4	3.3	43	21.8	2.6	117.4	5.4	121.4	6.0		50.0	2.9	51.9	3.1		62.3	3.6	65.4	3.9		26.5	1.5	26.3	1.0		20.6	1.3	19.5	1.2		58.1	2.8	56.8	2.2
8	62	21.8	4.0	39	24.2	3.6	120.8	4.9	126.2	5.7		51.2	2.6	53.5	3.3		64.2	3.8	68.0	3.3		27.2	1.4	26.8	1.3		21.1	1.5	20.1	1.3		59.5	4.1	57.9	2.9
9	60	25.4	4.7	52	26.3	4.1	125.5	6.2	130.1	4.6		53.9	3.4	55.0	2.8		67.7	6.2	71.0	3.1		28.3	1.7	27.9	1.3		21.8	1.3	20.8	1.7		60.6	3.3	59.7	3.7
10	68	29.1	4.4	46	31.5	5.8	132.4	6.5	137.1	6.4		57.2	3.6	58.7	4.0		72.3	4.4	75.3	4.5		29.7	1.5	29.5	1.5		23.3	1.7	21.8	1.5		63.0	4.0	64.7	6.4
11	65	30.9	5.5	48	34.2	7.1	137.1	7.7	141.4	6.2		59.9	4.1	61.0	4.4		76.0	5.2	78.5	4.0		30.4	1.8	30.1	1.7		24.5	2.0	22.5	1.5		64.9	4.8	66.9	7.5
12	76	35.2	7.2	43	39.7	7.6	144.4	7.4	149.3	6.6		62.9	4.2	65.0	4.0		78.9	5.4	82.6	4.1		31.9	2.2	31.7	2.3		25.8	2.7	24.2	1.7		69.7	5.5	72.3	7.6
13	64	38.1	5.8	72	43.5	7.6	149.8	7.1	154.8	7.1		65.6	3.6	67.3	4.0		82.4	4.1	84.3	4.7		33.1	1.9	33.3	1.9		27.4	1.8	25.4	1.7		72.1	4.5	77.0	7.3
14	65	44.9	7.1	81	47.8	7.2	155.3	6.1	156.3	4.7		67.9	3.2	68.1	3.2		85.1	4.2	85.3	3.7		34.7	1.7	34.0	1.5		29.6	1.9	26.3	1.7		76.7	5.2	80.6	7.2
15	56	49.0	6.3	73	51.7	7.1	157.8	4.9	158.6	4.9		68.3	3.5	69.7	3.0		85.5	4.0	86.9	3.4		35.5	1.6	34.9	1.7		30.5	1.8	27.4	1.6		78.0	5.0	83.9	6.9
16	31	49.0	6.1	45	52.2	7.4	157.6	5.7	159.6	5.9		69.2	3.4	70.3	3.4		84.5	4.0	87.1	3.7		35.7	1.5	34.6	1.4		30.5	1.7	27.3	1.6		78.9	3.3	83.3	6.5
17	22	51.4	6.4	65	52.7	6.9	161.5	5.1	159.2	5.5		71.0	2.9	70.2	3.7		87.1	3.1	86.2	3.8		36.5	1.7	35.1	1.8		31.1	2.0	27.8	1.4		81.5	4.7	85.1	5.2
18	24	54.0	2.9	34	55.6	6.5	160.3	5.8	158.9	5.3		69.5	3.8	69.4	3.6		86.5	4.6	86.2	4.1		36.8	1.7	34.9	1.7		31.5	1.9	28.1	1.6		81.5	4.9	87.0	5.5

$d = 3$ for weights and volumes of total body or any body part;
p is the prescribed value for the given item for the phantom (Eiben et al., 1976).

III. RESULTS AND DISCUSSION

A. Body Measurements

From the results, shown in Tables I and II, the following observations are possible.

1. *Weight.* The mean values in 1968 were, for boys 0.3—3.2 kg, and for girls 0.4—5.5 kg, higher than in 1958, except for children aged 3—4—5 years.

2. *Height.* The mean values for boys in 1968 were 1.2—5.5 cm higher than in 1958. The difference in the girls was 0.8—5.4 cm, except for girls of 17—18 years of age, who were shorter.

3. *Length of Upper Extremity.* The mean values in 1968 were, for boys 0.1—3.5 cm, and for girls 0.1—2.2 cm, higher than in 1958.

4. *Length of Lower Extremity.* The mean values for both boys and girls were higher in 1968 than in 1958: in boys up to the age of 13 years, in girls up to the age of 12. In the older groups, the 1958 values were higher.

5. *Biacromial Width.* In almost every case and in both sexes the values were smaller in 1968 than in 1958; the difference in the boys was 0.2—0.9 cm, in the girls 0.2—1.9 cm.

6. *Bicristal Width.* The above tendency was far more pronounced: the mean values were lower in 1968, in the boys by 0.9—3.4 cm, in the girls by 1.0—3.4 cm.

7. *Chest Circumference*, at steady breathing. In boys, no appreciable difference was found between the two investigations: in some cases the mean values of 1958 were higher (0.1—2.3 cm); in other cases they were lower (0.1—1.8 cm). In girls, the 1958 mean values for 3—9 year age groups were consequently higher by 0.6—2.7 cm, but for girls older than 10 years the 1968 mean values were higher by 1.7—5.9 cm. This phenomenon may be explained by earlier maturation,

Thus, we may conclude that the cross-sectional growth studies showed higher mean values for length measurements (and also for body weight), but lower values for trunk breadth, in 1968 than in 1958. It is therefore apparent that the physique of Körmend children between 1958 and 1968 became taller and narrower, i.e., more linear.

The reasons should be sought for in the changing environmental conditions. Industrialization is accompanied by urbanization; consequently, there is a tendency towards "leisure". Furthermore, the children no longer take part in agricultural occupations with their parents and they do not have to ride a bicycle daily (15—20 km) to school and back. In general, reduced physical activity unfavourably influences the development of body breadth. It is difficult to decide whether a change towards linearity is an "improvement" or rather a process of regression.

Another question that also arises is the degree of adaptation of the children's organism to a more linear physique. How will it resist dangers accompanying increasing urbanization? Will it be able to coordinate and reach equilibrium between body and mental development?

Perhaps we will have better chances to answer these questions more fully after the 1978 investigations.

B. Proportions

A total of 384 z values obtained from the six body measurements for each of the 16 age groups, both for boys and girls by the phantom formula, are displayed on the first lines (K-58) and on the second lines (K-68) for each designated body measurement in Tables III and IV. These values show how proportionally similar the obtained values are to phantom values. An obtained value of 0.00 indicates the mean value for the particular group is proportionally the same as the phantom. Positive values indicate the prototype is proportionally greater and the negative values that it is proportionally smaller than the phantom.

A second display of z values is shown on the third lines for each age group (Tables III and IV). The d values given here indicate the tendency to proportional changes between K-58 and K-68 values.

Comparison may be done either for one age group of K-58 and K-68 boys and girls for all their body measurements, or for one body measurement in all age groups. The latter is chosen, but reference will be given to differences between the two series of measurements in 1958 and 1968.

The z values of *body weight* during growth from the initial $+1.5$ to $+2.5$, by the age of 7 years decrease to -0.5, and by the age of 13 years to -1.0, while, upon reaching maturity, they fluctuate between -0.5 and $+0.5$ (Tables III and VI, Fig.1). These changes illustrate well the process from the squat physique of the children, through pubertal elongation, to the normal physique of the adult. The z values of the K-58 and K-68 investigations show no significant differences; since during the K-68 investigation the increase in body height and body weight mean values display a similar proportional tendency. The differences between the two sexes become apparent only after puberty.

The about -2.0 values of the *upper extremity length* as an initial value, increase by the age of 6–7 years, first rapidly then somewhat slowly, to $-0.5-0$ (Tables III and IV, Fig.2). The proportionally smaller length of the upper extremity of the young children approximates gradually the $z = 0$ value only when reaching puberty. There are no significant differences between the two series, but there are clear sex differences from the age of 8 years on.

During growth, the proportional changes of the *lower extremity length* show a similar trend to that of the upper extremity, but the z values change more significantly: from -2.5 initially, with an intensive change until the age of 8 years, to -0.5; subsequently, at least in the K-58 measurements, this value stabilizes at $-0.5-0$. On the other hand, in the K-68 data this value decreases slightly after puberty (Tables III and IV, Fig.3). The differences after puberty between the K-58 and K-68 measurements become more pronounced, especially in boys.

The *biacromial width* is great at early ages, then it gradually and proportionally becomes smaller, reaching $z = 0$ only in adult age (Tables III and IV, Fig.4). The difference between the z values $(1.0 - 1.5)$ of the two series measurements becomes moderate only following puberty. The boys and the girls change according to similar trends, but striking differences exist between the values of K-58 and K-68.

In the case of *bicristal width*, the most striking features are the proportional changes which accompany growth. The proportionally great bicristal width of the children gradually becomes smaller during prepuberty, scarcely changing at all in the time of puberty, and subsequently becomes proportionally greater in girls and reaches, and even surpasses, the z values of the small children. The pubertal z

Table III. The z values of Körmend boys.

Age (yr)		Weight	Length of the upper extremity	Length of the lower extremity	Biacromial width	Bicristal width	Chest circumference
3	K-58	+2.38	−2.03	−0.14	+1.26	+1.58	+1.15
	K-68	+1.52	−2.26	+0.11	+0.31	−0.44	+0.64
	d	−0.86	−0.23	+0.25	−0.95	−2.02	−0.49
4	K-58	+2.06	−1.90	+0.45	+1.14	+1.65	+0.96
	K-68	+1.00	−1.94	+0.58	+0.23	−0.10	+0.33
	d	−1.06	−0.04	+0.13	−0.91	−1.75	−0.63
5	K-58	+0.86	−1.56	+0.85	+0.62	+1.09	+0.13
	K-68	+0.38	−1.26	+2.15	+0.03	−0.45	−0.19
	d	−0.48	+0.30	+1.30	−0.59	−1.54	−0.32
6	K-58	+0.27	−0.94	+1.16	+0.24	+0.65	−0.42
	K-68	+0.19	−1.31	+1.64	−0.58	−0.66	−0.75
	d	−0.08	+0.37	+0.48	−0.82	−1.31	−0.33
7	K-58	−0.45	−0.78	+1.37	+0.26	+0.56	−0.70
	K-68	−0.41	−1.04	+1.83	−0.62	−1.04	−1.25
	d	+0.04	−0.26	+0.46	−0.88	−1.60	−0.55
8	K-58	−0.34	−0.80	+1.59	+0.02	+0.47	−0.98
	K-68	−0.55	−0.81	+2.16	−0.75	−1.00	−1.35
	d	−0.21	−0.01	+0.57	−0.77	−1.47	−0.37
9	K-58	−0.74	−0.47	+1.95	+0.26	+0.30	−1.09
	K-68	−0.64	−0.69	+2.39	−0.70	−0.97	−1.30
	d	+0.10	−0.22	+0.44	−0.96	−1.27	−0.21
10	K-58	−0.60	−0.20	+2.10	+0.03	+0.66	−1.13
	K-68	−0.70	−0.61	+2.30	−0.80	−1.06	−1.65
	d	−0.19	−0.41	+0.20	−0.83	−1.72	−0.52
11	K-58	−0.62	−0.22	+2.33	+0.08	+0.53	−1.04
	K-68	−0.86	−0.45	+2.80	−0.82	−0.92	−1.61
	d	−0.24	−0.23	+0.47	−0.90	−1.45	−0.57
12	K-58	−1.00	−0.28	+2.34	−0.26	+0.27	−1.19
	K-68	−0.89	−0.45	+2.69	−0.77	−1.01	−1.63
	d	+0.11	−0.17	+0.35	−0.51	−1.28	−0.44
13	K-58	−0.90	−0.11	+2.63	−0.10	+0.70	−1.01
	K-68	−1.06	−0.18	+2.78	−0.74	−1.09	−1.53
	d	−0.16	−0.07	+0.15	−0.64	−1.79	−0.52
14	K-58	−0.90	−0.21	+2.43	−0.06	+0.83	−0.93
	K-68	−0.90	−0.05	+2.58	−0.47	−1.02	−1.24
	d	0.00	+0.16	+0.15	−0.41	−1.85	−0.31
15	K-58	−0.70	−0.04	+2.45	+0.12	+1.13	−0.63
	K-68	−0.81	+0.03	+2.55	−0.33	−0.91	−1.16
	d	−0.11	+0.07	+0.10	−0.45	−2.04	−0.53
16	K-58	−0.55	+0.03	+2.33	+0.48	+1.00	−0.41
	K-68	−0.64	+0.11	+2.37	−0.34	−0.69	−0.72
	d	−0.09	+0.08	+0.04	−0.82	−1.69	−0.31
17	K-58	−0.40	−0.15	+2.19	+0.37	+1.47	−0.23
	K-68	−0.62	+0.25	+2.25	−0.08	−0.70	−0.56
	d	−0.22	+0.40	+0.06	−0.45	−2.17	−0.33
18	K-58	−0.29	+0.09	+2.15	+0.46	+1.63	+0.12
	K-68	−0.71	+0.02	+2.04	+0.20	−0.43	−0.45
	d	−0.42	−0.07	−0.11	−0.26	−2.06	−0.57

Table IV. The z values of Körmend girls.

Age (yr)		Weight	Length of the upper extremity	Length of the lower extremity	Biacromial width	Bicristal width	Chest circumference
3	K-58	+2.07	−2.07	−0.20	+1.90	+2.22	+1.43
	K-68	+1.64	−1.96	+0.30	+0.72	+0.09	+0.98
	d	−0.43	+0.11	+0.50	−1.18	−2.13	−0.45
4	K-58	+1.44	−2.06	+0.66	+1.04	+1.73	+0.55
	K-68	+0.91	−1.65	+1.22	+0.20	+0.01	−0.01
	d	−0.53	+0.41	+0.56	−0.84	−1.72	−0.54
5	K-58	+1.13	−1.43	+0.80	+0.43	+1.76	+0.40
	K-68	+0.10	−1.65	+1.38	−0.54	−0.72	−0.87
	d	−1.03	−0.22	+0.58	−0.97	−2.48	−1.27
6	K-58	+0.51	−1.13	+1.54	+0.43	+1.08	−0.90
	K-68	+0.31	−0.85	+1.79	−0.33	−0.73	−0.90
	d	−0.20	+0.28	+0.25	−0.76	−2.81	−0.60
7	K-58	−0.28	−0.95	+1.77	+0.18	+0.54	−0.71
	K-68	−0.58	−0.88	+2.10	−0.61	−0.89	−1.58
	d	−0.30	+0.07	+0.33	−0.79	−1.43	−0.87
8	K-58	−0.41	−1.05	+1.83	+0.15	+0.50	−0.77
	K-68	−0.60	−1.05	+2.13	−0.99	−1.00	−1.88
	d	−0.19	0.00	+0.30	−1.14	−1.50	−1.11
9	K-58	−0.14	−0.87	+2.15	+0.17	+0.42	−1.09
	K-68	−0.67	−1.11	+2.41	−0.80	−0.93	−1.88
	d	−0.53	−0.24	+0.26	−0.97	−1.35	−0.79
10	K-58	−0.33	−0.66	+2.43	+0.07	+0.65	−1.33
	K-68	−0.51	−0.84	+2.55	−0.75	−0.98	−1.45
	d	−0.18	−0.18	+0.12	−0.82	−1.63	−0.12
11	K-58	−0.73	−0.53	+2.65	−0.26	+0.80	−1.48
	K-68	−0.58	−0.70	+2.79	−0.95	−0.98	−1.41
	d	+0.15	−0.17	+0.14	−0.69	−1.78	+0.07
12	K-58	−0.80	−0.50	+2.45	−0.21	+0.93	−1.10
	K-68	−0.68	−0.53	+2.71	−1.01	−0.72	−1.05
	d	+0.12	−0.03	+0.26	−0.80	−1.65	+0.05
13	K-58	−1.02	−0.56	+2.57	−0.23	+1.27	−1.16
	K-68	−0.78	−0.54	+2.37	−0.75	−0.49	−0.62
	d	+0.24	+0.02	−0.20	−0.52	−1.76	+0.46
14	K-58	−0.64	−0.43	+2.48	−0.01	+2.08	−0.75
	K-68	−0.34	−0.49	+2.41	−0.54	−0.10	−0.03
	d	+0.30	−0.06	−0.07	−0.53	−2.18	+0.72
15	K-58	−0.36	−0.62	+2.26	+0.12	+2.31	−0.73
	K-68	−0.09	−0.32	+2.51	−0.32	+0.33	+0.41
	d	+0.27	+0.30	+0.25	−0.44	−1.98	+1.14
16	K-58	−0.33	−0.34	+2.02	+0.27	+2.33	−0.51
	K-68	−0.15	−0.28	+2.40	−0.60	+0.17	+0.19
	d	+0.18	+0.06	+0.38	−0.87	−2.16	+0.70
17	K-58	−0.51	−0.29	+2.15	+0.20	+2.27	−0.37
	K-68	−0.01	−0.25	+2.25	−0.26	+0.48	+0.60
	d	+0.50	+0.04	+0.10	−0.46	−1.79	+0.97
18	K-58	+0.01	−0.59	+2.16	+0.56	+2.61	−0.25
	K-68	+0.43	−0.44	+2.27	−0.32	+0.70	+1.03
	d	+0.42	+0.15	+0.11	−0.88	−1.91	+1.28

Changes in Body Measurements and Proportion of Children 195

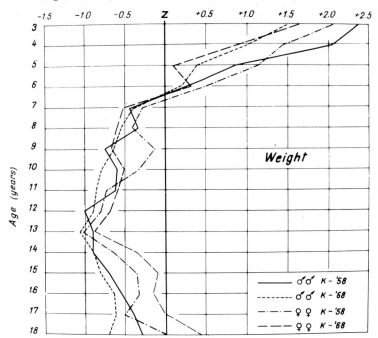

Fig. 1 Proportionality profile of weight.

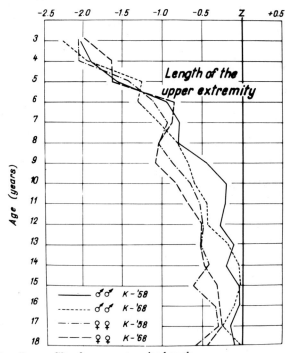

Fig. 2 Proportionality profile of upper extremity length.

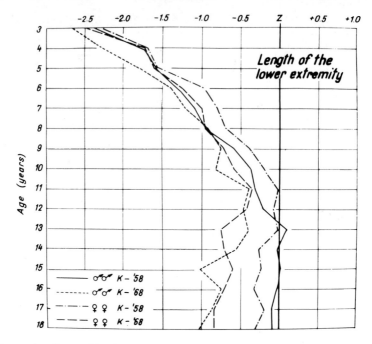

Fig.3 Proportionality profile of lower extremity length.

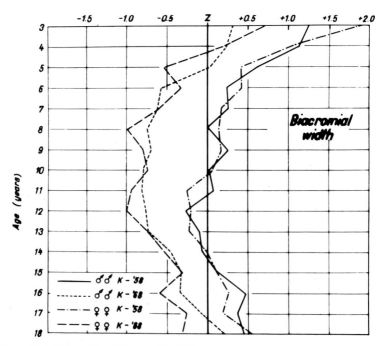

Fig.4 Proportionality profile of biacromial width.

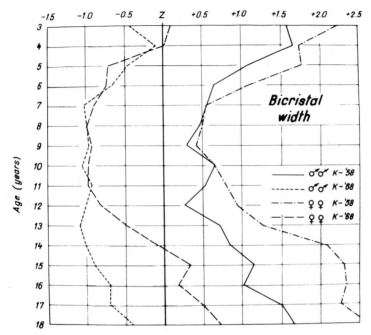

Fig. 5 Proportionality profile of bicristal width.

values of the boys increase at a more moderate rate (Tables III and IV, Fig.5). Significant differences are apparent in this feature between the values of the two series, K-58 and K-68: at 3 years of age, the difference between z values exceeds 2; between 7 and 10 years of age it is around 1.5, and subsequently it is about 1.5 in boys and 2 in girls.

The proportional changes are greatest in *chest circumference*. The z values were +0.5 to +1.5 at the age of 3 years in both series and in both sexes; at the age of 8—9 years, the values fluctuate between −2.0 and −1.0. From the time of prepuberty, chest circumference proportionally becomes greater, reaching a value between −0.5 and +1.0 at adult age (Tables III and IV, Fig.6).

Thus, the proportional changes during growth may readily be traced by the help of the unisex phantom. In our samples, from early childhood to adult age, the positive z values of body weight gradually change into negative values. The length growth of extremities is represented by the change of negative z values into positive ones. The breadth of the trunk — starting from a positive biacromial width and a nearly 0 value of z in bicristal width — during growth proportionally becomes small but only temporarily, then gradually increases. Chest circumference follows the same trend in broad outline.

The results of the two series of investigations demonstrate, by the help of the phantom, that body weight and length measurements of both boys and girls were proportionally greater in 1968 than in 1958, while the biacromial, and more especially bicristal, widths become proportionally smaller. The latter phenomenon became more pronounced with increasing age.

In the present work, I tried to analyse the proportional changes during growth. At the same time, I wish to stress the objective comparative possibilities

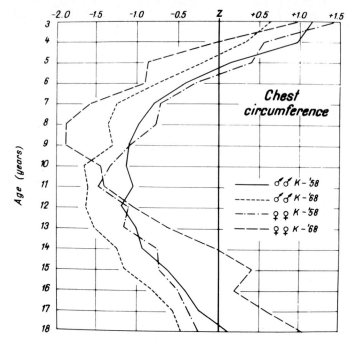

Fig.6 Proportionality profile of chest circumference.

we gain through the phantom model, both in growth investigations in general and especially in the analysis of proportional changes.

REFERENCES

Eiben, O.G. (1961). "Body Development of Körmend Youth" (in Hungarian). Thesis, Szombathely – Debrecen.
Eiben, O.G. (1963). *Anthropologie (Brno)* 1, 53-60.
Eiben, O.G. (1969a). Presentation at the Congress "Morphological, Physiological and Biochemical Problems of Growth and Development", Moscow 14–18, April 1969 (in Russian).
Eiben, O.G. (1969b). *Symp. Biol. Hung.* 9, 131-134.
Eiben, O.G., Ross, W.D., Christensen, W. and Faulkner, R.A. (1976). *Anthropol. Kozl.* 20, 55-67.
Ross, W.D. and Wilson, N.C. (1974). *In* "Children and Exercise" (J. Borms and M. Hebbelinck, eds). 6th International Symposium on Pediatric Work Physiology, den Haag, 1973. *Acta Paediatr. Begl.* **Suppl.** 169-182.

NUTRITION AND GROWTH PERFORMANCE IN AN ITALIAN RURAL ENVIRONMENT

A. Ferro-Luzzi, A. D'Amicis, A.M. Ferrini and G. Maiale

National Institute of Nutrition, Via Lancisi 29, Rome, Italy

SUMMARY

The use of anthropometric indicators for the diagnosis of marginal protein-energy malnutrition in the growing child is an accepted procedure. However, it is also recognized that many environmental factors, other than diet, may directly or indirectly influence the rate of growth. Evidence has been collected in Italy that indicates a close association between income and growth performance in school-age children. Very little information is available on this topic in children below 6 years of age. This chapter presents the results of a cross-sectional study on growth, nutrition, and associated environmental conditions, of a group of 514 preschool children (age 1—6 years) of rural Central Italy. Cases of growth retardation were evidenced in a small number of subjects; the nutrient intakes of the children who were considered malnourished were found to be similar to those of children with a satisfactory nutritional status, while children at risk of malnutrition because of unsatisfactory diets had a growth performance similar to that of the children with satisfactory diets. Growth retardation in this community may be better associated with other factors, such as genetic potential, family background or medical history, than with dietary intakes.

It is generally accepted that anthropometric indices are able to identify the presence and the degree of severity of protein-energy malnutrition at a community level (McLaren and Read, 1972; Waterlow, 1972; Jelliffe and Jelliffe, 1969; Keller, 1976; WHO, 1967; Gurney and Jelliffe, 1973) and they have been proposed as an essential step in the nutritional surveillance of population groups (WHO, 1976). A variety of these indices has been widely used by nutritionists in the field and has provided a concrete basis for programmes of nutrition intervention (Anderson, 1975). However, ample evidence has been collected throughout the world indicating that there are many factors other than nutrition — sociocultural, economic, and hygienic, to name a few — which, isolated or strictly interwoven, may come into play in the genesis of growth failure. Growth retardation has been shown to be related to socioeconomic conditions in African preschool children (Rea, 1971) and sociocultural factors have been identified as causing growth failure also in Indian children (Gokulanathan and Verghese, 1969). Striking anthropometric differences, possibly due to nutritional factors mediated by socioeconomic conditions, have been described in Italian schoolchildren (Ferro-Luzzi, 1967; Ferro-Luzzi and Proia, 1967). More recently, a longitudinal study conducted in Rome has confirmed the existence of a social gradient in the growth of Italian schoolchildren, and dietary factors were not considered to be the sole cause of the observed differences (Mariani et al., 1977).

Moving from these considerations, a cross-sectional study has been undertaken to assess the nutritional status, growth performance, and diet of a group of preschool children with the objective of identifying the environmental factors associated with the observed growth retardation in Italy, and to ascertain at what age this phenomenon makes its appearance.

A sample of 514 children, aged 1–6 years, representing over 80% of the population of a small rural town (Cori) in Central Italy, were studied. Nutritional status was assessed by a nutritionist and clinical signs of probable malnutrition were sought using the abbreviated method as described by Jelliffe and recommended by WHO (Jelliffe, 1966).

The results of this examination show that this community may be considered free of any major nutritional problem, with the possible exception of the recorded presence of clinical signs of iron deficiency in 26 children and of iodine deficiency in 18 children with an enlarged thyroid.

Table I. Mean percent deviation from reference height for age[a], weight for age[a], weight for height, and triceps for age[b], of 514 children from Cori (Central Italy).

Age (yr)	N	Height for age	Weight for age	Weight for height	Triceps for age
1	70	103	104	102	94
2	77	102	102	98	96
3	114	102	104	100	98
4	115	101	104	101	99
5	120	102	106	102	108
6	18	99	101	102	102

[a] British growth standards (Tanner et al., 1966). [b] British growth standards (Tanner and Whitehouse, 1975).

Body size and degree of leanness/fatness have been assessed by trained measurers using internationally approved methods and instruments (Weiner and Lourie, 1969).

Table I illustrates the somatic development of the subjects expressed as mean percent deviation from the chosen reference population (Tanner et al.,

1966; Tanner and Whitehouse, 1975). It may be observed that these children show, on the average, a satisfactory and smooth growth. They are as tall, or taller than, their British counterparts, and also slightly heavier. The average weight for height, which has the advantage of correcting the actual weight of the child for his own height, seems to suggest that the Italian children are also slightly heavier for each centimeter of stature than their British counterparts. Considering that the triceps skinfold (one index of adiposity) is slightly inferior to the reference standard, we are of the opinion that these Italian children are a little more muscular than the British ones.

Table II. Classification of Cori children (Central Italy) according to degrees of malnutrition and retardation.

			Growth retardation			
		Grade	0°	1°	2°	3°
		% height for age[a]	> 95	95–91	90–86	< 85
Malnutrition	Grade	% weight for height[a]				
	0°	> 90	446	11	—	—
	1°	90 – 81	45	5	—	—
	2°	80 – 71	2	1	—	—
	3°	< 70	—	—	—	—

[a] British growth standards (Tanner *et al.*, 1966).

The distribution of the children according to the combined classification of the weight for height and height for age, as proposed by Waterlow (1972), is shown in Table II. This combined classification has the advantage of providing a picture of the nutritional past, with the height for age distribution, as well as reflecting the current nutritional status, with the weight for height. The choice of the grouping is arbitrary and has been modified to suit the community under investigation. The overall picture thus obtained is one of a fairly well-nourished community. No child is severely malnourished. Only 9% (47 children) present a mild form of acute malnutrition (Wasting), without growth retardation. Eleven children (2%) have a slight degree (95–91%) of growth retardation without being malnourished. In this category one would expect to find grouped (if present) the "bonzai" children, or nutritional dwarfs, and children with a lesser genetic potential for growth. Finally, 6 children are slightly retarded in their growth while presenting a mild moderate degree of wasting.

Information about the 24-hour food intake of each child taking part in the survey was obtained by recall in an interview with the parents. Food samples, food models and sets of standardized household containers, specifically designed or selected to suit the local dietary practices, were used for the interviews.

The nutritive value of the diet, expressed as density of nutrients per 1000 kcal, is shown in Table III. It can be seen that the diet is very dense from the protein point of view, reaching values as high as 40–43 g of protein per 1000 kcal, as compared to the 23–22 of the recommended safe allowances (SINU,

Table III. Nutritive value (g/100 kcal) of the diet of boys and girls from Cori (Central Italy).

Age (yr)	Protein (g)	Carbohydrate (g)	Fat (g)	Calcium (mg)	Iron (mg)
		Boys			
1	40	142	34	677	4.9
2	40	139	35	555	4.7
3	35	143	37	398	4.5
4	37	142	36	417	5.1
5	35	144	35	370	4.7
6	34	146	35	402	4.3
		Girls			
1	43	141	33	656	5.0
2	41	133	38	522	4.9
3	38	134	38	477	4.4
4	36	143	36	439	4.4
5	34	133	37	351	3.9
6	35	135	38	356	3.9

1974). This reflects the large use of protein-rich baby foods during the 1st and 2nd year of life, commonly observed in the area. Calcium is also present in excess, while the 4–5 mg of iron per 1000 kcal do not reach the recommended Italian value of 5–7 mg.

Table IV. Percentage of energy supplied by protein, fat, and carbohydrates, in the diets of Cori children (Central Italy).

Age (yr)	Protein	Fat	CHO
1	16.6	30.2	53.1
2	16.1	32.9	51.0
3	14.7	33.5	51.9
4	14.6	32.1	53.3
5	14.5	31.4	54.4
6	13.8	32.8	52.8

In Table IV is shown that proteins contribute the very high value of 17% of the total energy intakes at 1 year of age, falling regularly with age to a more acceptable figure of 14%. The contribution of fat presents an inverse trend but appears to be contained in a more correct proportion (30–33%).

To investigate the possible existence of unequal distribution of food among the children and to identify the number of subjects at risk of malnutrition, the children were grouped according to the individual level of adequacy of energy and protein intakes.

Table V. Percentage distribution of Cori children (Central Italy) according to dietary adequacy.[a]

	Not protein deficient	Protein deficient
Not energy deficient	71	1
Energy deficient	25	4

[a] Percentage of subjects consuming diets deficient in energy, or protein, or both, calculated by comparing intakes with requirements or safe levels of intake (LARN, 1974). Deficient: less than 80% of requirement or safe level.

The percentage distribution in two classes of dietary adequacy, over and below 80% of the theoretical requirement, is shown in Table V. As much as 71% of the children had protein and energy intakes exceeding 80% of requirement. Less than 5% had protein deficits, but most of them had a simultaneous energy deficit. Another 25% had energy intakes less than 80% of requirement. The proportion of children whose diets were deficient in energy is thus rather unexpectedly high, whereas the percentage of protein deficits amounts only to 4–5%. A certain difference was present among the two sexes, girls showing evidence of a poorer diet than boys.

Table VI. Comparison of physical characteristics and dietary data of all the children *(a)*, only well-nourished children *(b)*, children classified as malnourished *(c)*, and children at risk of malnutrition *(d)*.

	N	% height for age	% weight for height	% body fat	Percent adequacy of intakes			
					Energy	Protein	Calcium	Iron
Boys									
a	259	101	101	19	99	161	133	88
b	175	101	104	19	113	183	139	106
c	28	100	87	16	100	174	147	85
d	60	102	101	19	69	115	102	66
Girls									
a	255	102	101	19	93	155	132	81
b	161	102	103	20	106	184	146	90
c	31	99	88	14	89	169	119	76
d	80	102	100	19	64	116	102	59

The dietary intakes of energy, protein, calcium, and iron, and the physical characteristics of these 140 children (60 boys and 80 girls) identified as being at risk of malnutrition because of their poor diets *(d)*, have been compared with the children identified as malnourished by anthropometric indices *(c)*, with the whole sample *(a)*, and with the group of selected normal, well-nourished children *(b)* (Table VI).

The energy and protein adequacies of the diets of the malnourished children do not differ substantially from those of the whole group. Compared with the children at risk, they have constantly higher values. Differences are found when physical characteristics are compared, as they appear to be the shortest of their age and the lightest for their height. On the other hand, the children at risk of malnutrition because of their poor diets show a growth performance as good as that of the whole group and of the well-nourished children.

The age at which the mother left school, as well as the occupations of both parents, were shown to be associated with the nutritive value of the diet as well as with the prevalence of overweight. In the general sample, 64% of the mothers were nonworking (housewives); in the group of children with a poor diet the percent of nonworking mothers fell to 53%. At the same time, the children of nonworking mothers showed a tendency towards a slightly higher prevalence of obesity (14% v. 11%). It was surprising to find that 75% of the malnourished children had nonworking mothers. Also, the age at which the parents had left school was shown to be associated, on the average, with the physical development of their children. Children of mothers who had left school at, or below, the age of 10 are more often lighter, smaller and leaner, and 51% of the malnourished children had less-educated fathers, as opposed to 43% of the well-nourished ones. As far as the father's occupation is concerned, those employed in farming differ from the other working categories in that a smaller proportion of their children are overweight.

In conclusion, in this fairly well-nourished community with an average satisfactory growth performance, we have evidenced a limited number of retarded growths and of poor diets. However, the children at risk of malnutrition because of their inadequate diets were not the same as those showing clinical and anthropometric signs of malnutrition.

The dietary intakes, as assessed in this survey, may not truly represent the usual food consumption, as it may and did happen that less or more food was eaten when a child had been ill or convalescent. However, it seems likely that food availability did not represent, in our subjects, the primary or only limiting factor, and that the onset of malnutrition was mediated through a legion of non-nutritional factors, either genetic or environmental, able to affect the child's genetic potential for physical development.

ACKNOWLEDGEMENTS

We are indebted to the parents for permission to make the measurements, to the Municipality of Cori for the provision of facilities and financial support. We are particularly grateful to Miss L. Attolini, sociologist, to Miss C. Bein, dietician, to Dr. L. Teodori, and to the undergraduate students, P. Colloridi, L. Lintas, P. Pagani, G. Perozzi, and F. Portoghesi, without whose splendid and dedicated cooperation this study would not have been possible.

REFERENCES

Anderson, M.A. (1975). *Am. J. Clin. Nutr.* 28, 775-781.
Ferro-Luzzi, A. (1967). *Quad. Nutr.* 17, 124-145.
Ferro-Luzzi, G. and Proia, M. (1967). *Quad. Nutr.* 17, 269-298.

Gokulanathan, K.S. and Verghese, K.P. (1969). *J. Trop. Ped.* **15**, 118-124.
Gurney, M.J. and Jelliffe, D.B. (1973). *Am. J. Clin. Nutr.* **26**, 912-915.
Jelliffe, D.B. (1966). WHO Monogr. Ser. 53.
Jelliffe, E.F.P. and Jelliffe, D.B. (1969). *J. Trop. Ped.* **15**, 177-260.
Keller, W., Donoso, G. and De Mayer, E.N. (1976). *Nutr. Abstr. Rev.* **46**, 591-609.
L.A.R.N.: Rapporto della Commissione ad hoc della VII Riunione Generale SINU. 1974 Riva del Garda.
Mariani, A., Lancia, B., Migliaccio, P.A. and Sorrentino, D. (1977). This volume, pp.49-61.
McLaren, D.S. and Read, W.W.C. (1972). *Lancet* **2**, 146-148.
Rea, J.N. (1971). *Hum. Biol.* **43**, 46-63.
Tanner, J.M., Whitehouse, R.M. and Takaishi, M. (1966). *Arch. Dis. Child.* **41**, 454-471; 613-635.
Tanner, J.M. and Whitehouse, R.M. (1975). *Arch. Dis. Child.* **50**, 142-145.
Waterlow, J.C. (1972). *Br. med. J.* **3**, 566-569.
Weiner, J.S. and Lourie, J.A. (1969). "Human Biology. A Guide to Field Methods". IBP Handbook No.9, Blackwell, Oxford.
WHO (1967). Doc. No. NUTR/67.1.
WHO (1976). WHO Tech. Rep. Ser. 593.

4. CLINICAL AUXOLOGY

GROWTH HORMONE AND TESTOSTERONE INTERACTIONS

M. Zachmann

*Department of Pediatrics, University of Zurich, Seinwiesstrasse 75
8023 Zurich, Switzerland*

SUMMARY

There are multiple interactions between GH and testosterone (T) both with respect to their secretion and their metabolic action. To elucidate these interrelations further, which are of physiological relevance in normal subjects, but also have clinical and therapeutic implications in patients with GH deficiency, three types of studies were undertaken. They include (1) the analysis of growth data of patients under T treatment with and without GH deficiency and with and without simultaneous hGH treatment; (2) the evaluation of the effect of T on the development of secondary sex characteristics in the presence and absence of GH; and (3) the study of nitrogen balance using the stable isotope ^{15}N.

(1) Growth and bone maturation were analyzed in 42 boys who had been treated with T; 16 had normal GH-secretion (7 gonadotropin deficiency, 9 congenital anorchia), 26 were GH- and gonadotropin-deficient. Of the GH-deficient patients, 12 received simultaneous hGH-treatment and 14 T only. In the patients with normal GH-secretion and in the GH-deficient patients with adequate hGH-replacement, the growth rate increased to normal with T treatment, but in the GH-deficient patients without hGH-replacement, T had almost no growth-stimulating effect.

(2) In GH-deficient patients, a much longer treatment period with T was necessary to induce axillary hair than in patients with normal GH-secretion.

(3) In the presence of GH-deficiency, there was only a weak T-induced nitrogen-retention. Under simultaneous hGH-treatment, the response to T became strikingly more positive. It is concluded that GH is necessary as a permissive factor to allow T to be fully effective with respect to protein anabolism, growth promotion, and androgenicity.

The interactions between the metabolic effects of growth hormone (GH) on the one hand, and of androgens, notably testosterone, on the other hand are multiple, and surprisingly few investigations have dealt with this interesting problem. The title has two main aspects: one is more practical and concerns treatment of short stature in general, with androgens or related anabolic compounds, as general and nonspecific stimulators of growth. The second aspect concerns true interactions between GH and testosterone, which are of physiological and clinical importance. Before hGH became available, androgens and anabolic steroids have been used as now obsolete treatment in patients with hypothalamo-pituitary dwarfism and in other types of short stature. Because of their stimulating effect on bone maturation, anabolic steroids have nowadays no place in the treatment of short stature.

Since short stature is a heterogeneous problem, there is no general growth-stimulating therapy, which would be independent of the diagnosis. Some years ago, it was thought that synthetic testosterone derivatives or other anabolic steroids could be used as general growth-stimulating agents. However, the deception followed quickly, when it became evident that the premature fusion of epiphyses induced by anabolic steroids results in a reduced or, at the most, equal adult height as would have been reached without any treatment.

Recent reports suggest even more caution with respect to the use of these drugs, since hepatomas and carcinomas of the liver have been reported after long-term treatment with 17-alkylated androgens.

The second point deals with the interactions between GH and testosterone. Many years ago, evidence has been gathered from animal studies, indicating that the effectiveness of exogenous testosterone depends on an intact pituitary function.

Already in 1944, Simpson *et al.* had shown that hypophysectomy reduces the effect of testosterone in rats. This fact has since been confirmed by many investigators. Scow and Hagan have shown — also in hypophysectomized rats — that testosterone alone did not increase the growth of any tissues except the accessory sexual muscles, and that it had no effect on body weight in these animals. When GH was given in addition to testosterone, the growth rate increased and was greater than with GH alone. This finding suggests that GH is necessary for the growth-promoting action of testosterone. On the other hand, GH production and release in intact animals are also influenced by testosterone. The mature male pituitary gland of the rat contains more than the female gland. By contrast, the plasma levels are higher in female animals. During male puberty, in animals as well as in the human, plasma GH levels increase.

The next point concerns the interactions between testosterone and GH secretion in normal human subjects; fasting plasma GH levels do not appear to be influenced by sex hormones under physiological conditions. However, the maximum GH levels after various stimuli show some sex differences. It has been shown by various authors that these maximum levels are under certain conditions higher in women than in men. From this observation, however, it should not be concluded that testosterone blunts the GH response to stimuli. In fact, both estrogens and testosterone stimulate the release of GH from the pituitary, but testosterone to a lesser extent than estrogens. Illig and Prader (1970) have studied the GH response to insulin in four patients with anorchia, one with delayed puberty, and one hypopituitary dwarf, before and on the second or third day after a single injection of a long-acting testosterone preparation. In instances, the test was repeated after three months of treatment with testosterone. With the exception of the GH-deficient patient, who never had any response, the GH levels on androgen treatment were higher in all cases compared to the levels before treatment, indicating that the maximum GH-secretory capacity is indeed stimulated by testosterone.

In the last section, some of our own studies on testosterone secretion and testosterone effect in GH-deficient patients are discussed. The effect of hGH on the testosterone secretory capacity of the Leydig cells is only based on the experience with one single case. Although these results should be regarded as preliminary and should be interpreted with caution, they are quite striking. We have seen an adult patient with familial isolated GH deficiency. His basal plasma and urinary testosterone levels were those of a normal man. When given a single dose of 5000 U of HCG per m^2, his urinary testosterone increased from 49 to 68 $\mu g/24$ h, which is a response in the lower normal range. We have then treated this patient with hGH for 3 months and repeated the HCG stimulation on hGH. Interestingly, this time, the response to HCG was much more marked than in the first test, and urinary testosterone increased from 38 to 138 $\mu g/24$ h. This finding requires confirmation, but it seems to indicate that, somehow, the maximum testosterone secretory capacity of the Leydig cells depends on GH.

While this single observation is mainly of physiological interest, we have recently been mainly interested — also from a practical point of view — in the opposite, namely, the alteration of the effect of testosterone by a lack of GH. For several years, we had the clinical impression that testosterone and anabolic steroids are less active — both with respect to growth stimulation and to the development of secondary sex characteristics — in boys with severe pituitary insufficiency than in boys with normal pituitary function.

The growth response to androgens in hypopituitary patients is different individually. If only the growth velocity, but not bone maturation, is considered, some patients do respond markedly to androgens, while others do not at all. In the majority, however, the growth response to anabolic steroids alone is much smaller than that to hGH.

To check this impression in a more precise and up-to-date way, we selected a number of patients for growth analysis, who were treated with replacement doses of the same long-acting testosterone preparation. In this way, we could compare the growth response in patients with normal GH production with that in patients with GH deficiency on and off simultaneous hGH treatment.

The growth-stimulating effect of testosterone was studied in a total of 56 patients and subjects. Among these, 15 patients had a normal GH secretion, 6 suffered from isolated gonadotropin deficiency, and 9 from congenital anorchia. The results in these patients were compared with those in 26 GH-deficient patients. Among these, 12 were on simultaneous hGH treatment, and 14 on testosterone alone. The bone age at the start of treatment was similar in all groups and ranged from 12.5 to 13.5 years. For comparison, growth velocity in normal subjects from the Zurich longitudinal growth study was analyzed. In order to be able to compare mean values of the different groups statistically, we have calculated the ratio between observed and expected growth velocity for bone age using Tanner's standards as normal values.

This ratio should be 1 or near 1 in normal subjects and also in patients who lack spontaneous puberty, but have normal GH secretion and adequate testo-

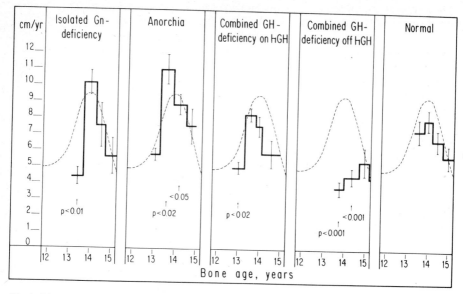

Fig. 1 Mean height velocity plotted against bone age (cm/year, $\bar{x} \pm$ Se) before and during three treatment periods with testosterone. For details, see Aynsley-Green et al. (1976). P values indicate significance of difference against the respective period in the 15 normal subjects.

sterone replacement therapy. If our hypothesis of a diminished testosterone response in GH-deficient patients without hGH treatment is correct, the ratio should be below 1 in patients treated with the same testosterone dose alone, and it should be 1 or more in patients simultaneously treated with hGH and testosterone. This is in fact what we have found.

In *isolated gonadotropin deficiency*, the growth velocity for bone age was too slow before testosterone treatment, with an observed-to-expected ratio of 0.52. During the first year of treatment, the ratio increased to 1.3. Later on, it levelled off at 1.

In *congenital anorchia*, pretreatment and prepubertal growth velocity was normal with a ratio of 0.96, because, in these patients, the anorchia was known and treatment was started immediately at the appropriate bone age, when the pubertal growth spurt should have started. Again, the ratio increased during the first year of treatment to 1.47 and levelled off at 1 thereafter.

In the *GH-deficient patients* off hGH treatment, the ratio before treatment was 0.53. When they were given testosterone alone, there was practically no stimulation of growth, with a ratio remaining considerably below 1, quite in contrast to the two previous groups of patients who had received the same or lower testosterone doses.

In the patients *on hGH* before testosterone, when they received hGH alone, the ratio was only 0.7, because all these patients had also a lack of gonadotropins and therefore no spontaneous pubertal development. However, when testosterone was added to the continued hGH therapy, a normal growth spurt took place and the ratio increased to 1.19.

In summary, these results show that, in GH deficiency, testosterone stimulates growth to a much lesser extent than in subjects with normal GH secretion.

They are summarized in Fig.1, showing the growth velocity in cm per year for bone age in the different groups. The pubertal growth spurt is normal in the first 3 groups with normal GH secretion or adequate hGH replacement, but not in the GH-deficient patients off hGH in the 4th group.

A practical conclusion is that, in GH-deficient patients at a pubertal bone age, hGH treatment should not be stopped and replaced by testosterone, but both hormones should be given simultaneously until the final height has been reached.

Surprisingly, not only the testosterone-induced growth spurt is lacking in GH-deficient patients, but also secondary sex characteristics develop more slowly or not at all, if testosterone is given without hGH.

We have computed the total testosterone dose (in mg/m^2) that is required to produce axillary hair in patients with normal GH secretion and in GH-deficient patients. This total dose is quite homogeneous in the first group, the mean value being about 2500 mg/m^2. By contrast, in the GH-deficient patients not treated with hGH, the mean dose to have this effect is above 9000 mg/m^2 and there is a much wider scatter. The wide scatter in the GH-deficient patients may indicate that some of the cases still have some endogenous GH production, which is insufficient for normal growth, but sufficient to enhance the activity of testosterone.

The mechanism of the interaction between GH and testosterone with respect to the development of secondary sex characteristics is entirely unknown. That this reduced effect is indeed due to the lack of GH and not to that of ACTH, which was also present in some of the cases, is shown by the fact that axillary hair develops normally, when the patients are given hGH and testosterone simultaneously, but not ACTH.

Recently, we have been able to confirm by metabolic studies that also biochemically the effect of testosterone is reduced in GH-deficient patients. We were interested to see whether the testosterone-induced nitrogen retention would be less marked in GH-deficient patients off hGH and whether it could be enhanced by hGH. Because the classic nitrogen balance studies are time-consuming, we have devised a test without a diet, using the stable isotope ^{15}N. The patients were given 50% ^{15}N-labelled ammonium chloride before and after testosterone and 24-hour urines were collected.

The total N was determined and the percentage of ^{15}N was estimated on a ^{15}N analyzer. From this, the excreted amount of ^{15}N was calculated and together with the ingested quantities, a true urinary ^{15}N balance could be determined. By this technique, we have studied the metabolic response to testosterone in a GH- and gonadotropin-deficient patient who did no longer respond to hGH alone, but who had a normal pubertal growth spurt with hGH *and* testosterone. In this patient, the ^{15}N balance study before and after testosterone was done twice, once without any additional treatment and once under simultaneous treatment with hGH for 3 months. The untreated GH-deficient patient had a definite, but weak and subnormal, response to exogenous testosterone. By contrast, his response was strikingly more marked when the test was repeated with the same testosterone dose, but under hGH treatment. These results are preliminary, but they confirm our clinical impression and the results from the growth analysis.

REFERENCES

Aynsley-Green, A., Zachmann, M. and Prader, A. (1976). *J. Pediatr.* **89**, 992.
Illig, R. and Prader, A. (1970). *J. clin. Endocrinol.* **30**, 615.
Scow, R.O. and Hagan, S.N. (1965). *Endocrinology* **77**, 852.
Simpson, M.E., Marx, W., Beck, H. and Evans, H.M. (1944). *Endocrinology* **35**, 309.
Zachmann, M. and Prader, A. (1970). *J. clin. Endocrinol.* **30**, 615.

HORMONAL AND METABOLIC RESPONSES TO FAST: INFLUENCE OF GROWTH IN NORMAL CHILDREN AND IN PATHOLOGICAL CONDITIONS

J.L. Chaussain, P. Georges, P. Olive and J.C. Job

Research Centre on Growth and Child Endocrinology
St. Vincent de Paul Hospital, 75014 Paris, France

SUMMARY

The ability of the organism to sustain a normal blood glucose (BG) level during fasting periods is fundamental for a normal brain maturation. During gast, BG level is regulated by hepatic glycogen stores, gluconeogenesis substrates, and enzymatic or hormonal systems responsible for their mobilization.

In normal children under 10 years of age, the major characteristic is the inability to sustain a normal BG value during a prolonged fast. In 56 normal children, aged 1 to 9 years, a 24-hour fast induced a highly significant ($p < 0.001$) drop of BG (from 86 ± 15 to 52 ± 14 mg/100 ml). After fast, individual values were distributed according to a Gaussian curve, and 22% of the children had values below 40 mg/100 ml. This sensitivity to fast may be related to a reduction of hepatic glycogen stores, demonstrated by the glucagon tolerance test, and to a physiological deficiency of gluconeogenesis substrates: serum alanine levels are strongly reduced during fast (from 36 to 17.8 μM/dl) and correlate to BG ($r = 0.68, p < 0.02$). The fast-induced rise of plasma cortisol and GH, the FFA mobilization, the ketosis with the parallel reduction of plasma insulin, fail to prevent the fall of BG. This sensitivity to fast appears to improve with growth. In a group of 28 normal children aged 2 to 7 years, after a 24-hour fast, BG values were correlated to age ($r = 0.68, p < 0.001$). An age-correlated increase of plasma alanine ($r = 0.86, p < 0.001$) demonstrated the improvement of gluconeogenesis. The consequence is an age-related decrease of the fasting level of plasma cortisol, GH, and FFA.

In children with substrate-limited hypoglycemia, BG during fast drops more than in controls ($p < 0.01$), with a parallel decrease of plasma alanine. A fast-induced hypoinsulinism induces in these children a rise of branched-chain aminoacids ($p < 0.01$). In obese children, fast-induced variations of BG and alanine are identical to those of controls. However, fat mobilization appears to be reduced ($p < 0.05$) in obese children, in parallel with fasting hyperinsulinism and low levels of GH. In hypopituitary children, results are identical to those observed in children with substrate-limited hypoglycemia; however, the GH deficiency reduces the fast-mobilization of FFA and the degree of ketosis.

These results indicate that the maintenance of a normal BG during fast and its improvement with age are primarily related to gluconeogenesis capacities. Hormonal responses, fat mobilization, and ketosis, appear to be secondary in this process.

I. INTRODUCTION

The brain requires a constant perfusion of glucose to maintain a normal metabolism. For this reason, any period of fast and especially when prolonged, appears to become critical in the process of blood glucose (BG) regulation. During fast, this regulation depends on the mobilization of glycogenic, protein, and lipid reserves, by hormonal and enzyme systems.

Hypoglycemic syndromes are relatively frequent in young children, and mostly induced by fast. By contrast, in adults, hypoglycemic syndromes are less frequent and mainly related to an excess of insulin. This discrepancy in the frequency of the fast-related hypoglycemic syndromes between children and adults, leads to the idea that the ability of the organism to sustain a normal BG level during fast improves with age.

The aim of the present work is the study of the age-related variations of the sensitivity to fast in normal children and adolescents, and also in pathological conditions which emphasize the particular role of metabolic and/or hormonal factors.

II. METHODS

Children and adolescents were submitted to a 24-hour fast (Chaussain, 1973; Chaussain et al., 1974). The fast lasted from noon on the first day until noon the second day. During this time, the subjects were kept in bed and water was permitted as desired.

Blood samples were collected at time 0 and 24 hours of the test, and urines were collected at the end of the fast for determination of ketonuria. Additionally, in some cases, a glucagon tolerance test (0.03 mg/kg IV) was performed, both before and after fast.

BG was measured with an autoanalyzer by the ferricyanide method (Hugget and Nixon, 1957). Plasma insulin and GH were measured by radioimmunoassay (Sizonenko et al., 1968), and plasma cortisol by a method derived from Braunsberg and James (1961). Serum FFA were determined by a colorimetric method (Duncombe, 1964), and serum aminoacids by colum chromatography (Moore et al., 1958). The extent of ketonuria was estimated by Ketostix (Ames Co., Inc.).

III. NORMAL CHILDREN

The first step of this study concerned a group of 56 normal children aged 1 to 9 years (Chaussain, 1973; Chaussain et al., 1974). The major fact in this population was the highly significant fall of BG induced by fast. At 24 hours, the mean BG value was 52 ± SD 14 mg/100 ml and the individual values were distributed according to a Gaussian curve. One third of the studied children exhibited values in the hypoglycemic range. The BG response to glucagon was also highly significantly diminished by fast, and completely suppressed in 5% of the cases, indicating insufficient glycogenolytic possibilities of the liver. The insulin response to glucagon is also deeply reduced by fast.

This fast-induced drop of BG resulted in an increase of plasma cortisol, GH, and FFA. At the end of the fast, plasma cortisol and FFA were significantly negatively correlated ($p < 0.01$) to BG. The degree of ketonuria was also grossly correlated to the BG value (Chaussain, 1973). According to Pagliara et al. (1972), protein gluconeogenesis is the most important factor in the maintenance of BG during fast. Alanine, which represents the key aminoacid of gluconeogenesis, offers a valuable index to appreciate this phenomenon. In fact, fast induced in normal children a constant and highly significant fall of serum alanine levels ($p < 0.001$) and, at the end of the test, BG and alanine levels were significantly correlated (Chaussain et al., 1974).

The second step of this study concerned the age-related variations of the sensitivity to fast (Chaussain et al., 1977). For this purpose, 27 children and adolescents of both sexes, aged 1 to 18 years, were studied. In this group, the fasting BG values increased progressively with age, and a significant correlation ($p < 0.01$) existed between the two parameters. No fasting values below 50 mg/100 ml were observed after the age of 10 years. This age-related improvement of the glycemic response to fast has for consequence a decrease of plasma cortisol, GH, and FFA fasting levels, and these three parameters were significantly negatively correlated with age. By contrast, alanine fasting levels increased progressively with age, indicating that the improvement of the resistance to fast is related to increasing stores of alanine.

IV. KETOTIC HYPOGLYCEMIA

Ketotic hypoglycemia, as defined by Colle and Ulstrom (1974), represents the most common form of hypoglycemia in childhood. Insufficient stores of alanine have been demonstrated by Pagliara et al. (1972) in children with ketotic hypoglycemia. In fact, in our own study concerning 14 children with ketotic hypoglycemia, the drop of alanine was more accentuated than in controls. The terminal value was 12.9 ± 1.3 µM/dl in ketotic hypoglycemic children and 17.8 ± 1.8 µM/dl in controls (Chaussain et al., 1974). The consequence of this acute degree of hypoalaninaemia was a greater fall of BG during the 24-hour fast, and the mean fasting value of ketotic hypoglycemic children was only 27 mg/dl, which represents the second inferior standard deviation of controls.

V. OBESE CHILDREN

The 24-hour fast was performed in a group of 21 nondiabetic obese children, 7 boys and 14 girls, aged 7 to 16 years (Chaussain et al., 1976). In this group, the fasting BG value (63 ± 2 mg/dl) was not significantly different from that of controls of similar ages (54 ± 4 mg/dl). Serum alanine concentrations decreased significantly ($p < 0.05$) and similarly in the two groups. These data established that, even in the presence of an excess of lipid reserves, the role of alanine remains primordial.

One of the numerous consequences of obesity is a decrease of GH levels. In fact, in the present study, the GH response to glucagon was significantly lower in obese children than in controls, both before and after fast. The lower degree of FFA mobilization induced by fast in obese children of this group may be related to the decreased levels of plasma GH.

VI. HYPOPITUITARY CHILDREN

Our results in hypopituitary children are still preliminary. However, at this time, 11 children aged 4 to 16 years were studied. All were idiopathic hypopituitary dwarfs with at least GH and ACTH deficiencies. In these children, the major facts demonstrated by the 24-hour fast were low fasting BG values which did not increase with age, low alanine and FFA fasting levels. These results demonstrate the major role played by hormones in the improvement of the BG regulation during fast with age, and explain the higher risk of hypoglycemic seizures in these children.

VII. CONCLUSIONS

Fasting tests demonstrate in normal children a very critical period in BG homeostasis between 1 and 9 years of age. At this period, the regulation of BG during fast appears to be deficient in a large proportion of the population. The risk of symptomatic or asymptomatic hypoglycemia is high, and increased by an eventual hormonal deficiency. For all these reasons, prolonged period of fast must be avoided in this period of age and in cases of fasting hypoglycemic seizures, the normalization of alanine stores by frequent feedings, must be part of the therapy.

REFERENCES

Braunsberg, M. and James, W.M.T. (1961). *J. clin. Endocrinol.* **21**, 1146.
Chaussain, J.L. (1973). *J. Pediatr.* **82**, 438.
Chaussain, J.L., Georges, P., Olive, G. and Job, J.C. (1974). *J. Pediatr.* **85**, 776.
Chaussain, J.L., Georges, P., Olive, G. and Job, J.C. (1976). *Biomedicine* **25**, 229.
Chaussain, J.L., Georges, P., Olive, G. and Job, J.C. (1977). *J. Pediatr. (in press).*
Colle, E. and Ulstrom, R.A. (1964). *J. Pediatr.* **64**, 632.
Duncombe, W.G. (1964). *Clin. Chim. Acta* **9**, 122.
Hugget, A.S.G. and Nixon, D.A. (1957). *Biochem. J.* **66**, 12.
Moore, S., Spackman, D.H. and Stein, W.H. (1958). *Ann. Chem.* **30**, 1185.
Pagliara, A.S., Karl, I.E., De Vivo, D.C., Feigin, R.D. and Kipnis, D.M. (1972). *J. clin. Invest.* **51**, 1440.
Sizonenko, P.C., Job, J.C. and Michel, R. (1968). *Rev. Fr. Étud. Clin. Biol.* **13**, 406.

EVIDENCE FOR THE SOCIAL INHERITANCE OF OBESITY IN CHILDHOOD AND ADOLESCENCE

S. Garn, S. Bailey, P. Cole and I.T.T. Higgins

*Center for Human Growth and Development and the School of Public Health
University of Michigan, Ann Arbor, Michigan 48109, USA*

SUMMARY

As shown in 6,372 biological parent-child pairs, 425 adoptive parent-child pairs, 3,256 biological sibling pairs, and 196 adoptive sibling pairs, genetically unrelated individuals living together tend to the same degree of fatness similarity as serologically-verified biological parent-child and sibling pairs.

I. INTRODUCTION

Although hundreds of "explanations" have been advanced for human obesity, few of them have been derived in population context and by the study of family lines. Rather, many of the explanations have been derived from unique animal models, such as genetically-obese rats and mice, rodents with hypothalamic lesions, tube-fed animals and those on unique dietary regimens. Many other attempted explanations have invoked psychoanalytic theory, "set-point" theory, the early use of cow's milk, induction of increased fat cell number and the extensive use of sucrose, but without documentation.

The most compelling evidence, however, is that fatness follows family line. This has been shown by the anecdotal method in Angel's study (1949) of obese women, in Wither's analysis (1964) of weight data on parents and siblings reported by his students, and — in particular — by the extensive data of the Ten-State Nutrition Survey of the United States (Garn and Clark, 1976). The question arises, however, as to whether parent-child and sibling similarities in fatness (as commonly observed) have a genetic basis, or whether fatness similarities among relatives are socially rather than genetically inherited (Steinberg, 1960; Garn et al., 1975).

The answer to this question bears both on the cause and the prevention of obesity in man, and is important to the private practitioner and to public-health practice as well. If obesity is genetic, then we must search for the metabolic error or the enzyme defect in the families of the obese and in the obese themselves. If the level of fatness is attained in family lines by learned attitudes toward food and eating, then we must discover how to modify these attitudes in the context where they are learned.

II. MATERIALS AND METHODS

This study is based upon fatfold measurements of biological parents and their children, siblings of like and unlike sex, of adoptive parents and their adopted children, and of genetically unrelated "siblings", all of them participants in the Tecumseh (Michigan) study of the University of Michigan School of Public Health (Napier *et al.*, 1971).

In making the comparisons, the correlational method was employed, as well as the use of parental fatness combinations — Lean × Lean, Lean × Medium, through to Obese × Obese. All correlations were based on age-specific, sex-specific, normalized Z-scores for triceps and subscapular fatness (although residuals from segmented regressions yield very nearly the same results). Parental mating combinations (Lean × Lean, Lean × Medium, Lean × Obese, Medium × Medium, Medium × Obese, and Obese × Obese) were based on the 15th and 85th percentiles, respectively for fatness, for age and for sex. Thus, a "lean" father was in the lower 15% for fatness for his age, while an "obese" mother was in the upper 15% for age and sex.

Biological parent-child and sibling pairings were serologically verified. Adoptive (or "social") parent-child pairs rigorously excluded children adopted by close relatives. Further data included the age at adoption, and (in a later part of the study) comparison of parents and adult children, including those living apart. Parent-child pairs were separately analyzed in this study so as to encompass (1) the singly-adopted, i.e., the biological child of one parent and the adoptive child of the other parent, and (2) the double-adopted — i.e., children who were genetically unrelated to both adoptive parents. Only doubly-adopted children were included in the analyses using parental fatness combinations.

In short, the basic design involved fatness comparisons of genetically related individuals living together and genetically unrelated individuals living together, thus providing an exact test of the genetic hypothesis.

III. RESULTS

As shown in Table I, where 6,372 single-parent v. child fatfold correlations are set forth, parent-child fatfold correlations are positive at all age groupings, increasing somewhat through to adolescence. Moreover, with an overall r value of 0.20, parent-child fatfold correlations are the order of magnitude commonly found for genetically-determined traits. Indeed, these correlations (grouped in 4-year age intervals) involving both like-sexed and cross-sexed one-parent v. child fatfold fatness similarities, quite resemble parent-child fatfold correlations from the Ten-State Nutrition Survey of 1968-1970, involving a separate sample of some 4,874 pairs (see Table I).

Table I. Biological parent-child correlations for triceps fatness.

Age	Father-son		Mother-son		Father-daughter		Mother-daughter		All pairs	
	N	r	N	r	N	r	N	r	N	r
0–4	402	0.23	432	0.09	387	0.08	426	0.03	1647	0.11
5–9	503	0.18	555	0.19	452	0.25	514.	0.20	2024	0.21
10–14	429	0.29	492	0.22	413	0.29	472	0.23	1806	0.25
15–18	206	0.41	240	0.22	202	0.29	247	0.29	895	0.30
Mean r:										
This study	1540	0.25	1719	0.18	1454	0.22	1659	0.18	6372	0.21
Ten-State Study	1379	0.21	1185	0.18	1234	0.25	1076	0.22	4874	0.22

Mean values of r from the mean z transform of r.

Table II. Biological sibling correlations for fatness.

Fatfold	Brother-brother		Sister-sister		Brother-sister		All pairs[a]	
	N	r	N	r	N	r	N	r
Triceps	866	0.32	791	0.33	1599	0.31	3256	0.32
Subscapular	864	0.31	785	0.26	1590	0.29	3239	0.29

[a] From age- and sex-specific normalized z transforms as in Table I.

In similar fashion, sibling similarities in fatness are significant overall and positive, with no marked influence of sex. As shown in Table II, both for triceps fatfold and the subscapular fatfold, brother-brother, sister-sister, and brother-sister correlations for these fatfold values are very much the same and closely approximate 0.31 overall. (Again, this is what might be expected for genetically-determined traits, since siblings share more genes than do single parent-child pairs.) And, for a total of 3,256 pairs of siblings, the r of 0.31 provides a very good indication of the magnitude of similarity for the triceps and subscapular fatfolds (shown in Table II) and for the iliac and abdominal fatfolds as well.

Now, the single parent-child correlations given in Table I (i.e., mother-son, mother-daughter, father-son, father-daughter) are much improved in meaning when the fatness level of *both* parents is taken into account. Setting up the parental matings as Lean X Lean, Lean X Medium, Lean X Obese, etc., it is then possible to depict the fatness level of the doubly-adopted offspring after correction for sex and correction for age — using the sex-specific, age-specific z transforms. When this is done, as in Fig.1, the results are dramatic indeed. In simple stepwise fashion, the fatness level of the biological children increases according to the parental fatness combination. Thus, for 2,845 serologically-verified biological children of different parental fatness combinations, the fatness level of the children is seen to rise from approximately −0.50 for the children of two lean parents (Lean X Lean) to +0.50 for the children of two obese parents (Obese X Obese). These results, here pictured only for the abdominal fatfold, are closely paralleled by results for the triceps, subscapular, and iliac fatfolds as well.

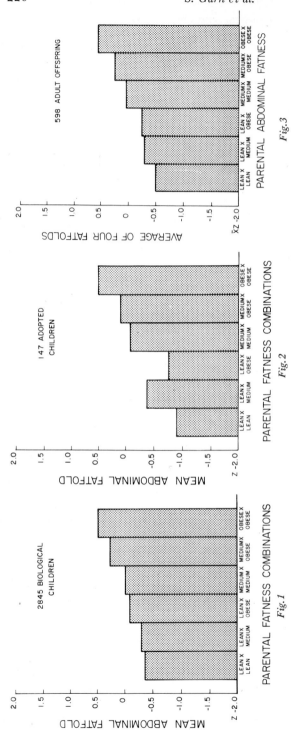

Fig.1 Relationship between parental fatness combination (Lean × Lean, Lean × Obese, etc.) and normalized age- and sex-corrected fatfold thicknesses of their offspring, shown for the abdominal fatfold in 2,845 serologically-verified biological children. In general, there is a stepwise relationship such that the children of two lean parents (i.e., Lean × Lean) tend to have fatfold values well below average for age and sex, while children of obese parents (Obese × Obese) tend to have the greater fatfold values.

Fig.2 Relationship between adoptive parental fatness combination and abdominal fatfold thicknesses of their adopted offspring. As shown here, using the same parental fatness pairings (Lean × Lean, Lean × Medium, Lean × Obese), fatness of the adopted children closely parallels the fatness combinations of the adoptive parents to a degree closely resembling that in biological parent-child pairs.

Fig.3 Relationship between parental fatness level (Lean × Lean, Lean × Medium, etc.) and average thickness of four different fatfolds for 598 serologically-verified adult offspring. Note that the relationship between parental fatness combination (Lean × Lean, etc.) and fatfold thickness of the offspring extends even into adulthood (> 18.5), and using an average of all four fatfolds (triceps, subscapular, iliac and abdominal).

Data based on examination round 2 of the Tecumseh Study.

Table III. Comparisons of biological and adoptive parent-child correlations in triceps and subscapular fatness.

Comparison	Triceps		Subscapular	
	N	r	N	r
Biological parent v. child	6372	0.20	6256	0.19
Adoptive parent v. child	425	0.19	329	0.23

Values of r calculated from mean z transforms of age-specific rs, as in Table I.

Now, the biological parent-child similarities in fatness, shown as product-moment correlations in Table I and as z-scores for fatness of children of different parental fatness combinations in Fig.1, rather surprisingly and closely resemble correlations for adoptive parent-child pairs. This is shown simply in Table III, where biological parent-child correlations (for 6,372 pairs) and adoptive parent-child correlations (for 425 pairs) are contrasted for the triceps and subscapular fatfolds. Overall, the biological parent-child pairs (involving single parent-child comparisons) and the adoptive parent-child pairs (again involving single parent-child comparisons) are of the same order of magnitude, i.e., approximately 0.20 overall. Similar results were obtained for two other fatfold data from the same (Tecumseh) study.

Table IV. Comparisons of biological and adoptive sibling correlations in triceps fatness.

Type of relationship	Sibling combination	Triceps fatfold	
		N	r
Biological	Brother-brother	866	0.32
Adoptive	Brother-brother	45	0.42
Biological	Sister-sister	791	0.33
Adoptive	Sister-sister	54	0.31
Biological	Brother-sister	1599	0.31
Adoptive	Brother-sister	97	0.21
Biological	All siblings	3256	0.32
Adoptive	All siblings	196	0.29

All data from Tecumseh examination round 1, calculated from age- and sex-specific normalized z-scores, with mean values of r from the mean z transform of r.

In comparable fashion as shown in Fig.2, adoptive parents grouped according to fatness level (Lean X Lean, Lean X Medium, etc.) are closely paralleled by the fatness level of their adoptive offspring. Here shown for 147 doubly-adopted children (i.e., involving two lean adoptive parents, two medium adoptive parents, two obese adoptive parents, etc.) the fatness level of the children also increases according to the parent fatness combination. Thus, Fig.2 (showing adoptive parent-child fatfold similarities) may be contrasted with Fig.1, where the fatness level of biological doubly-adopted parent v. child pairs is shown.

Finally, and as shown in Table IV, adoptive sibling pairs (involving biologically unrelated boys and girls living together) and serologically-verified biological sibling pairs (involving true siblings) evidence closely comparable similarities in their fatfold levels. Accordingly, and as also shown in Table III and Fig.2, genetically unrelated individuals living together and genetically related individuals living together tend towards similarity in the various fatfolds measured.

These data, intentionally abridged to conserve space, and including only children below 19 years of age, clearly indicate the extent to which living together brings about fatfold similarities between parents and "children" and between siblings of both sexes.

IV. DISCUSSION

This study, based on an unusually large single-population sample (N ⩾ 8,000) and with serological verification of 13,309 parent-child and sibling pairings, makes two obvious points. First, it shows that fatness follows family line, with the leanest parents producing the leanest children and the fattest patents producing the fattest sons and daughters (Garn et al., 1976a). Indeed, the progression is so regular from the parental fatness combination Lean X Lean through Obese X Obese as to resemble a hypothetical model rather than actual findings. Second, the present study shows that adopted children also resemble their adoptive parents in the level of fatness, such that two lean adoptive parents also tend to generate lean "children" and two obese adoptive parents tend to engender obese adopted sons and daughters, much as with the biological parent-child pairs just described (Garn et al., 1976a, 1976b).

It should be understood that the families involved in this data analysis come from a total-community study, thus largely eliminating ordinary sampling biases. Furthermore, serological data were used to exclude questionable parent-child and sibling pairs. In addition, children adopted by known relatives were also excluded from the adopted child/adoptive parent and adopted sibling comparisons. In this way, we avoided attenuating fatness correlations involving biological relatives or exaggerating the correlations involving genetically unrelated pairs. Nevertheless, biologically related and biologically unrelated individuals living together tend to very comparable levels of fatness similarity, the more so with increasing years together (cf. Fig.3).

Limitations of space do not allow us to describe the influence of parental sex in detail, except to note that (somewhat to our surprise) paternal fatness level tends to have slightly more influence on the fatness level of the children than does maternal fatness. Nor is there space to discuss age effects, except to note increasing parent-child and sibling similarities in fatness level through adolescence (Tables I and II), in both the biologically related and the biologically unrelated groupings. Taking all the data together, it is highly probable that the level of fatness is attained in family-line context through learned attitudes toward food and eating, energy expenditure, and energy conservation (and even the programmed stimuli that call for eating). For example, we have very considerable evidence (from the Ten-State Survey of 1968-1970 and still-earlier studies of ours) that parents and children resemble each other in caloric and nutrient intake to the same extent that they resemble each other in fatness (Garn and Rohmann, 1966; Garn and Clark, 1975).

While we had designed the data analysis so as to include (1) biologically related individuals living together and (2) biologically unrelated individuals living together, thus extending the strategies commonly employed in genetic analyses, the findings apply to more than fatness. True, the findings show that obese parents tend to produce obese children, whether the children are the products of their loins or obtained from adoption agencies. In this respect, we can surely say that level of fatness is acquired in family-line context rather than genetically inherited. And we have thereby achieved our objective, given the sample size, analytical design, and remarkably clear-cut nature of the results.

To return to the central problem, it is now clear that the level of fatness (and not just "obesity") is attained within the context of the family. Second, it is clear that the familial nature of fatness involves more than shared genes, since biologically unrelated individuals living together come to resemble each other in fatness, even if they come to live with each other only in adult life. Under these circumstances, the prevention of obesity can be directed to the individuals at particular "risk", i.e., children living with obese individuals, whether they are biological relatives or not.

REFERENCES

Angel, J.L. (1949). *Am. J. Phys. Anthropol.* 7, 433-472.
Garn, S.M. and Clark, D.C. (1975). *Pediatrics* 56, 306.
Garn, S.M. and Clark, D.C. (1976). *In* "Report on the Second Wyeth Nutrition Symposium". Wyeth Laboratories, New York.
Garn, S.M. and Rohmann, C.G. (1966). *Pediatr. Clin. North Am.* 13, 353-379.
Garn, S.M., Clark, D.C. and Ullman, B.M. (1975). *Ecol. Food Nutr.* 4, 57-60.
Garn, S.M., Bailey, S.M. and Cole, P.E. (1976a). *Am. J. Phys. Anthropol.* 45, 539-543.
Garn, S.M., Cole, P.E. and Bailey, S.M. (1976b). *Ecol. Food Nutr.* 6, 1-3.
Napier, J.A., Johnson, B.C. and Epstein, F.H. (1971). *In* "Case Book of Community Study". Johns Hopkins Press, Baltimore.
Steinberg, A.G. (1960). *Am. J. Clin. Nutr.* 8, 752-759.
Withers, R.J. (1964). *Eugen. Rev.* 56, 81-90.

PRIMARY GROWTH DEFICIENCIES: DIAGNOSTIC PROBLEMS

G. Segni and P. Mastroiacovo

National Research Council, Rome: Program of Preventive Medicine
(Perinatal Preventive Medicine Subproject MPP-4)
and
Department of Pediatrics, Catholic University, Rome, Italy

SUMMARY

The diagnostic problems of the syndromes with primary growth deficiency, not depending on chromosomal abnormalities or inborn errors of metabolism, are discussed. Many of these anomalies have a genetic etiology, few have an environmental origin, and some are of unknown etiology. A specific diagnosis is necessary for an adequate genetic counselling. The greatest problems come either from the fact that the diagnosis is based only on the phenotypic features of the patients, and therefore only on clinical criteria, or from the large number and variety of the recognized disorders, many of which are very rare. Some points are emphasized, such as the importance, during physical examination, to search for minor malformations and, whenever possible, to take measurements in a standard way. A classification of syndromes with growth deficiency, based on the major clinical-morphological features, is proposed as a first help to arrive at a specific diagnosis. On account of the increasing number of new syndromes, a Computer-Assisted Congenital Syndrome Identification Program (CACSIP) has been developed and its way of functioning is illustrated.

I. INTRODUCTION

In general, growth deficiencies can be grouped into two general categories: (1) primary growth deficiencies, resulting from intracellular disturbances which affect the growth of skeletal cells, and (2) secondary growth deficiencies, stemming from extrinsic problems, such as hormonal defects or altered nutrient

PRIMARY AND SECONDARY GROWTH DEFICIENCY

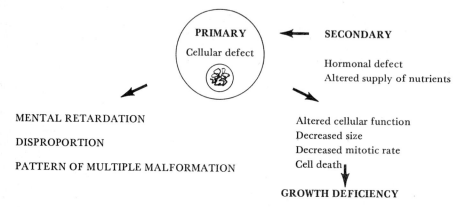

Fig. 1 General distinction of growth deficiency disorders. (From Smith, 1976, modified).

supply, which limit the capacity for growth of skeletal cells (Smith, 1976). This general distinction is illustrated in Fig.1.

In primary growth deficiency, various defects are present, individually or in combination, the most important of which are an alteration of skeletal cell function (e.g., by an inborn metabolic error), a decrease in cellular size, and a slowing down of mitotic rate. This type of growth deficiency may often be accompanied by malproportionment or defects in skeletal molding (as in skeletal dysplasias). The morphogenesis of other tissues may be affected as well. Thus, the patient often shows multiple malformations and/or mental retardation.

The general modes of etiology in primary growth deficiency are illustrated in Table I (Smith, 1976). These include chromosomal abnormality syndromes, mutant gene syndromes, and a number of syndromes of unknown etiology. In a few of these conditions, a specific metabolic error is known, as in mucopolysaccharidosis, but in the majority a biochemical defect has yet to be pinpointed.

Table I. General modes of etiology of primary growth deficiencies.

1. Chromosome abnormalities
2. Single mutant genes (Mendelian inheritance)
   A. Known biochemical defect
   B. NOT known biochemical defect
3. Etiology not established
4. Maternal infections, medications and other

When a Mendelian genetic etiology is found, it is appropriate to regard these syndromes as inborn errors of early morphogenesis. Such analogy with hereditary metabolic diseases might be useful in throwing light on hereditary errors in morphogenesis. In a number of syndromes, however, etiology is completely unknown, either as a consequence of their rare occurrence, or as a result of insufficient study in the morphogenetic process.

Primary Growth Deficiencies: Diagnostic Problems

Other growth deficiencies may be considered the consequence of adverse prenatal environmental factors, such as infectious diseases contracted by the mother, or drugs taken in pregnancy. In reality, however, the inadequate knowledge of the growth pattern of these conditions makes it difficult to regard them as true primary growth deficiencies.

II. SYNDROMES WITH PRIMARY GROWTH DEFICIENCY: DIAGNOSTIC APPROACH

The most important clinical features of syndromes with primary growth retardation are summarized in Table II. Short stature is frequently evident from birth, because it is of prenatal onset, but exceptions do exist, and the retardation may only become evident at 2 or 3 years of age.

Table II. Clinical features of primary growth deficiency.

1. Usually prenatal onset
2. Variable rate of osseous maturation, usually normal
3. Frequent malproportionment of body size
4. Frequent association with major or minor congenital anomalies and mental retardation
5. No specific therapy available. Genetic counselling, psychologic management and prognostic advice may be planned

The malproportionment of the body is not usually associated with the presence of major or minor congenital anomalies or mental retardation, and these syndromes are divided into two well-defined groups, skeletal dysplasias and dysmorphic syndromes.

Although there does not yet exist a specific therapy, an accurate diagnosis is absolutely necessary for genetic counselling and the formulation of a precise prognosis. This, in turn, favors both the psychological equilibrium and a possible social integration of the patient.

A primary diagnostic difficulty stems from the fact that, if one excludes those conditions arising from chromosomal defects and those few conditions for which a particular metabolic defect is known, the diagnosis depends largely on phenotypic observation not supported by any laboratory testing, as is especially true for the dysmorphic syndromes. Finally, signs such as the minor malformations are difficult to detect, although they are of diagnostic importance.

Table III. Evaluation of the short child with primary growth deficiency.

1. Family history
   Patient history
2. Physical examination
   Searching for and measurement of minor malformations
3. X-ray evaluation
4. Standard photographs

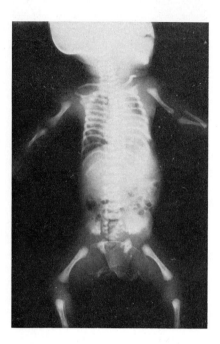

Fig.2 X-ray of a newborn with chondroectodermal dysplasia. Retrospective diagnosis was placed on the basis of this X-ray and the description of anomalies in the clinical record.

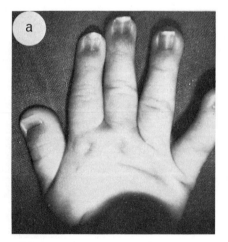
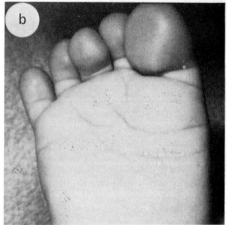

Fig.3 Hand and foot of a child with Rubinstein-Taybi syndrome. Note the enlarged thumb and hallux.

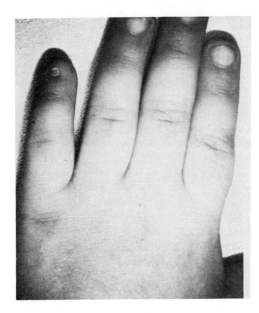

Fig.4 Hand of a child with Coffin-Siris syndrome. Note the pronounced hypoplasia of the nail and terminal phalanx of the fifth finger.

Consequently, it is necessary to apply a precise diagnostic method (see Table III), which attempts to transcend the subjective character of a diagnosis based on personal clinical sense and experience. The following points are proposed as a guide for the examination of a child with primary growth deficiency.

(1) The family history is of fundamental importance. A systematic survey of the data must be performed to assure that no information is overlooked. Family members must be directly examined; retrospective diagnosis can be done on the basis of radiography and photography (Feingold, 1975). Figure 2 shows an infant suffering from a chondroectodermal dysplasia which was diagnosed after death.

(2) The physical examination must not only be very accurate, but also specifically directed to the search for minor malformations of the hands, genitalia, skin, and face. Figure 3 shows the hand and foot of a child suffering from Rubinstein-Taybi syndrome. The enlarged thumb and hallux point towards diagnosis. In Fig.4, the pronounced hypoplasia of the nail and terminal phalanx of the fifth finger indicates Coffin-Siris syndrome (Mastroiacovo *et al.*, 1977).

The facies is often very typical, so that it may immediately indicate the diagnosis (Figs 5, 6, and 7).

The simple physical observation is, however, often insufficient. It is incorrect to define minor malformations as large ear, small anterior fontanel, or wide-set eyes, without appropriate measurements. The boy in Fig.8, suffering from Aarskog-facial-digital-genital syndrome (Mastroiacovo *et al.*, 1976), showed hypertelorism, but this could be defined only after measuring the canthal distance. There exist precise data in standard measurement; as for weight, height, and head circumference, percentile grids are available for the most important anatomical structures (Feingold, 1975).

(3) Standard X-rays and systematic measurement of traits are also of value in the definition of skeletal defects.

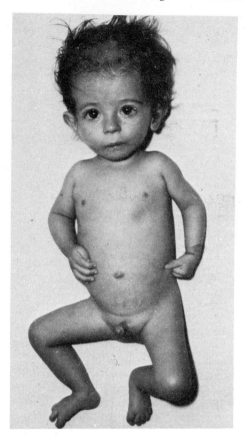

Fig.5 Silver-Russel syndrome. Note small triangular facies, craniofacial disproportion, and slight facial asymmetry.

(4) Complete documentation of a subject cannot be done without photographic record. One picture is worth a thousand words, and a photograph may convey much more information than an extensive and detailed physical examination. The importance of these methodological suggestions will become clearer after consideration of the successive stages in the diagnostic process, consisting in making a diagnosis on the basis of the observed phenotypic signs.

III. A CLINICAL CLASSIFICATION OF RECOGNIZABLE SYNDROMES WITH SHORT STATURE WITHOUT KNOWN BIOCHEMICAL OR CHROMOSOMAL DEFECT

In order to simplify the clinical diagnostic process, we have worked out a classification of primary growth deficiencies ordered according to the evidence of the most important features (Table IV).

This classification omits, however, the primary growth deficiencies resulting from chromosomal or known biochemical defects, for which a 'laboratory' diagnosis is possible. The clinical parameters which may help the physician in the diagnostic procedure are: evidence of signs in the neonatal period, presence of body size malproportionment, and severity of the condition. Among the severe

Primary Growth Deficiencies: Diagnostic Problems 231

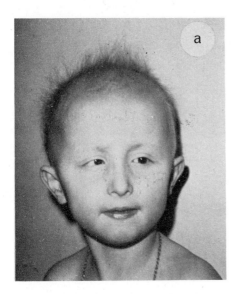
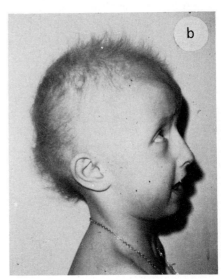

Fig.6 Hallermann-Streiff syndrome. Note hypotrichosis, thin nose, malar hypoplasia, bilateral microphthalmia, and low-set slanted ear.

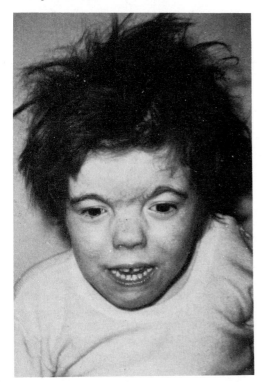

Fig.7 De Lange syndrome. Note bushy eyebrows and synophrys, small nose with anteverted nostrils and low nasal bridge; the facies "gestalt" is very characteristic.

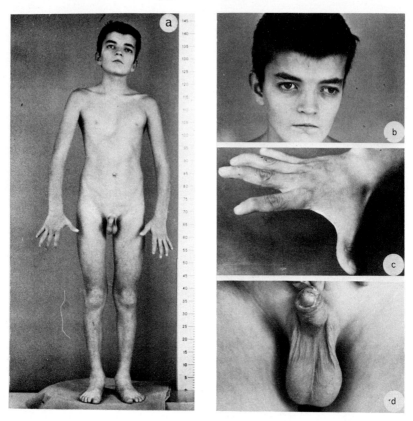

Fig. 8 Fourteen-year-old boy with Aarskog facial-digital-genital syndrome. Note: (a) short stature, pectus excavatum, and anomalies of toes; (b) hypertelorism (inner canthal distance, 38 mm; outer canthal distance, 102 mm), strabismus, myopia, left palpebral ptosis, widow's peak, malar hypoplasia, tiny upper lip, down-turned mouth, and semilunar crease in the chin; (c) short, broad hand (hand length, 143 mm; middle finger, 37.7% of total hand length), mild interdigital webbing, and peculiar finger extension; (d) congenital penile torsion of 90°; IVP showed bilateral idiopathic megaureter.

types, examples are achondrogenesis, thanatophoric dwarfism, and camptomelic syndrome. Among the less severe types, examples are achondroplasia and chondroectodermal dysplasia. Less severe types with localized manifestations in the limbs include various types of mesomelic dwarfism.

Among the relatively proportionate types, one should consider the presence of major or minor malformations and in which structure they are most frequently evident. Among the types with major malformations, Roberts, Holt-Oram, and De Lange syndromes are examples of cases showing limb malformations. Syndromes showing minor malformations include Rubinstein-Taybi, Seckel, Smith-Lemli-Opitz, and Prader-Willi syndromes.

In this classification, however, one group has been separated from the rest, because growth retardation is isolated or associated in an inconstant mode with minor malformations. This group includes small-for-date newborns with a slow

Table IV. A clinical classification of recognizable syndromes with short stature without known biochemical or chromosomal defects.

I. **Syndromes recognizable in the newborn**

  A. Disproportionate short stature
    1. Severe forms, short-limbed, usually lethal in newborn or in the first year of life
    2. Not severe forms, only sometimes lethal, associated with:
      a. Generalized manifestations: limbs, spine, pelvis, facies
      b. Relatively localized manifestations: limbs +/− facies
  B. Relatively proportionate short stature and/or low birth weight
    1. With major malformations
    2. Without major malformations *but with* minor malformations:
      a. Facies
      b. Hands
      c. Skin
      d. Other
    3. Without major and minor malformations

II. **Syndromes not usually recognizable in the newborn, only aspecific minor anomalies present**

III. **Syndromes recognizable after infancy**

  A. Disproportionate short stature
    1. Short limbs
    2. Short trunk
  B. Relatively proportionate short stature with minor malformations
    1. Facies
    2. Hands
    3. Skin
    4. Other

rate of growth and the following characteristics: (1) sporadic occurrence in the family; (2) occasional presence of one or more associated minor malformations, such as clinodactyly, syndactyly of 2nd-3rd toes, or other; and (3) no recognizable pattern of known syndrome. This category should provide a large amount of material for investigation; it is likely that a detailed study may lead in the future to the definition of new syndromes in this area.

    Another group has been kept separated, in which either growth deficiency or other manifestations are not so evident at birth, or affected newborns usually escape detection. Diagnosis is only possible when a complete family history is available or an in-depth clinical examination has been performed. A follow-up is needed in such cases to confirm the diagnosis. A classical example is hypochondroplasia, in which affected newborns usually escape detection, but a careful examination may disclose a sudden block in full elbow extension and sometimes a failure of fingertips to reach the mid thigh. This condition, which is increasingly recognized, is however more evident towards the second or third year of life.

The same clinical parameters have been used in the classification of the primary growth deficiencies that come into evidence only after the second or third year of life. These conditions include some types of skeletal dysplasias characterized by a malproportioned short stature due to the shortening of the limbs or of the trunk and a great number of dysmorphic syndromes. Obviously, in these cases we are only talking of minor malformations, the major ones being easily recognizable at birth.

IV. COMPUTER-ASSISTED CONGENITAL SYNDROMES IDENTIFICATION PROGRAM

Applying this diagnostic approach, which we extend to the dysmorphic syndromes not characterized by growth deficiency, we have obtained good results.

Out of 94 patients, excluding those affected with Down's syndrome, in 66 (70%) it was possible to make a diagnosis based only on clinical data (Table V);

Table V. Syndrome identification in 94 cases.

	Nonchromosomal	Chromosomal	Total
Identified	52	14	66 (70.6%)
Nonidentified	21	4(+3)*	28 (29.4%)
Total	73	21	94

* Definition of chromosomal defect under study.

in 14 of these, the diagnosis was confirmed by a chromosomal test, which shows that it was possible to identify 2/3 of the chromosomal syndromes only on the basis of clinical data. Limiting ourselves to the patients with growth deficiency, but no chromosomal or metabolic anomalies, out of 40 cases, 28 (70%) were accurately diagnosed on clinical grounds (Table VI). Dysmorphic syndromes (18 cases) were more frequent than skeletal dysplasias (10 cases). The majority of patients were found among those hospitalized in our clinic, only a few of whom had been referred to us for diagnosis.

In spite of these results, it is our opinion that it becomes more and more difficult, even for a specialist, to keep in mind the large and continually growing quantity of information about malformative syndromes. For this reason, we have prepared a Computer-Assisted Congential Syndrome Identification Program (CACSIP), particularly aiming to facilitate the diagnosis of rare and private syndromes.

The IBM STAIRS program was chosen to provide the data bank, which includes all the information about the various syndromes. The data bank is based on information derived from 450 previously known malformative syndromes. The STAIRS program (Storage and Information Retrieval System) offers particular advantages for the kind of work we are doing. First, it allows the use of words instead of code numbers. Second, it is not necessary to prepare previously a dictionary of all the words-symptoms which would be used in the future, the so-called thesaurus, since this dictionary, synonyms included, can be progressively

Table VI. Syndrome identification in 40 patients with primary growth retardation syndromes. (Excluding chromosomal and metabolic syndrome.)

1.	**IDENTIFIED**		28 (70%)
	A.	Syndromes	
		Silver-Russel	3
		Hallerman-Streiff	2
		Prader-Willi	2
		Rubinstein-Taybi	2
		Ataxia-Teleangectasia	1
		Coffin-Siris	1
		De Lange	1
		Facio-digito-genital	1
		Fanconi pancytopenia	1
		Bird-headed Dwarfism, Boscherini type	1
		Rapp-Hodgkin Ectodermal Dysplasia	1
	B.	Skeletal Dysplasias	
		Chondroectodermal dysplasia	3
		Achondroplasia	2
		Camptomelic syndrome	2
		Multiple exostosis	2
		Osteogenesis imperfecta	1
	C.	Anomalad	
		Klippel-Feil with CHD	1
	D.	Provisional Syndromes	
		Coloboma-CHD-other anomalies	1
2.	**NONIDENTIFIED**		12 (30%)

made and enriched as one goes along. Finally, since the computer program uses a language and not a code, one can rapidly ask questions and, on the basis of the answers, ask further questions without time loss.

Figure 9 presents a general outline of the program. Of course, the patients who are to be introduced in the program should present the phenotypic pattern of a malformative syndrome. They can be seen by the doctor himself, or referred to us through special forms provided by us.

Our team performs a complete physical examination or analyzes the data and photographs submitted. At this point, there are three possibilities:

(1) The data are incomplete: further data are requested.

(2) The patient presents a specific syndrome readily recognizable through observation: the diagnosis and the data are sent back together with selected bibliographic notes.

(3) The patient presents a rare, uncertain or unknown clinical picture: the signs are translated into the standardized language and critically listed according to their presumed importance. Then, the data are introduced into the program in order to find equal or similar clinical pictures on the basis of the given signs.

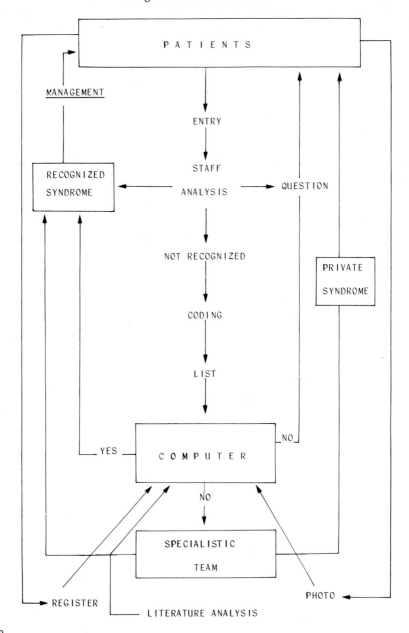

Fig. 9

Obviously, the greater the number of data, the fewer the clinical pictures sharing the same symptoms, so that their number becomes progressively very small, and the possible syndromes may thus be singularly analyzed by the doctor. Two different results may be obtained. The computer gives an affirmative answer, which means that patients with the same associated symptoms have already been described; the inquirer receives one or more diagnoses with a detailed note of when, where, and by whom, similar patients have been described, and may then take the appropriate contacts. In the second case, the computer does not point out any previous case: this may happen when were are in the presence of a sporadic, not yet described clinical form.

REFERENCES

Feingold, M. and Bossert, W.H. (1974). *Birth Defects Orig. Artic. Ser.* 13.
Feingold, M. (1977). *Obstet. Gynec.* 45, 237.
Mastroiacovo, P., Mastrangelo, R. and Segni, G. (1976). *Riv. Ital. Pediatr.* 2, 405.
Mastroiacovo, P., Salvaggio, E. and Parenti, D. (1977). *Minerva Pediatr.* 29, 773.
Smith, D.W. (1976). "Recognizable Patterns of Human Malformation", 2nd Ed. W.B. Saunders, London.

GROWTH DISTURBANCES IN HYPOTHALAMIC PITUITARY DISEASES

Z. Laron[*], Z. Zadik, A. Pertzelan and A. Roitman

*Institute of Pediatric and Adolescent Endocrinology
Beilinson Medical Center, Petah Tikva
and
Sackler School of Medicine, Tel Aviv University, Israel*

SUMMARY

The growth pattern of 151 children and adolescents with hypothalamic-pituitary insufficiency of one or more hormones were analyzed at the time of referral and during prolonged follow-up. The following types of growth patterns were distinguished: (a) growth retardation in patients lacking GH and/or TSH; (b) slowing of growth at puberty when gonadotropins are lacking; (c) accelerated growth associated with tumours secreting GH or with tumours causing precocious puberty; (d) postoperative accelerated catch-up growth of undetermined origin in children with craniopharyngioma, in the absence of IR-GH.

I. INTRODUCTION

The etiology of diseases of the hypothalamus and pituitary in childhood is various, some disturbances being hereditary or congenital and others acquired (Laron, 1969). While it is possible to distinguish between a number of these entities, with the exception of tumors, the exact etiology and cellular pathology is uncertain in most patients, even in those with an isolated hereditary hormone deficiency. Recent advices in laboratory methods have made possible the evaluation of the secretory capacity of individual pituitary hormones and thus a more exact differentiation between isolated and multiple pituitary hormone deficiencies (Laron, 1977). It has even proved possible to detect a deficiency of gonadotropins prior to puberty (Dickerman et al., 1976; Laron et al., 1977a), and it has also

---

[*] Established investigator of the Chief Scientist's Bureau, Ministry of Health.

become apparent that hormone deficiencies can be partial as well as total, and sometimes are transitory. Further, it has been suggested that, in some cases, the pituitary hormone secreted may be biologically inactive, though immunologically active (Laron, 1974), and that there may be a failure of synthesis of an intermediate substance or a defect in receptors (Jacobs et al., 1976). One of the remaining difficulties is to distinguish more easily between purely hypothalamic and pituitary lesions.

Considering that normal growth of a child, i.e., growth along the percentile of his genetic potential, depends upon hormonal harmony, it is generally accepted that either under- or oversecretion of any of the pituitary hormones will affect the normal growth pattern. The latter may be affected by any of the above enumerated defects in secretion, depending upon the genetic background, time of onset of the disturbance, the number of hormones involved, the degree of hormone deficiency, and duration of the disease.

Table I. Classification of 151 children and adolescents with hypothalamic pituitary hormone deficiencies.

Diagnosis		No. of patients		Sex distribution (% values)	
		Males	Females	Males	Females
Isolated Hormone Deficiency					
Group I	Isolated GH deficiency	20	13	61	39
Group II	Isolated GH deficiency, partial	2	4	33	67
Group III	Isolated gonadotropins deficiency	18	5	77	23
Group IV	Isolated ADH deficiency (diabetes insipidus)	3	9	25	75
Group V	High IR-GH, Laron-type dwarfism	12	16	43	57
Multiple Pituitary Hormone Deficiency					
Group VI	Craniopharyngioma	11	7	61	39
Group VII	Idiopathic multiple pituitary hormone deficiency	17	14	55	45

Having analyzed a large group of patients for their hypothalamic or pituitary hormone deficiencies, we have attempted to evaluate the influence of these hormones on the linear growth pattern. This survey, performed last year, included 151 children and adolescents with a hypothalamic pituitary hormone insufficiency of one or more hormones. A multifactorial analysis was performed from the data obtained at referral and during prolonged follow-up. The distribution of the patients, in accordance with the various diagnostic groups, is shown in Table I. It is noteworthy that more males than females have an isolated deficiency of GH or gonadotropins, idiopathic multiple pituitary hormone deficiencies, and craniopharyngiomas. It was also evident that more females have diabetes insipidus, pituitary dwarfism with high plasma IR-GH (Laron-type dwarfism), and possibly isolated partial GH deficiency, although, with the latter entity, the numbers are too small to be conclusive.

We shall now discuss a few problems relating to the growth pattern of children with hypothalamic or pituitary hormone deficiencies which came forth during the survey.

One of the basic questions it whether or not there are characteristic growth patterns for each type of pituitary hormone insufficiency or for certain combinations of hormone deficiencies. We must also ask how heredity affects the growth pattern of a hypopituitary patient: (1) in hereditary hormone deficiency, will all generations be short? and (2) in sporadic cases, will the height of the child depend upon that of the parents? That is to say, will children with the same hormone deficiency and tall parents be taller than similar children with short parents? Finally, what are the limits of the classical description for hypopituitarism (Root, 1972), i.e., growth in a channel which progressively deviates from the previously recorded channel accompanied by a markedly retarded bone age.

II. INFLUENCE OF PARENT HEIGHT ON THE STATURE OF THE HYPOPITUITARY CHILD

Table II presents the mean stature of the parents for each diagnostic group as expressed by mean deviation from the median adult height (Tanner *et al.*, 1966). When this is compared to that of the general population (Laron, 1976), it is evident that the mean height of both father and mother are all below the mean. In groups III, IV, VI, and VII, all disorders considered not to be hereditary, the deviation was small, while in the groups of isolated GH deficiency (IGHD) and the syndrome of pituitary dwarfism and high plasma IR-GH, which are largely hereditary in nature, the parents were shorter and the number of other family members with short stature was also greater than in the other groups.

Table II. Mean height of parents of children with hypothalamic pituitary hormone deficiencies.

Diagnosis		N	Fathers Deviation below median	N	Mothers Deviation below median
Group I	Isolated GH deficiency	22	1.4	24	0.9
Group II	Isolated GH deficiency, partial	6	1.8	6	2.4
Group III	Isolated gonadotropins deficiency	18	0.8	17	0.7
Group IV	Isolated ADH deficiency	9	0.6	6	0.5
Group V	High IR-GH, Laron-type dwarfism	18	1.8	18	1.6
Group VI	Multiple pituitary hormone deficiency, craniopharyngioma	11	0.7	11	0.3
Group VII	Idiopathic multiple pituitary hormone deficiency	19	0.3	19	0.8

Height standards according to Tanner *et al.* (1966).

Since there is now a means of treating GH deficiency, and treatment is begun at a very young age in these patients, it is now very difficult to relate parental height to the final stature of the child. The syndrome of pituitary dwarfism and high plasma IR-GH, while resembling IGHD clinically and in many metabolic respects (Laron *et al.*, 1972), is as yet untreatable (Laron *et al.*, 1971b) however, and an answer to the above question was therefore sought in this group. This proved somewhat difficult, since not all the parental heights are available for the 18 families of these 28 patients.

From what data was available, it would appear that parental height has little influence, despite the fact that all of these parents are relatively short. A comparison of final height in individual families revealed that, even within the same sex, the growth channel and final height varied. This was true even in a pair of twins, probably nonidentical, with IGHD, one of whom had a final height of 123.6 cm, while the other was only 112.5 cm tall.

III. INTRAUTERINE GROWTH IN THE ABSENCE OF GROWTH HORMONE

The observation that many anencephalic babies are born with body weights within normal range led to the assumption that pituitary GH is not essential for prenatal growth (Talbot and Sobel, 1947; Reid, 1960; Seckel, 1960; Dibbern, 1962; Kind, 1962). A review of additional data from the literature (Haworth and McRae, 1967; Laron et al., 1971a), as well as that collected in our own clinic (Laron and Pertzelan, 1969), reveal that many infants with developmental abnormalities of the pituitary and hypothalamus were short at birth (Table III). Table IV presents the available birth lengths of the children included in this survey in accordance with diagnostic groups. On the basis of standards established for our population (Zaizov and Laron, 1966), we divided these findings in both sexes into subnormal (A) and normal (B) ranges. With the exception of group V (the patients with the syndrome of pituitary dwarfism and high plasma IR-GH), the numbers involved are quite small, but nevertheless it would appear that most of the patients with this syndrome, as well as those with IGHD, are short at birth. Among the groups in which the hormone deficiency is an acquired one, the birth length tends to be within normal limits. Thus, it would seem that the role of the pituitary hormones in determining birth length is relatively limited, and that shortness is due only to GH deficiency. TSH deficiency at birth is very rare and, when present, its effect is probably mediated only through the thyroid hormone,

Table III. Birth length (cm) of children with hypothalamic pituitary hormone deficiencies.

	Diagnosis	Males A	Males B	Females A	Females B
Group I	Isolated GH deficiency	46, 48	51	46	
Group II	Isolated GH deficiency, partial		51, 51	37, 44	
Group III	Isolated gonadotropins deficiency	48	52, 54		50
Group IV	Isolated ADH deficiency				49
Group V	High IR-GH Laron-type dwarfism	44, 45 46, 47 47, 47 47	49	34, 46 46, 46 47	49
Group VI	Multiple pituitary hormone deficiency - craniopharyngioma		52		
Group VII	Idiopathic multiple pituitary hormone deficiency	48	50, 51		

A = Subnormal; B = Within normal range.

Table IV. Birth length (cm) of full-term infants with hypothalamic pituitary hormone deficiencies.

Diagnosis		Males A	Males B	Females A	Females B
Group I	Isolated GH deficiency	46, 48	51	46	
Group II	Isolated GH deficiency, partial		51, 51	37, 44	
Group III	Isolated gonadotropins deficiency	48	52, 54		50
Group IV	Isolated ADH deficiency				49
Group V	High IR-GH Laron-type dwarfism	44, 45 46, 47 47, 47 47	49	34, 46 46, 46 47	49
Group VI	Multiple pituitary hormone deficiency - craniopharyngioma		52		
Group VII	Idiopathic multiple pituitary hormone deficiency	48	50, 51		

A = Subnormal; B = Within normal range.

IV. INFANT GROWTH IN THE ABSENCE OF ONE OR MORE PITUITARY HORMONES

It is a matter of controversy whether children deficient in pituitary hormones, particularly in GH, show a decrease in growth rate immediately after birth or only after the age of 6 to 12 months (Blizzard, 1968). It is possible that the variability of patterns reported is a result of the lack of evidence for a total absence of GH secretion at birth. The fact that measurement from birth is not a routine procedure detracts from any retrospective studies made.

The study of hereditary IGHD constitutes a most interesting model. Figure 1 presents the growth pattern of a girl (S.K.) with isolated GH deficiency. This girl belongs to a sibship containing several adult dwarfed patients, all of whom were proved to have a monohormone deficiency. The father of this girl was 179 cm in height, and the mother 150 cm. The birth length of the child is not known, but she was said to have been small. At the age of one month, her supine length was 48 cm. For 4 months, she grew along a line below, but possibly parallel to the 3rd percentile, and subsequently a progressive decline in growth velocity was observed. Between age $1^{1}/3$ and 3 years (Fig.2), when hGH therapy was instituted, the growth rate varied from slow growth to short periods of arrested growth.

Another two children with familial IGHD showed marked retardation of growth and bone age as early as at the age of 6 months (Figs 3,4). Furthermore, it is noteworthy that patient E.B., between the age of 6 months to 1 year, had a growth rate which was slow, but parallel to the 3rd percentile. It may be stated with certainty that, even in the complete absence of GH, there is a slow growth at times (Fig.5: L.I.; Fig.6: E.I.), in some cases with complete growth arrest later in childhood. What it is that makes these children grow, albeit slowly, is not yet known. Somatomedin is not only low in early infancy, but is also GH dependent (Van den Brande, 1976). Insulin secretion is also low in infants lacking GH (Laron et al., 1977b). It may be concluded that, *in utero* and for several months postnatally, these infants are able to grow despite a reduced growth potential. *In utero*, the nutrients passing from the mother through the placenta may be in-

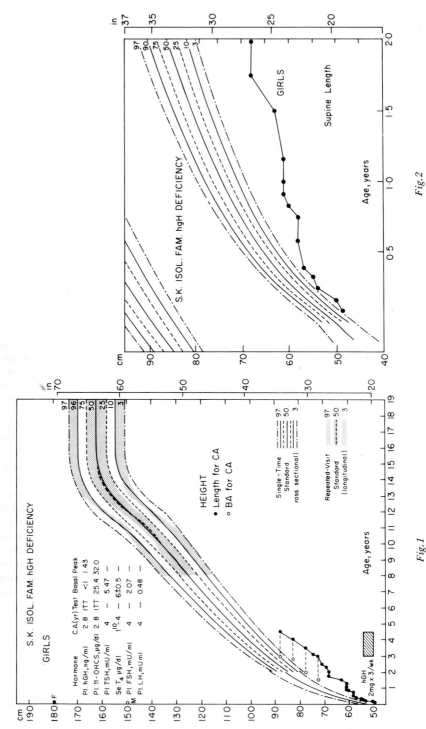

Fig. 1

Fig. 2

Growth Disturbances in Hypothalamic Pituitary Diseases

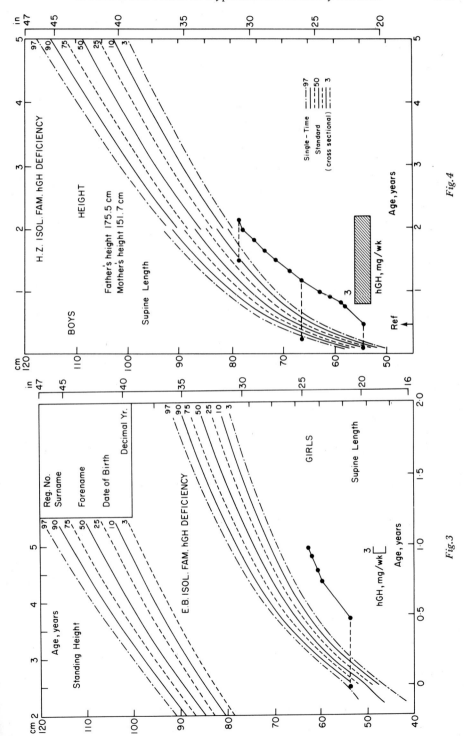

Fig. 4

Fig. 3

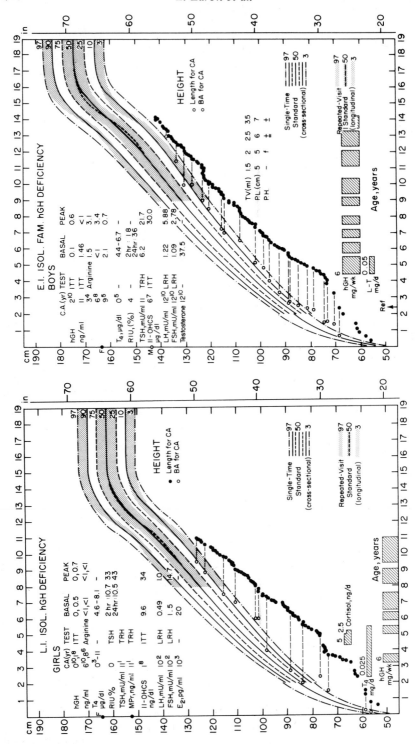

Fig. 5

Fig. 6

Growth Disturbances in Hypothalamic Pituitary Diseases

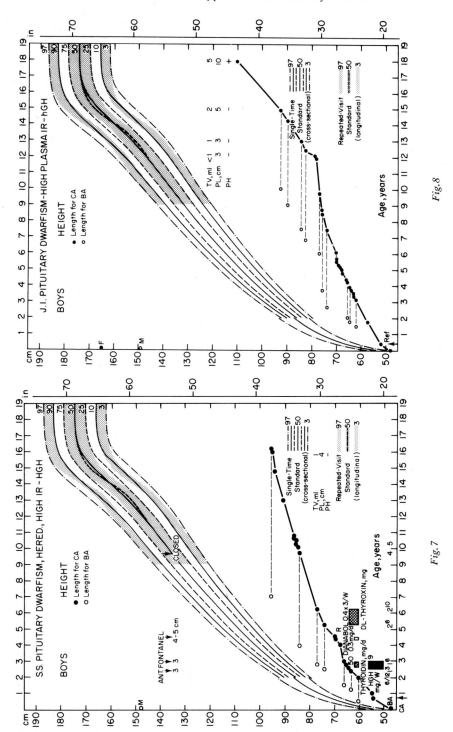

Fig. 7

Fig. 8

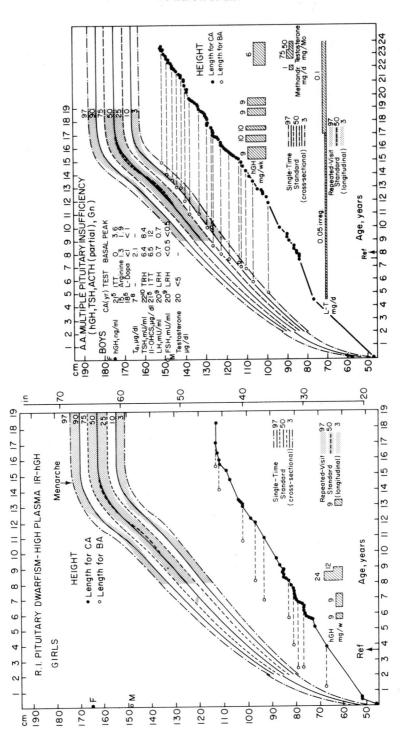

Fig. 9

Fig. 10

Growth Disturbances in Hypothalamic Pituitary Diseases

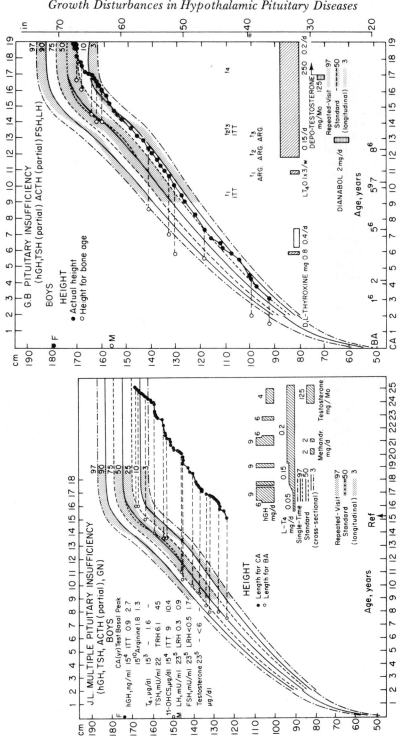

Fig. 11

Fig. 12

strumental in maintaining growth, but postnatally, the instigation for growth must be contained within the infant himself.

If the child lacking GH is not treated from the age of one year, there is a definite progressive slowing in growth, with subsequent short periods of actual growth arrest. Substitutional therapy, whether continuous (Braunstein et al., 1975) or intermittent (Pertzelan et al., 1976), serves to improve the growth rate dramatically. The optimal dose for the achievement of normal growth has yet to be determined and remains the subject of controversy. It may be asked whether increasing the dose above 6 mg/week would benefit patient L.I. (Fig.5), who still has a bone age retardation of 1½ years, but this would not necessarily be true. Patient E.I. (Fig.6), who has a very short mother (144 cm) and a relatively short father (163.5 cm), at the age of 14 has attained a corresponding bone age, but is still below the 3rd percentile. Thus, genetic factors not necessarily related to the extent of his GH deficiency, and the rate of puberty, which may also be genetically determined, will all influence his final height. Whether or not higher doses of hGH and shorter intervals of treatment interruption at the ages of 8 and 10½ (the longer interruptions having been caused by temporary shortages in hGH) would have led to a higher growth channel is another question.

V. THE SYNDROME OF PITUITARY DWARFISM AND HIGH PLASMA IR-GH (LARON-TYPE DWARFISM)

Since this syndrome and hereditary IGHD are so similar, the growth during the first year of life should be noted. We have accurate data on several patients for this period (in Figs 7–9: S.S., J.I., R.I.). All were short at birth, and, during the first half-year of life, already showed a deviation from the normal growth channel (Fig.7). With advancing age, both the deviation from the normal and the retardation in bone age progressed. At puberty, the boys showed a growth spurt (Fig.8). The girls had a similar growth pattern, but without the pubertal spurt (Fig.9). This would appear to be the classical growth pattern of boys and girls with hereditary IGHD who have not received any substitutional therapy with exogenous hGH.

VI. MULTIPLE PITUITARY HORMONE DEFICIENCY (MPHD) OR PANHYPOPITUITARISM

It is to be expected that, in the absence of several pituitary hormones, growth would be more impaired than in the absence of only one hormone. The type and degree of hormone insufficiency in the individual constitute the major limiting factors. Thus, a total absence of ACTH is not compatible with life and a total absence of TSH would lead to the severe growth retardation seen in hypothyroidism. Only scant data are available on the perinatal growth of these children, who, if they survive, are usually diagnosed at a later age. There is also no certainty that the hormonal insufficiency remains constant over the years. This is particularly true in cases of craniopharyngioma or other types of tumor.

Figure 10 presents the growth pattern of A.A., a boy with MPHD in whom measurements had been made from birth. His birth length was short (47 cm). He began to receive thyroxine at age 4½ years and, in general, his growth pattern up to the age of 14 years is similar to that seen in IGHD, with the exception of a greater bone age retardation. The initiation of exogenous hGH administration at the age of 15 years led to an improvement in growth velocity, but to a lesser extent than that which would be expected in IGHD. The continuous growth in the absence of all known growth hormones is astonishing. Even hGH therapy given with long interruptions appears to suffice for growth (Fig.11: J.L.), even though occurring at an unusual age. The addition of androgens was found to induce a marked spurt: patient J.L. has reached a height of 168 cm and has a bone age of 16 years at the chronological age of 24 years, still having the capability for further growth.

Our experience indicates that many patients with MPHD, with or without minimal therapy, have a growth potential which, in some cases, has appeared to be greater than that noted in IGHD. Certain basic conditions have to be fulfilled, however. Thyroid hormone is necessary for growth and for hGH to be effective (Roitman and Laron, 1977). We hypothesize that extensive damage to the area of the hypothalamic pituitary deprives the organism not only of growth-stimulating hormones, but also of growth-inhibiting hormones (e.g., somatostatin and possibly others), thus allowing a slow continuous growth adapted to the still present and now inhibited growth factors. This theory is compatible with the growth seen in patients who grow without measurable immunoreactive GH (Laron et al., 1976). It is of note that many of these are patients with craniopharyngioma (Karp and Laron, 1967; Kenny et al., 1968) who show catch-up growth after total or subtotal extirpation of the tumor and/or pituitary.

Less easy to explain is the growth within normal channels of a few patients who, upon repeated testing, are found to lack endogenous GH as well as gonadotropins. One such patients (G.B.) is shown in Fig.12. It is of note that his bone age is less retarded than in patients with MPHD and marked growth retardation. In contradistinction to the latter, G.B. has a normal insulin response to the usual stimuli, despite his not being obese. It is worth mentioning that most of these patients are males (Laron et al., 1976).

In conclusion, it may be stated that normal growth is a highly complex process which is largely dependent upon the integrity of the hypothalamic pituitary hormones, whether stimulators or inhibitors. The fact that growth at a low velocity is still present in the absence of the hypothalamic pituitary hormones compels us to search for other growth hormones or factors not controlled by the hypothalamic pituitary hormones. It has been suggested that insulin is such a hormone (Karp and Laron, 1967), but it is quite likely that there are others as well.

REFERENCES

Blizzard, R.M. (1968). *In* "Human Growth" (D.B. Cheek, ed), pp.41-59. Lea & Febiger, Philadelphia.
Braunstein, G.D., Raiti, S., Hansen, J.W. and Kohler, P.O. (1975). *N. Engl. J. Med.* 292, 332.
Dibbern, H.H. (1962). Inaugural dissertation for M.D. degree, Medical, pp.1-15. Faculty of Zurich.
Dickerman, Z., Prager-Lewin, R. and Laron, Z. (1976). *Am. J. Dis. Child.* 130, 634-638.
Haworth, J.C. and McRae, K.N. (1967). *Lancet* 87, 41-46.
Jacobs, L.S., Sneid, D.S., Garland, J.T., Laron, Z. and Daughaday, W.H. (1976). *J. clin. Endocr. Metab.* 42, 403-406.
Karp, M. and Laron, Z. (1967). *Harefuah* 73, 41-45.
Kind, C. (1962). *Helv. Paediatr. Acta* 17, 244-258.
Kenny, F.M., Iturzaeta, N.F., Mintz, D., Drash, A., Garces, L.Y., Susen, A. and Askari, H.A. (1968). *J. Pediatr.* 72, 766.

Laron, Z. (1969). *In* "Paediatric Endocrinology" (D.W. Hubble, ed), pp.35-111. Blackwell Sci. Publ., Oxford.
Laron, Z. (1974). *Israel J. Med. Sci.* 10, 1247-1253; and *in* "Heterogeneity of Polypeptide Hormones" (D. Rabinowitz and J. Roth, eds), pp.65-71. Academic Press, New York.
Laron, Z. (1976). *Nordic Council Arc. Med. Res. Rep.* 14, 15-24.
Laron, Z. (1977). *In* "Hypothalamic and Pituitary Diseases in Childhood". C.C. Thomas, Springfield. *(In preparation.)*
Laron, Z. and Pertzelan, A. (1969). *Lancet* 1, 680-681.
Laron, Z., Pertzelan, A. and Frenkel, J. (1971a). *In* "Hormones in Development" (M. Hamburg and E.J.W. Barrington, eds), pp.573-585. Appleton Century Crofts, Meredith Corp., New York.
Laron, Z., Pertzelan, A., Karp, M., Kowadlo-Silbergeld, A. and Daughaday, W.H. (1971b). *J. clin. Endocrinol.* 33, 332-342.
Laron, Z., Karp, M., Pertzelan, A., Kauli, R., Keret, R. and Doron, M. (1972). *In* "Growth and Growth Hormone" (A. Pecile and E.E. Muller, eds), pp.458-482. Excerpta Medica, Amsterdam.
Laron, Z., Pertzelan, A., Kiwity, S., Livneh-Zirinsky, M. and Keret, R. (1976). *In* "Convegno Internazionale di Endocrinologia Pediatrica" (Serono Symposia, Bologna), pp.131-152. Pacini, Pisa.
Laron, Z., Kaushanski, A. and Josefsberg, Z. (1977a). *Clin. Endocrinol.* 6, 265-270.
Laron, Z., Mimouni, M., Josefsberg, Z., Zadik, Z. and Doron, M. (1977b). *Diabetologia (in press).*
Pertzelan, A., Kauli, R., Assa, S., Greenberg, D. and Larzon, Z. (1976). *Clin. Endocrinol.* 5, 15-24.
Reid, J.D. (1960). *J. Pediatr.* 56, 658-664.
Roitman, A. and Laron, Z. (1977). *Submitted for publication.*
Root, A.W. (1972). *In* "Human Growth Hormone", pp. 108. C.C. Thomas, Springfield.
Seckel, H.P.G. (1960). *Am. J. Dis. Child.* 99, 349-379.
Talbot, N.B. and Sobel, E.H. (1947). *Adv. Pediatr.* 2, 238.
Tanner, J.M., Whitehouse, R.H. and Takaishi, M. (1966). *Arch. Dis. Child.* 41, 454 and 613.
Van den Brande, I.L. (1976). *In* "Growth Hormone and Related Peptides" (A. Pecile and E.E. Muller, eds), pp.271-285. Excerpta Medica, Amsterdam.
Zaizov, R. and Laron, Z. (1966). *Acta Paediatr. Scand.* 55, 524-528.

HUMAN GROWTH HORMONE, IMMUNOREACTIVE INSULIN, AND SOMATOMEDIN ACTIVITY IN ACHONDROPLASIA

G. Giordano, E. Foppiani, F. Minuto, R. Cocco, M. Di Cicco, A. Barreca
F. Caiazza and F. Morabito

*Department of Endocrinology and Constitutional Pathology
University of Genoa
and
Italian Auxologic Center, Piancavallo (Novara), Italy*

SUMMARY

GH, IRI, and ASm (somatomedin activity) patterns were studied in 20 achondroplastic subjects. Arginine and L-Dopa administration produced a normal rise in GH concentration, while somatomedin was, in basal conditions, statistically higher than control subjects. The administration of hGH produced a significant increase in ASm. Furthermore, during the course of the oral glucose tolerance test, a normal blood glucose pattern was observed, while insulin secretion was significantly lower. A similar behaviour was observed after tolbutamide administration. These results are in agreement with the hypothesis of a peripheral resistance to the Sm action and suggest that these patients should present a noticeable predisposition to diabetes mellitus.

I. INTRODUCTION

Achondroplasia is the most frequent and well-known example of growth osteochondrodystrophy. The disease is caused by a mutant gene, which, by means of biochemical mediators, influences the development of the epiphyseal-metaphyseal complex, with disruption of morphogenesis, calcification, ossification, and growth (Boni et al., 1970; Lamy and Maroteaux, 1961).

Endocrine causes are, by definition, excluded from the primary pathogenesis of this group of disorders, particularly achondroplasia. The endocrine aspects of this disease have been poorly studied and have yielded only equivocal results. Of particular interest should be the behaviour of the Growth Hormone (GH) — somatomedin (Sm) axis, due to its overall influence on growth.

Basal values of GH within normal limits in achondroplasia have been observed by several authors (Frasier, 1967; Isidori et al., 1974), while others (Collip et al., 1972) have reported, with some frequency, consistently low GH levels. Still other researchers (Crisalli and Chiossi, 1974), referring to rare cases of osteochondroplasia, note variable GH responses to stimulation which could be due to disregulation of the secretion.

Attempts at therapy with hGH, however, have proved either negative (Gershberg et al., 1964), doubtful (Escamilla et al., 1974), or, with high doses of hGH, moderately positive (Escamilla, 1974).

With regard to somatomedin activity (ASm), surprisingly high levels were found in two cases (Van den Brande and Du Caju, 1974) in which samples of blood were taken in the afternoon. This contrasts with studies performed after a night's fast in 7 other achondroplastic patients in which the ASm levels were within normal limits (Van den Brande, 1975) and with the results observed in another group of 16 achondroplastic patients, in which radioreceptor assayable Sm was again within the normal range (Horton et al., 1976).

Such data seem to indicate that the slow growth rate of achondroplasia is accompanied by at least normal ASm and is probably caused by reduced sensitivity of the target tissues (Van den Brande, 1975).

Histochemical research performed *in vitro* on achondroplastic cartilage (Shepard, 1971; Shepard et al., 1968; Stanesio et al., 1966) has demonstrated, besides a reduction of glycogen levels in the cells of the proliferative stratum, alterations in the enzymatic activity of glucose metabolism. Such observations lead to the hypothesis of a metabolic defect in the pathogenesis of chondrogenic disturbances characteristic of achondroplasia. Furthermore the oral glucose tolerance test (OGTT), performed in 24 achondroplastic patients (4 months to 2 years of age), showed an evident glucose intolerance in 16 cases (Collin et al., 1972). Such observations seem to suggest that the metabolic defect of achondroplasia is not restricted to the cartilage, but may involve the whole organism (Collip et al., 1972).

For these reasons, we have further investigated the behaviour of GH, ASm and immunoreactive insulin (IRI) in a group of achondroplastic patients.

II. MATERIALS AND METHODS

The study was performed on 21 achondroplastic subjects (11 males and 10 females aged 5 to 25 years), inpatients at the Italian Auxologic Center at Piancavallo. In none of these patients did the family history reveal evidence of diabetes mellitus in parents, grandparents, or relatives; during the course of the study, none of these subjects presented manifest signs of diabetes mellitus; in all, T3, T4 and FT4 were within the normal range.

The patients were allowed a free diet with normal carbohydrate intake during the days preceding the study. The tests were performed during the morning in fasting conditions. One hour prior to the administration of the test substance, an i.v. drip of physiological saline was applied into the cubital vein, in order to avoid repeated venipunctures and the thrombosis of the needle. Samples were taken at 0, 30, 60, 90 min after the beginning of arginine infusion; similarly, at 30 min intervals up to 180 min after L-Dopa administration and oral glucose load; finally, at -15, 0, 2, 5, 20, 30, 40 and 60 min after infusion of tolbutamide.

The arginine test was carried out in 19 patients, infusing 0.5 g/kg body weight of arginine chloridrate in 50 ml of solvent over a period of 30 min. L-Dopa testing was performed in 19 patients with oral administration of 250-500 mg at time 0. The oral glucose tolerance test (OGTT) was performed in 14 subjects with the administration of 1.75 g/kg ideal body weight over a period of 5—10 min. Tolbutamide testing was carried out in 14 patients by i.v. injection, over a period of 3 min, of 1 g of tolbutamide in 20 ml of solvent. ASm was studied in 20 patients in basal conditions and in 14 patients also at 8, 24, 48 h after i.m. injection of 4 mg of hGH (Grorm, Serono).

The control group for the study of the behaviour of GH secretion was made up of 24 subjects (7 to 15 years of age) for arginine testing, and of 7 subjects (8 to 13 years) for L-Dopa testing. An ambisexual control group for evaluation of ASm was made up of 57 subjects of corresponding age. The control group for blood glucose (BG) and IRI during the course of the OGTT and tolbutamide test was composed of 7 subjects aged 9 to 16 years.

BG was assayed by glucose-oxidase method (Keston, 1956); GH and IRI were determined by radioimmunoassay (Kits CEA-IRE-Sorin); ASm was determined by radiobioassay, according to Yde (1968) partially modified (Giordano et al., 1976).

III. RESULTS

During the course of the arginine test, serum levels of GH rose from 1.1 ± 1.5 to 9.2 ± 5.2 ng/ml (mean of peak values $\pm$ SD); these values are not statistically different from those observed in the control group (2.6 ± 2.4 and 10.6 ± 7.0 ng/ml, respectively). After L-Dopa administration, GH levels varied from 0.3 ± 0.4 to 10.7 ± 5.0 ng/ml; again, the mean of the peak values was not statistically different from those of controls (14.6 ± 7.0 ng/ml) (Fig.1).

ASm was 1.3 ± 0.4 U/ml, significantly higher ($p < 0.01$) than that observed in a group of 57 age-matched subjects (0.92 ± 0.3 U/ml). Considered case by case, it appears evident that 10 of 20 patients showed basal values of ASm in the high normal, or even above normal, range for adult controls (1.02 ± 0.18 U/ml). After administration of hGH, the ASm varied from 1.3 ± 0.4 to 1.5 ± 0.5 U/ml between 24 and 48 h, with a variation not statistically significant, while, in normal controls, we recorded a significant increase, from 0.92 ± 0.08 to 1.61 ± 0.21 U/ml at 48 h (Fig.2). The mean of the peak values, however (observed at 24 h in 5, and at 48 h in 9 patients) was 1.7 ± 0.5 U/ml, significantly different ($p < 0.05$) with respect to the basal values.

In the course of OGTT, the BG pattern of the achondroplastic subjects behaved similarly to that of the control group (Fig.3). The IRI level, however, was significantly lower at all points of the curve, with the exception of the 60 min value. The insulinogenic index, calculated according to Seltzer and Smith (1959) in all points of the curve, was in these subejcts absolutely flat.

During the tolbutamide test, the BG values were not significantly different from controls, while IRI concentrations were significantly lower at all points, with the exception of the 60 min value (Fig.4).

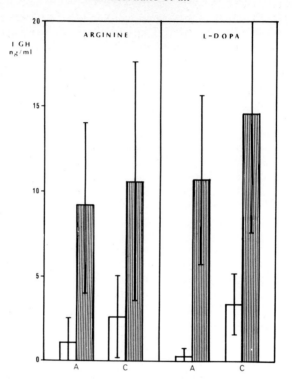

Fig. 1 Effect of arginine and L-Dopa on GH secretion in patients affected with achondroplasia (A) and in control subjects (C). Open columns represent basal values ± SD. Dark columns represent peak values ± SD.

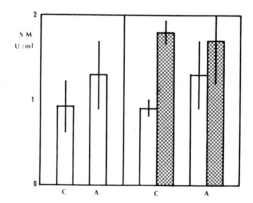

Fig. 2 Somatomedin activity in control subjects (C) and in patients affected with achondroplasia (A). Open columns: basal values ± SD; dark columns: values observed after hGH administration ± SD.

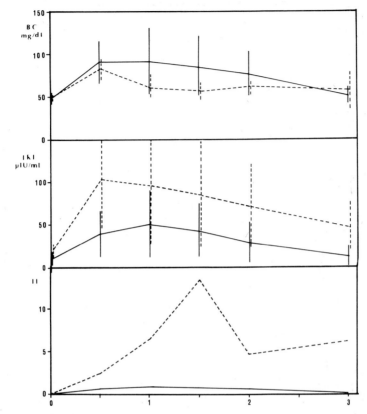

Fig. 3 Behaviour of blood glucose (BG), immunoreactive insulin (IRI), and insulinogenic index (II) during oral glucose tolerance test in control subjects (dashed lines) and in patients affected with achondroplasia (full lines) (mean ± SD).

IV. DISCUSSION

In basal conditions, GH appeared normal, although individual values were frequently at the lower limits, in agreement with the findings of other researchers (Collip *et al.*, 1972; Frasier, 1967; Isidori *et al.*, 1974). The response of GH to stimulation was similar to that of controls, making evident the lack of impairment of GH secretion in achondroplasia.

Our results demonstrate an elevated ASm in these patients. Such an elevation appears more evident when individual values are considered; in fact, the level of ASm was higher than normal in 50% of the cases studied. A similar pattern was observed by Van den Brande and Du Caju (1974) in 2 cases studied in the afternoon, but was not confirmed by the same authors in 7 other patients whose serum was taken in the morning after a night's fast (Van den Brande *et al.*, 1975). Such behaviour might be interpreted in terms of circadian variations of ASm, which have been observed in normal children by the same authors (Van den Brande *et al.*, 1975). Our results show ASm to be elevated in achondroplasia after a night's fast.

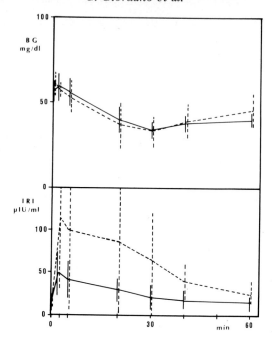

Fig.4 Effect of tolbutamide on blood glucose (BG) and on immunoreactive insulin (IRI) concentration in control subjects (dashed lines) and in patients affected with achondroplasia (full lines) (mean ± SD).

The Sm-generating system did not appear seriously impaired in our subjects, as we observed an increase in ASm after i.m. administration of hGH.

On the basis of these results, one might conclude that achondroplastic subjects present a negative correlation between elevated or relatively elevated ASm values and a slow growth rate.

This would indicate a hyporesponsiveness by the target tissues to Sm action. Such hyporesponsiveness can probably be explained in terms of a defective binding to specific receptors or of structural and enzymatic anomalies of the cartilagineous substrate (Shepard, 1971; Shepard et al., 1968; Stanesio et al., 1966).

Particularly interesting is the impairment of insulin secretion in these subjects. Even without a clear intolerance to carbohydrate, the insulin response was significantly lower than control group and the insulinogenic index appeared flat. This fact seems to support the hypothesis of an abnormality of glucose metabolism involving the whole organism and evidenced by the impairment of insulin secretion.

We are not aware of statistical studies on the incidence of diabetes in adult achondroplastic patients; however, if our results will be confirmed, they would suggest that such patients should present a noticeable predisposition to diabetes mellitus.

REFERENCES

Boni, M., Ceciliani, L., Ghisellini, F. and Lenzi, L. (1970). In "Atti 55° Congr. Soc. It. Ortopedia e Traumatologia", pp.93-115. Pozzi, Rome.

Collip, P.J., Sharma, R.K., Thomas, J., Vaddanahally, T.M. and Sliang, Y.C. (1972). *Am. J. Dis. Child.* **124**, 682-685.
Crisalli, M. and Chiossi, F.M. (1974). *In* "Proc. Intern. Symp. on Management of Height Retardations". *Acta Med. Auxol. Suppl.* **6**, 125-129.
Escamilla, R.F. (1974). *In* "Advances in Human Growth Hormone Research" (S. Raiti, ed), pp.765-781. Dhew. Publ., Washington.
Escamilla, R.F., Hutchings, J.J., Li, C.H. and Forsham, P. (1966). *Calif. Med.* **105**, 104-110.
Frasier, S.D. (1967). *J. Pediatr.* **71**, 625-638.
Gershberg, H., Mari, S., Hulse, M. and St. Paul, H. (1964). *Metabolism* **13**, 152-160.
Giordano, G., Foppiani, E., Minuto, F. and Perroni, D. (1976). *Acta endocr. (Copenh.)* **81**, 449-454.
Giordano, G., Foppiani, E., Perroni, D. and Minuto, F. (1976). *In* "Atti 16° Congr. Soc. It. Endocrinologia", pp.175-199. Serono Symposia, Rome.
Horton, W.A., Rimoin, D.L., Underwood, L.E. and Van Wyk, J.J. (1976). *N. Engl. J. Med.* **295**, 453.
Isidori, A., Cianfriglia, F. and Dondero, F. (1974). *Folia Endocrinol.* **27**, 179-184.
Keston, A. (1956). Abstr. 129th Meet. Am. Chem. Soc. S.31 c.
Lamy, M. and Maroteaux, P. (1961). "Les Chondrodystrophies Genotypiques". L'Expansion Scientifique, Paris.
Seltzer, H.S. and Smith, W.L. (1959). *Diabetes* **8**, 417-424.
Shepard, T.H. (1971). *J. Embriol. Exp. Morphol.* **25**, 347-363.
Shepard, T.H., Maffet, B.C. and Fry, L.H. (1968). *Soc. Pediatr. Res.* **38**, 21.
Stanesio, V., Bona, C., Jonesco, V. and Gottlieb, F. (1966). *Rev. Fr. Étud. Clin. Biol.* **11**, 600-610.
Van den Brande, J.L. (1975). *In* "Growth Hormone and Related Peptides" (A. Pecile and E.E. Muller, eds), pp.271-285. ICS 381. Excerpta Medica, Amsterdam.
Van den Brande, J.L. and Du Caju, M.V.L. (1974). *In* "Advances in Human Growth Hormone Research" (S. Raiti, ed), pp.85-115. Dhew Publ., Washington.
Van den Brande, J.L., van Buul, S., Heinrich, V., van Roon, F., Zurcher, T. and van Steirtegem, A.C. (1975). *Adv. Metab. Disord.* **8**, 171-181.
Yde, H. (1968). *Acta endocr. (Copenh.)* **57**, 557-564.

HUMAN GROWTH HORMONE, IMMUNOREACTIVE INSULIN, AND SOMATOMEDIN ACTIVITY IN TURNER'S SYNDROME

F. Morabito, F. Caiazza, A. Barreca, M. Di Cicco, F. Minuto, E. Foppiani and G. Giordano

Italian Auxologic Center, Piancavallo (Novara)
and
Department of Endocrinology and Constitutional Pathology
University of Genoa, Italy

SUMMARY

GH, IRI, and ASm (somatomedin activity) patterns were studied in 19 subjects affected with Turner's syndrome. GH secretion in basal conditions and in response to arginine and L-Dopa was not statistically different from control subjects. ASm was within the normal values, in basal conditions, in 10 patients, while in 2 other cases, it was lower than in the corresponding control group. After hGH administration, similarly to normal adults, ASm varied significantly at 48 h. During the course of oral glucose tolerance test, blood glucose was considerably, but not statistically, higher than controls, and IRI appeared constantly lower. Consequently, the insulinogenic index was absolutely flat. These results seem to confirm a hypofunction of the insular-pancreatic system, while GH secretion and ASm appeared normal, in agreement with the hypothesis of a peripheral resistance to the action of GH and Sm.

I. INTRODUCTION

Short stature in Turner's syndrome does not appear to be due to an impairment of growth hormone (GH) secretion or to the absence of somatomedin activity (ASm).

GH secretion, in basal conditions and after stimulation, has been reported as normal or elevated (Daughaday *et al.*, 1969; Kaplan *et al.*, 1968; Lindsten et al., 1967; Peake et al., 1972; Saenger et al., 1976). In addition, the observation of an early paradoxical GH increment during the oral glucose tolerance test (OGTT) (Lindsten et al., 1967; Marrama et al., 1973; Nielson et al., 1969) points out the possibility of a disregulation in GH secretion.

ASm in basal conditions is normal or even elevated (Almqvist et al., 1963; Daughaday et al., 1969; Daughaday and Parker, 1963).

While the metabolic responses of these subjects to administration of hGH appears normal (with regard to increase in FFA levels, nitrogen retention, and insulin response to glucose (Daughaday et al., 1969; Merimee et al., 1969)) no variations were observed in the ASm (Daughaday, 1971; Daughaday et al., 1969). Such observation does not exclude, however, the possibility of a partial resistance of the somatomedin-generating system to the GH action.

For a long time the frequent association between gonadal dysgenesis, Turner's syndrome, and diabetes mellitus has been emphasized. According to several authors (Engel and Forbes, 1965; Forbes and Engel, 1963; Freychet et al., 1967; van Campenhout et al., 1973), Turner's syndrome is frequently correlated with diabetes mellitus which presents the features of the maturity-onset diabetes (Nielsen et al., 1969); other authors evidenced the frequent association with chemical diabetes (Jackson et al., 1966; Nielsen et al., 1969; Rasio et al., 1976; van Campenhout et al., 1973), or with latent diabetes, evidenced by means of the cortisone-glucose tolerance test (Andreani et al., 1966; Menzinger et al., 1966).

Insulin secretion, however, is frequently impaired, even when intolerance to carbohydrates is not evident (Jackson et al., 1966; Lindsten et al., 1967; Soyka et al., 1964; van Campenhout et al., 1973), with consequent lowered insulinogenic index (Lindsten et al., 1967; Nielsen et al., 1969; van Campenhout et al., 1973). Insulin pattern may present a response delayed or reduced, similar to that observed in maturity-onset diabetes, or, alternatively, it may present an abrupt increase, with persistence of high values at plateau (Nielsen et al., 1969). In addition, while the response of insulin to tolbutamide is said to be normal (Nielsen et al., 1969; van Campenhout et al., 1973), according to Andreani et al. (1966), the blood glucose (BG) behaviour is similar to that of hyperinsulinism.

Such conflicting reports have motivated a study of the behaviour of GH, immunoreactive insulin (IRI), and ASm in a group of subjects affected with gonadal dysgenesis.

II. MATERIALS AND METHODS

The study was performed on 19 inpatients of gonadal dysgenesis (aged 9 to 23 years) at the Italian Auxologic Center at Piancavallo. In none of these patients did the family history reveal evidence of diabetes mellitus in parents, grandparents, or relatives. During the course of the study, none of these subjects presented manifest signs of diabetes mellitus; in all, T3, T4 and FT4 were within the normal range and none were undergoing estrogen therapy.

The patients were allowed a free diet with normal carbohydrate intake during the days preceding the study. The tests were performed during the morning in fasting conditions. One hour prior to administration of the test substance, an i.v. drip of physiological saline was applied into the cubital vein, in order to avoid repeated venipunctures and the thrombosis of the needle. Samples were taken at

0, 30, 60, 90 min after the beginning of arginine infusion; similarly, at 30 min intervals up to 180 min after L-Dopa administration and glucose oral load; finally, at -15, 0, 2, 5, 20, 30, 40 and 60 min after infusion of tolbutamide.

The arginine test was carried out in 19 patients, infusing 0.5 g/kg body weight of arginine chloridrate in 50 ml of solvent over a period of 30 min. L-Dopa testing was performed in 11 patients with oral administration of 250-500 mg at time 0. The OGTT was performed in 17 subjects with the administration of 1.75 k/kg ideal body weight over a period of 5-10 min. Tolbutamide testing was carried out in 17 patients by i.v. injection, over a period of 3 min, of 1 g of tolbutamide in 20 ml of solvent. ASm was studied in 12 patients under basal conditions and in 6 patients also at 8, 24 and 48 h after i.m. injection of 4 mg of hGH (Grorm, Serono).

The control group for the study of the behaviour of GH secretion was made up of 24 subjects (7 to 15 years of age) for arginine testing and of 7 subjects (8 to 13 years) for L-Dopa testing. An ambisexual control group of evaluation of ASm was made up of 45 subjects of corresponding age. The control group for study of the behaviour of BG and IRI during the course of the OGTT and tolbutamide test was composed of 7 subjects (9 to 16 years).

BG was assayed by glucose-oxidase method (Keston, 1956); GH and IRI were measured by radioimmunoassay (Kits CEA-IRE-Sorin). ASm was measured by radiobioassay, according to Yde (1968) partially modified (Giordano et al., 1976).

III. RESULTS

In subjects affected with gonadal dysgenesis, arginine produced a rise of GH from 1.2 ± 1.1 to 9.7 ± 8.7 ng/ml (mean of peak values $\pm$ SD). These values were not statistically different from those observed in the control group (2.6 ± 2.4 and 10.6 ± 7.0 ng/ml, respectively). Similarly, after L-Dopa administration, the GH concentration varied from 1.4 ± 1.3 to 10.7 ± 5.0 ng/ml; these values are not statistically different from those found in control subjects (3.4 ± 1.8 and 14.6 ± 7.0 ng/ml, respectively) (Fig.1).

In 10 patients aged 9-14 years, ASm was 0.8 ± 0.2 U/ml, which is not statistically different from the value found in 16 control subjects of corresponding age (0.9 ± 0.3 U/ml) (Fig.2). In 2 other cases (22.4 and 23.4 years of age), ASm was 0.6 and 0.86 U/ml, respectively, which is lower than the value found in a group of 29 adult controls (20-50 years of age) (1.02 ± 0.18 U/ml). After hGH administration, ASm varied significantly ($p < 0.05$), from 0.7 ± 0.2 U/ml to 1.5 ± 0.7 U/ml at 48 h. This behaviour is similar to that observed in 4 normal adult subjects aged 20-35 years (Fig.2).

During the course of OGTT, BG was considerably, but not statistically, higher than in controls (Fig.3), and IRI appeared constantly lower, with a statistically significant difference only at 30 min ($p < 0.05$). The insulinogenic index, calculated according to Seltzer and Smith (1959) in all points of the curve, was in these subjects absolutely flat.

During the tolbutamide test, the BG concentration was statistically higher than normal only at 2 min ($p < 0.05$); similarly, IRI behaviour was generally lower with respect to that recorded in normal subjects, but significantly different only at 20 min ($p < 0.05$) (Fig.4).

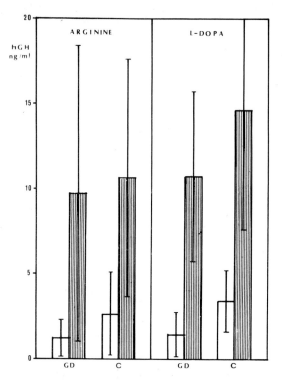

Fig.1 Effect of arginine and of L-Dopa on hGH secretion in patients affected with gonadal dysgenesis (GD) and in control subjects (C). Open columns represent basal values ± SD. Dark columns represent peak values ± SD.

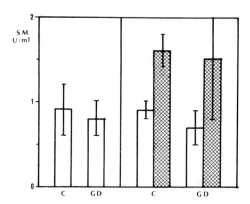

Fig.2 Somatomedin activity in control subjects (C) and in patients affected with gonadal dysgenesis (GD). Open columns represent basal values ± SD. Dark columns represent values observed after hGH administration ± SD.

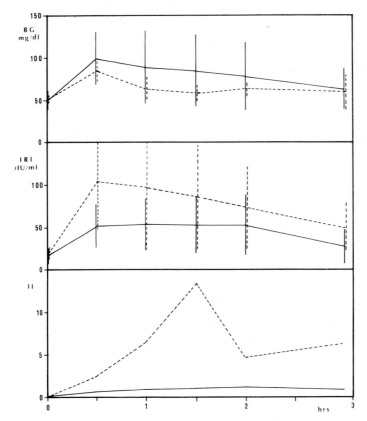

Fig.3 Behaviour of blood glucose (BG), immunoreactive insulin (IRI), and insulinogenic index (II) during oral glucose tolerance test in control subjects (dashed lines) and in patients of gonadal dysgenesis (full lines). (Mean values ± SD.)

IV. DISCUSSION

In agreement with some authors (Kaplan *et al.*, 1968; Marrama *et al.*, 1973; Saenger *et al.*, 1976), and in contrast with others (Lindsten *et al.*, 1967; Peake *et al.*, 1972), who have observed elevated basal values of GH, our results confirm normal GH concentration in basal conditions.

Furthermore, the response to L-Dopa and to arginine appeared normal, in agreement with the results obtained with arginine (Franchimont and Burger, 1975; Marrama *et al.*, 1973; Tzagournis, 1969) and with insulin-induced hypoglycemia (Franchimont and Burger, 1975; Lindsten *et al.*, 1967; Marrama *et al.*, 1973; Meadow *et al.*, 1968). Nevertheless, individual cases provided varying responses, sometimes elevated, other times absent (Rasio *et al.*, 1976).

The disagreement of the results may be due to the heterogeneity of the cases studied. Considering, however, the paradoxical response to glucose load (Lindsten *et al.*, 1967; Marrama *et al.*, 1973; Nielsen *et al.*, 1969), such variable behaviour of GH secretion can be explained by disturbances in secretory regu-

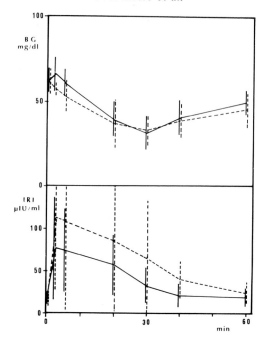

Fig.4 Effect of tolbutamide on blood glucose (BG) and on immunoreactive insulin (IRI) concentration in control subjects (dashed lines) and in patients affected with gonadal dysgenesis (full lines). (Mean ± SD.)

lation in all subjects affected with Turner's syndrome.

With the exception of the two adult cases, in which surprisingly low values of ASm are recorded, ASm seems to be within normal limits (Fig.2). The literature reports both normal (Saenger *et al.*, 1976; Wiedemann *et al.*, 1976) and elevated (Almqvist *et al.*, 1963; Daughaday *et al.*, 1969) levels of ASm. Such differences can be at least partially explained by the difference in the subjects studied and the assay used by the different groups.

The responsiveness of the Sm-generating system to hGH administration is controversial. Daughaday *et al.* (1971) could not observe a significant ASm increase after hGH chronic administration for several days, while other investigators (Peake *et al.*, 1972), after a short treatment, have recorded an increase similar to that observed in normal subjects. Our results agree with the hypothesis of a Sm-generating system responsive to the hGH stimulation.

It has been observed (Saenger *et al.*, 1976) that treatment with ethinyl estradiol inhibits Sm generation in subjects with Turner's syndrome as well as in normal subjects (Wiedemann *et al.*, 1975). On the basis of these observations, it seems possible to exclude a pathological involvement of Sm generation in subjects with gonadal dysgenesis. The normal or elevated ASm observed in the presence of normal levels of GH are consistent with such a supposition.

These results lead us to hypothesize a peripheral resistance to the action of GH and Sm in subjects with gonadal dysgenesis. The resistance to GH is evident at least as regards the metabolic actions of the hormone. The resistance to GH, and mainly to Sm, appears evident from the very poor results observed after massive and prolonged hGH treatment in these subjects (Soyka *et al.*, 1964; Tanner *et al.*, 1971; Tzagournis, 1969). If the increase of the growth rate (Escamilla, 1974; Soyka *et al.*, 1964; Tanner *et al.*, 1971; Tzagournis, 1969) seems to confirm the increase of ASm, the very poor results observed seem to be due to a hyporesponsiveness of the cartilagineous substrate, at least in part justified by the osteochondrodystrophic alterations (Borghi, 1974).

The OGTT in patients with gonadal dysgenesis evidenced a relative intolerance to carbohydrates, although the BG pattern was not statistically significant with respect to normal (Szczepski *et al.*, 1972) (Fig.3).

Particularly interesting is the reduced insulin response to the glucose load and the behaviour of the insulinogenic index (Fig.3). These seem to evidence a relative insular-pancreatic hypofunction, which can represent a sign of potential diabetes in subjects with gonadal dysgenesis. Impairment of insulin secretion is frequently evident, even when intolerance to carbohydrates is not present (Jackson *et al.*, 1966; Lindsten *et al.*, 1967; Nielsen *et al.*, 1969; van Campenhout *et al.*, 1973) and precedes, in these subjects, the appearance of a pathological OGTT and of symptoms of diabetes mellitus. In fact, an intolerance to carbohydrates is more frequently observed in older patients (Nielsen *et al.*, 1969; van Campenhout *et al.*, 1973), whereas in younger subjects, a diminished insulin secretion was found, without evident alterations of BG pattern (Jackson *et al.*, 1966; van Campenhout *et al.*, 1973). These observations provide a possible explanation for the conflicting results obtained in various case studies, and agree with the findings of the present study.

The presence of a hypofunction of the insular-pancreatic system is confirmed by the behaviour of the tolbutamide test. In contrast with other authors (Rasio *et al.*, 1976; van Campenhout *et al.*, 1973), we observed an insulin response constantly lower than normal, with a statistically significant difference at 20 min (Fig.4).

None of our subjects presented a family history positive for diabetes mellitus, in agreement with the observations of Lindsten *et al.* (1967) and in contrast with other authors (Nielsen *et al.*, 1969; van Campenhout *et al.*, 1973). This, in our opinion, seems to indicate that impairment in insulin secretion and in carbohydrate tolerance are directly related to the disease (Lindsten *et al.*, 1967), while family history positive for diabetes mellitus represents perhaps a genetic predisposition necessary for the manifestation of clinical diabetes.

As to the nature of the genetic error, it seems (Jackson *et al.*, 1966; Nielsen *et al.*, 1969) that the absence of an X chromosome produces the loss of a genetic locus responsible for the induction of an enzyme or enzymes important in glucose metabolism. According to other researchers, the anomalous insulin response is specifically correlated to the chromosomal aberration of Turner's syndrome (Lindsten *et al.*, 1967).

In conclusion, our results seem to confirm the presence, in subjects affected with gonadal dysgenesis, of an insular-pancreatic hypofunction, while the secretion of GH and Sm appears normal. In consideration of the role of insulin in growth processes, it would appear of interest to clarify whether the relative lack of this hormone can be considered one of the factors involved in the reduced growth rate and, therefore, in the short stature of these subjects.

REFERENCES

Almqvist, J., Linsten, J. and Lindvall, N. (1963). *Acta endocr. (Copenh.)* **42**, 168-186.
Andreani, D., Menzinger, G., Carratù, R. and Cohen, G. (1966). *In* "Atti 11º Congr. Naz. Soc. It. Endocrinologia", pp.229-291. Soc. Tipografica Piemontese, Torino.
Borghi, A. (1974). *In* "Proc. Inter. Symp. on Management on Height Retardations". *Acta Med. Auxol. Suppl.* **6**, 235-250.
Daughaday, W.H. (1971). *Adv. Intern. Med.* **17**, 237-263.
Daughaday, W.H., Laron, Z., Pertzelan, A. and Heins, J. (1969). *Trans. Assoc. Am. Physicians* **82**, 129-140.
Daughaday, W.H. and Parker, M.L. (1963). *J. clin. Endocr. Metab.* **23**, 638-650.
Engel, E. and Forbes, A.P. (1965). *Medicine* **44**, 135-164.
Escamilla, R.F. (1974). *In* "Advances in Human Growth Hormone Research" (S. Raiti, ed), pp.765-781. Dhew Publ., Washington.
Forbes, A.P. and Engel, E. (1963). *Metabolism* **12**, 428-439.
Franchimont, P. and Burger, H. (1975). *In* "Human Growth Hormone and Gonadotropins in Health and Disease", p.339. North Holland Publ. Co., Amsterdam.
Freychet, P., Rosselin, G., Assan, R., Tchoubroutsky, G., Delais, J. and Deret, M. (1967). *Presse Méd.* **75**, 2381-2385.
Giordano, G., Foppiani, E., Minuto, F. and Perroni, D. (1976). *Acta endocr. (Copenh.)* **81**, 449-454.
Giordano, G., Foppiani, E., Perroni, D. and Minuto, F. (1976). *In* "Proc. 16º Congr. Soc. It. Endocrinologia", pp.175-199. Serono Symposia, Rome.
Jackson, I.M.D., Buchanan, K.D., McKiddie, M.T. and Prentice, C.R.M. (1966). *J. Endocrinol.* **34**, 289-298.
Kaplan, S.L., Abrams, C.A.L., Bell, Y.Y., Conte, F.A. and Grumbach, M.M. (1968). *Pediat. Res.* **2**, 43.
Keston, A. (1956). *Abstr. 129th Meet. Am. Chem. Soc.* S.31 c.
Lindsten, J., Cerasi, E., Luft, R. and Hultquist, G. (1967). *Acta endocr. (Copenh.)* **56**, 107-131.
Marrma, P., Della Casa, L., Carani, C., Remaggi, F. and Bonati, B. (1973). *Recenti Prog. Med.* **54**, 197-212.
Meadow, S.R., Boucher, B.J., Mashiter, K., King, M.N.R. and Stimmler, L. (1968). *Arch. Dis. Child.* **43**, 595-597.
Merimee, T.J., Rimoin, D.L., Hall, J.D. and McKusick, V.A. (1969). *Lancet* **1**, 963.
Menzinger, G., Falluca, F. and Andreani, D. (1966). *Lancet* **1**, 1269.
Nielsen, J., Johansen, K. and Yde, H. (1969). *Acta endocr. (Copenh.)* **62**, 251-269.
Peake, G.T., Rimoin, D.L., Paekman, S. and Daughaday, W.H. (1972). *Clin. Res.* **20**, 436.
Rasio, E., Antaki, A. and van Campenhout, J. (1976). *Eur. J. clin. Invest.* **6**, 59.
Saenger, P., Schwartz, E., Wiedemann, E., Levine, L.S., Tsai, M. and New, M. (1976). *Acta endocr. (Copenh.)* **81**, 9-18.
Seltzer, H.S. and Smith, W.L. (1959). *Diabetes* **8**, 417-424.
Soyka, L.F., Ziskind, A. and Crawford, J.D. (1964). *N. Engl. J. Med.* **271**, 754-764.
Szczepski, O., Walczak, B., Pawlaczyk, B. and Socha, J. (1972). *Pediatrie* **27**, 149-152.
Tanner, J.M., Whitehouse, R.H., Hughes, P.C.R. and Vince, F.P. (1971). *Arch. Dis. Child.* **46**, 745-782.
Tzagournis, M. (1969). *J. Am. Med. Assoc.* **210**, 2373-2376.
van Campenhout, J., Antaki, A. and Rasio, E. (1973). *Fertil. Steril.* **24**, 1-9.
Wiedemann, E., Levine, L., Lewy, J., Schwartz, E. and New, M. (1975). *In* "Inter. Symp. on Growth Hormone and Related Peptides", p.43. Ricerca Scientifica ed Educazione Permanente, Suppl.1, University of Milan.
Wiedemann, E., Schwartz, E. and Frantz, A.G. (1976). *J. clin. Endocr. Metab.* **42**, 942-952.
Yde, H. (1968). *Acta endocr. (Copenh.)* **57**, 557-564.

GONADOTROPINS, PROLACTIN AND THYROTROPIN (BASELINE VALUES AND PITUITARY RESPONSES) IN TURNER'S SYNDROME

G. Valenti, G.P. Ceda, S. Bernasconi, E. Tarditi, P. Chiodera,
P.P. Vescovi, A. Banchini and U. Butturini

*First Institute of Medical Pathology, and Department of Pediatrics
University of Parma, Italy*

SUMMARY

GnRH and TRH tests were performed in 14 subjects with Turner's syndrome as compared with normal subjects from prepubertal to adult age.

Gonadotropins. LH basal values were clearly higher in patients than controls since prepubertal age and progressively increased till adult age. FSH values were also higher since prepubertal age, but went on steadily from prepubertal to adult age. This pattern is much less evident when incretory areas are considered. The results support the hypothesis of two different feedback mechanisms for LH and FSH, and that of a greater sensitivity of FSH to negative feedback which seems to be present since early prepubertal age. Thereafter, in Turner's syndrome, the different behaviour of LH/FSH ratio in prepubertal and postpubertal age of normal subjects is likewise found.

PRL and TSH. When the subjects with Turner's syndrome in postpubertal age are compared with controls of the same age, overlapping basal values and peaks of prolactin are found. Nevertheless, the fall of prolactin levels is much faster and consequently a significant decrease of incretory area is found. TSH, on the contrary, shows basal values and incretory areas similar to the controls in the different ages.

Estrogen deficiency, though able to modulate prolactin secretion, does not seem to be able to interfere with TSH, thus supporting the independence of the two systems.

I. INTRODUCTION

Turner's syndrome is a dysgenetic disease in which the pathogenetic importance of the hypothalamus-pituitary axis is not quite elucidated. The present study was carried out with the purpose of examining the function of pituitary sections active on the secretion of LH, FSH, PRL, and TSH, in a group of such patients.

II. MATERIALS AND METHOD

A total of 14 patients (aged 9 months to 24 years) suffering from Turner's syndrome was studied. A 45,XO caryotype was shown by all but 3 patients (1 XXp— and 2 XX/XO), and laparoscopic control indicated complete gonadal dysgenesis.

The controls were subdivided as follows:
— 6 girls (aged 6 to 10 years, stage G1P1 (Tanner, 1962)) and 10 women in the early follicular phase of the menstrual cycle (aged 18 to 32 years) for the gonadotropin study;
— 15 girls (aged 3 to 11 years; stage G1P1) and 17 women partly in the early follicular and partly in the luteinic phase of the menstrual cycle (aged 18 to 36 years) for the prolactin study;
— 7 girls (aged 3 to 8 years; stage G1P1) and 12 women in the early follicular phase of the menstrual cycle (aged 18 to 32 years) for the thryotropin study.

LH and FSH were radioimmunologically assayed, using a double antibody method, and expressed in mIU of the 2nd IRP-HMG standard. PRL was radioimmunologically assayed, using a double antibody method, and expressed in ng of the 71/222 WHO standard for human prolactin. TSH was radioimmunologically assayed, using a double antibody method, and expressed in μU of the 68/38 WHO standard for human thyroid stimulating hormone.

These hormones were evaluated as baseline values (at least two determinations) and as pituitary responses to GnRH (100 μg i.v. acute injection with assays, in heparinized samples at times 15, 30, 45, 60, 90, 120, 150, and 180 min) and to TRH (200 μg i.v. acute injection with assays in heparinized samples at times 10, 20, 30, 45, 60, 90, and 120 min). The responses were evaluated as area (triangulation technique) circumscribed by serum hormonal curves and by line of basal values, and expressed in mIU/3h for gonadotropins, ng/2h for PRL, and μU/2h for TSH.

III. RESULTS

A. Gonadotropins

1. The results are shown firstly according to a longitudinal study in correlation with age.

Baseline values (Fig.1). LH, constantly well above the normal range, shows a progressive increase from prepubertal to adult age; FSH, decidedly higher than in controls, persists evenly elevated until adult age. It follows that FSH prevails over LH in the prepubertal and pubertal ages. This pattern magnifies that of the gonadotropins observed in physiological conditions. The predominance of FSH is progressively lower in the adult age, when the LH/FSH ratio is sometimes reversed.

Gonadotropins, Prolactin and Thyrotropin in Turner's Syndrome

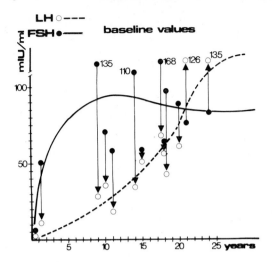

Fig. 1 Behaviour of baseline values of gonadotropins in subjects with Turner's syndrome: correlation with age.

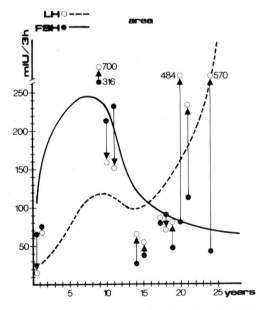

Fig. 2 Behaviour of pituitary gonadotropin responses (areas) to GnRH in subjects with Turner's syndrome: correlation with age.

Areas (Fig. 2). Considering the areas of pituitary hormonal responses, the values of LH in the prepubertal age constantly higher than the controls are documented: they are blunted in the early postpubertal age and show a huge increase in the adult age. FSH levels, well above normal in the prepubertal age, show on the contrary a progressive decrease until adult age. It follows that FSH

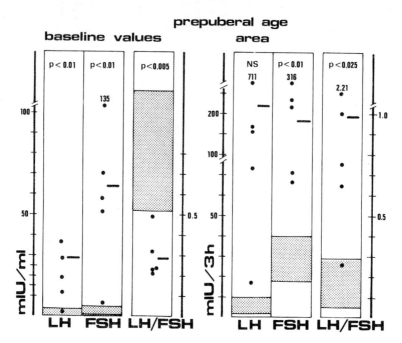

Fig.3 Behaviour of gonadotropins (baseline values and areas) in subjects with Turner's syndrome in prepubertal age as compared with controls.

is clearly predominant over LH in the prepubertal age. The behaviour is quite similar to that shown by baseline values, the only difference being that the reverse of LH/FSH ratio develops more precociously.

2. Secondly, the subjects with Turner's syndrome, subdivided into two groups (9 months to 11 years and 14 to 24 years of age) were compared with two equivalent groups of controls.

In the first comparison, relevant to the *prepubertal age* (Fig.3), the baseline values of the subjects with Turner's syndrome were found to be significantly higher than in controls for both LH ($p < 0.001$) and FSH ($p < 0.001$); the LH/FSH ratio shows a significant decrease ($p < 0.005$). Considering the areas of pituitary responses, a significant increase of FSH ($p < 0.01$) is documented; that of LH is not statistically significant, but only because of the dispersion of the data; nevertheless, both are constantly higher than in controls. In fact, the huge increase of LH shows its evidence when we consider that the LH/FSH ratio is significantly higher ($p < 0.025$) than in controls.

In the second comparison, relevant to the *adult age* (Fig.4), baseline values are significantly higher than in controls for both LH ($p < 0.001$) and FSH ($p < 0.001$); the increase is even more evident in comparison with prepubertal subjects. The LH/FSH ratio, owing to higher increase of FSH than LH, is significantly blunted ($p < 0.02$). As for the areas of pituitary hormonal responses, the increase, in comparison with controls, is significant for both LH ($p < 0.05$) and FSH ($p < 0.001$); consequently, the ratio of the areas is blunted, but not significantly.

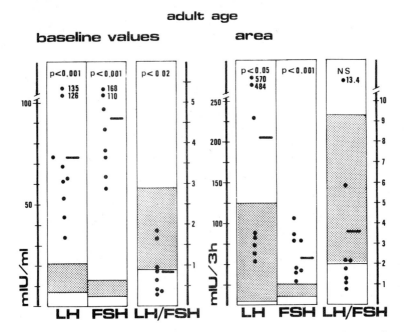

Fig.4 Behaviour of gonadotropins (baseline values and areas) in subjects with Turner's syndrome in adult age as compared with controls.

B. Prolactin

The subjects with Turner's syndrome, subdivided into two groups (1 to 11 and 14 to 24 years of age) were compared with two equivalent groups of controls (Fig.5). No significant difference is observed for baseline values, while the pituitary responses to TRH are significantly lower ($p < 0.05$) in the adult age. In the prepubertal age, owing to the scarcity of data, a statistical analysis is not possible, but the behaviour, especially for baseline values, seems to be different for the frequent finding of basal values above the normal range.

Considering the pattern as a whole, a significant linear negative coefficient is documented between the baseline values of PRL and the ages of the patients ($r = -0.76, p < 0.01$) and between the baseline values of PRL and of LH ($r = -0.59, p < 0.05$).

C. Thyrotropin

The subjects with Turner's syndrome, subdivided into two groups (9 to 11 and 14 to 24 years of age), were compared with two equivalent groups of control (Fig.6). In the adult age, the baseline values are significantly higher ($p < 0.001$) than in controls; the pituitary responses to TRH, on the contrary, appear to be substantially similar in the two groups. Owing to the scarcity of data, a statistical analysis in the prepubertal age is impossible. Nevertheless, the results seem to show a similar behaviour. When examining the single values, it can be seen that, out of twelve observations, only one shows a quite pathological behaviour, i.e., basal values higher than in controls and pituitary response to TRH higher and prolonged.

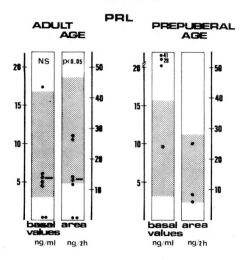

Fig. 5 Behaviour of prolactin (baseline values and areas) in subjects with Turner's syndrome in prepubertal and adult age as compared with controls.

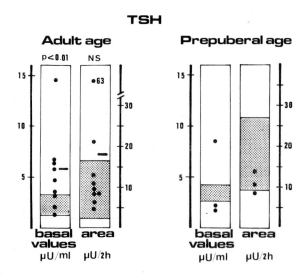

Fig. 6 Behaviour of thyrotropin (baseline values and areas) in subjects with Turner's syndrome in prepubertal and adult age as compared with controls.

All the results are summarized in Table I.

IV. DISCUSSION

A. *Gonadotropins*

LH and FSH in patients with Turner's syndrome, studied at different ages, essentially follow the same qualitative pattern of gonadotropin secretion found during normal sexual maturation. The variation of the LH/FSH ratio, passing

Table I. Gonadotropins (LH and FSH), prolactin (PRL), and thyrotropin (TSH) in 14 patients with Turner's syndrome.

Case no.	Age	Caryotype	LH Baseline values (mIU/ml)	LH Area (mIU/3h)	FSH Baseline values (mIU/ml)	FSH Area (mIU/3h)	PRL Baseline values (ng/ml)	PRL Area (ng/2h)	TSH Baseline values (µU/ml)	TSH Area (µU/2h)
1	9 mo	XO	1.5	17.6	6.1	66.5	—	—	—	—
2	14 mo	XO	12.0	73.2	50.8	70.4	41.0	—	—	—
3	9 yr	XO	28.5	700.0	135.0	316.0	21.0	23.5	8.5	10.3
4	10 yr	XO	36.0	159.2	70.0	211.8	28.0	5.4	1.9	13.4
5	11 yr	XO	18.5	151.1	58.0	231.5	9.4	8.2	2.1	8.6
6	14 yr	XXp—	34.0	61.7	110.0	28.7	17.5	0.5	4.5	21.1
7	15 yr	XO	53.0	52.6	58.0	39.0	—	—	5.6	9.2
8	16 yr	XO	—	—	—	—	0.6	12.6	3.4	8.9
9	18 yr	XO	61.5	72.7	64.0	89.7	—	—	6.1	12.8
10	18 yr	XO	44.0	82.9	97.0	43.9	5.4	0.6	2.0	4.2
11	18 yr	XO	69.0	85.1	168.0	82.5	6.0	11.9	6.6	8.9
12	20 yr	XO/XX	62.0	483.7	89.0	81.7	0.5	14.4	14.7	63.1
13	21 yr	XO/XX	126.0	232.0	76.0	113.0	5.0	27.7	1.2	6.1
14	24 yr	XO	135.0	571.0	74.0	42.5	4.3	26.5	3.1	10.8

from prepubertal to adult age, should be correlated with a different sensitivity of pituitary cells active on FSH secretion, and might be the consequence of a progressive maturation of the hypothalamus-pituitary axis, which is promoted independently on endocrine activity of the ovary, according to the observations of other authors (Conte et al., 1975). These findings agree with the results of Boyar et al. (1973) who, comparing LH concentrations during sleep and wake, documented the same gonadotropin pattern found in physiological conditions. Job et al. (1974) and Winter and Faiman (1972) also demonstrated, in agonadal children and adolescents, a gonadotropin behaviour independent of gonadal steroids.

A second important finding seems to be the behaviour of the LH/FSH ratio relevant to baseline values, that is found to be significantly lower than in controls both in prepubertal and adult age. This is the consequence of a higher increase of FSH than LH, as has been well documented in the literature (Job et al., 1974; Suwa et al., 1974; Winter and Faiman, 1972; Roth et al., 1973; Ko-en Huang, 1975). Such behaviour suggests the hypothesis that, especially in prepubertal age, the FSH pituitary secretion may be more sensitive than LH secretion to the negative ovarian feedback, in agreement with other reports (Donald and Espiner, 1974; Valenti et al., 1976), who documented this gonadotropin pattern in male and female patients with primary hypogonadism. When the areas of pituitary responses are considered, the phenomenon loses its evidence, owing to a huge variability of LH responses to GnRH, as also documented by others (Suwa et al., 1974; Ko-en Huang, 1975).

From the gonadotropin pattern found in subjects with Turner's syndrome in prepubertal age, one can also postulate that the negative feedback of gonadal steroids is active since the early stages of prepubertal age, when the ovary is still far from its germinal and endocrine maturation (Illig et al., 1974; Winter and Faiman, 1972).

Finally, another important observation should be stressed. Recent experimental reports suggest that attainment of a critical body weight is the trigger of the onset of puberty (Frish and Revelle, 1969; Kennedy and Mitra, 1963; Visser, 1973); these authors find a close correlation between the maturation of the body as a whole and that of the brain areas on which the diencephalic-gonadotropic axis is dependent. Since the subjects with Turner's syndrome, characterized by a typical impairment of body size, show a gonadotropin qualitative pattern similar to that found in physiological conditions, we can exclude the hypothesis that body weight may be the determinant factor in the maturation of the hypothalamus-pituitary-gonadal axis.

B. Prolactin

As for impaired pituitary PRL response to TRH, this pattern could be explained on the basis of the typical decrease of estrogen levels in the patients with ovarian dysgenesis: the lack of the positive feedback of estrogens on the hypothalamus-pituitary axis (Frohman and Stachura, 1975; Marchandise, 1973) might be the answer to the problem. The negative significant correlation between basal PRL and LH levels could be the consequence of a displacement of pituitary function to the preferential and almost exclusive secretion of gonadotropins. The interpretation of the negative significant correlation between PRL and the age of the patients seems to be much more difficult, if one considers that, in our normal controls, we find a positive correlation close to statistical significance.

C. Thyrotropin

As can be seen from the results, a slight correlation between Turner's syndrome and thyroid hypofunction might be suggested, in agreement with reports of association of gonadal dysgenesis with clinical signs of hypothyroidism (McHardy-Young et al., 1970; Mirouze et al., 1966; Williams et al., 1964).

REFERENCES

Boyar, R.M., Finkelstein, J.W., Roffwarg, H., Kapen, S., Weitzman, D.E. and Hellman, L. (1973). *J. clin. Endocr. Metab.* **37**, 521-525.
Conte, F.A., Grumbach, M. and Kaplan, S.L. (1975). *J. clin. Endocr. Metab.* **40**, 670-674.
Donald, R.A. and Espiner, E.A. (1974). *J. clin. Endocr. Metab.* **39**, 364-369.
Frisch, R.E. and Revelle, R. (1969). *Hum. Biol.* **41**, 536-542.
Frohman, L.A. and Stachura, M.E. (1975). *Metabolism* **24**, 211-234.
Illig, R., Tolksdorf, M., Mürset, G. and Prader, A. (1974). *Acta endocr. (Copenh.)* **Suppl. 184**, 26.
Job, J.C., Garnier, P.E., Chaussain, J.L., Scholler, R., Toublanc, J.E. and Canlorbe, P. (1974). *J. clin. Endocr. Metab.* **38**, 1109-1114.
Kennedy, G.C. and Mitra, J. (1963). *J. Physiol.* **166**, 408-418.
Ko-en Huang (1975). *J. clin. Endocr. Metab.* **41**, 771-776.
Marchandise, B. (1973). *Ann. Endocrinol.* **34**, 308-310.
McHardy-Young, S., Doniach, D. and Polani, P.E. (1970). *Lancet* **2**, 1161-1164.
Mirouze, J., Jaffiol, C.L., Emberger, J.M., Badach, L. and Jambon, F. (1966). *Sem. Hop.* **42**, 731-739.
Roth, J.C., Kelch, R.P., Kaplan, S.L. and Grumbach, M.M. (1973). *In* "Hypothalamic Hypophysiotropic Hormones", Sec.3, pp.236. Excerpta Medica, Amsterdam.
Suwa, S., Maesaka, H. and Matsui, I. (1974). *Pediatrics* **54**, 470-475.
Tanner, J.M. (1962). *In* "Growth and Adolescence", pp.433-469. Blackwell Scientific Publications, Oxford.
Valenti, G., Tarditi, E., Ceda, G.P., Banchini, A., Vescovi, P.P., Chiodera, P. and Butturini, U. (1976). *Ann. Endocrinol.* **37**, 63-73.
Visser, H.K.A. (1973). *Arch. Dis. Child.* **48**, 169-182.
Williams, E.D., Engel, E. and Forbes, A.P. (1964). *N. Engl. J. Med.* **270**, 805-810.
Winter, J.S.D. and Faiman, C. (1972). *J. clin. Endocr. Metab.* **35**, 561-564.

RESULTS OF THE TREATMENT WITH GROWTH HORMONE IN A CASE OF GOLDENHAR-LIKE SYNDROME WITH DELETION OF THE SHORT ARM OF CHROMOSOME 18

G. Aicardi, L. Buffoni, A. Naselli and A. Tarateta

*Department of Child Health and Human Growth
University of Genoa School of Medicine
and
Admission and Observation Service, Children's Hospital "G. Gaslini"
Genoa, Italy*

SUMMARY

The case is reported of a male patient aged about 3 years, affected by polymalformative Goldenhar-like syndrome, showing deletion of the short arms of one chromosome 18 and a remarkable retardation of stature, weight, and bone maturation. Having observed selective pituitary deficiency of GH and a normal response to GH secretion induction, the patient was treated with hGH in weekly doses of 20-25 IU/m^2. In contrast to what occurs in almost all the syndromes due to chromosomal aberration, the statural growth of the subject, which was below 2 cm/year before the treatment, achieved 11 cm in the first year of therapy and 10 cm during the second. The treatment did not appear to lead to unfavourable modifications in the chronological age:bone age ratio.

I. CASE REPORT

In February 1974, a male subject aged 2 years and 9 months came to our observation. His morphological features were quite peculiar: *facies sui generis*; slight dolicocephaly with frontal bossing; antimongoloid slant; hypertelorism; epiphora with stagnation of tears in the medial canthus of the eye due to lacrimal canaliculus impotency; coloboma iris; piriform pupils with horizontally-positioned

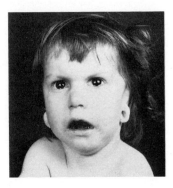

Fig.1 Quite clear antimongoloid slant, hypertelorism, epiphora with tear stagnation in the medial canthus, coloboma iris, macrostomy, overfolded auricles.

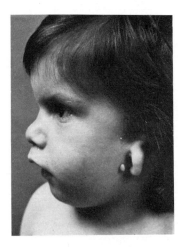

Fig.2 Prominent forehead, nose root depression, retro- and micrognathia, left overfolded, almost hanging, auricle with small tag in the preauricular region.

apexes; slightly depressed nose root; macrostomy; ogival palate; retro- and micrognathia; short neck (Fig.1). The auricular malformations were particularly interesting: low-set overfolded rudimentary auricles with external auditory canal agenesis and, on the left side, a small tubercle-shaped tag located in the preauricular region (Fig.2).

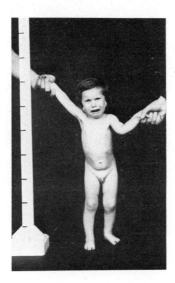

Fig.3 Metrical photo of the patient on admission. Chronological age 2 years and 9 months, height 71.5 cm, weight 7.4 kg.

Nothing else was found at the physical examination, except for deafness of moderate degree, clearly correlated to the absence of auditory canals, and moderate mental retardation, partly due to deafness. As to motor function, in spite of a moderate muscle hypotonia, the child kept a standing posture with no aid, and his gait handling was quite normal.

From the auxological point of view, the subject showed a remarkable retardation in growth and development. His height was 71.5 cm, his weight 7.4 kg, both far lower than Tanner's 3rd centile (Tanner *et al.*, 1975). Bone age, evaluated by Greulich and Pyle's method (1966), was about 7 or 8 months, also much lower than the 3rd centile (Fig.3).

Nothing particularly significant was found in either the family or the personal history, except for a rather low birth-weight (full-term delivery, 2.8 kg, 48 cm) and a progressively slower growth in stature and weight, starting from the 2nd semester of life.

Quite numerous clinical and laboratory examinations were carried out, for which we refer to the full case report (Buffoni *et al.*, 1976), only the main pathological findings being mentioned in the following.

Urinary Tract: an i.v. pyelogram and an isotope nephrogram documented the complete absence of the left kidney.

Chromosome Analysis: carried out with G banding, showed the deletion of the short arm of one chromosome 18. The deletion appeared almost total (Fig.4),

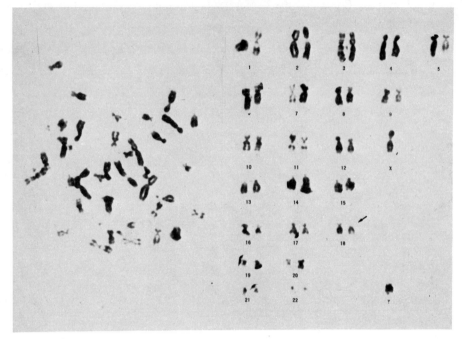

Fig.4 Karyotype of the patient. The arrow shows the deletion.

although a quantitative determination was impossible.

Pituitary Function: Insulinic hypoglycemia, L-Dopa *per os*, and i.v. arginine hydrochloride, showed a serious deficiency of growth hormone (GH) incretion, with a top peak of 2.5 ng/ml after this last stimulation.

II. RESULTS OF TREATMENT AND DISCUSSION

On the whole, our subject could be clinically classified in an intermediate position between Goldenhar's syndrome (Feingold, 1973) and chromosome 18p-syndrome (Lurie and Lazjuk, 1972). To the former can be attributed the typical ocular and auricular malformations (the absence of vertebral anomalies and of epibulbar dermoids being interpreted as a reduced expressivity of the syndrome); to the latter can be attributed the renal anomaly, short neck, micrognathia, hypertelorism, ogival palate, growth deficiency, and finally, the endocrinopathy.

The hormonal disorder was well documented, in our case, as a serious deficiency in GH incretion, and was represented by diabetes mellitus in the case of Van Dyke *et al.* (1964), by hypothyroidism in the case of Battin *et al.* (1973), and by low plasma GH levels in the case of Leisti *et al.* (1973) that, besides, proved nonsusceptible to replacement therapy. In their report, Leisti *et al.* did not exclude the possibility of a random association between GH deficiency and chromosomal aberration, but emphasized the possibility of a cause-effect relationship, an interpretation which seems to be supported in our case.

Along with the low plasma-GH levels, we found in our patient a normal somatomedinic response to exogenous GH and a spontaneous statural growth of

1 cm in 9 months. Consequently, in November 1974, we decided to start the replacement therapy in weekly doses of 20-25 IU/m^2.

Table I. Main anthropometric values and bone age of the patient, before therapy and after its beginning.

Age (yr)	Height (cm)	Weight (kg)	Sitting height (cm)	Head circ. (cm)	Thorax circ. (cm)	Subscap. skinfold (mm)	Bone age (TW2) (yr)
Before therapy							
2.77	71.5	7.40	40.3	45.5	40.5	5.3	0.7[a]
3.15	72.0	7.60	40.7	46.0	41.0	5.2	0.9[a]
3.60	72.5	7.70	41.0	46.0	42.0	5.4	1.1
After therapy							
3.94	77.5	9.20	45.0	46.5	45.0	6.4	1.6
4.53	83.0	9.50	47.2	47.0	46.5	5.8	2.1
4.89	86.5	10.80	48.5	47.0	47.0	6.6	2.4
5.24	89.0	10.90	49.0	47.5	47.5	5.8	2.6
5.69	92.5	11.80	50.0	47.5	47.5	6.6	2.9

[a] Evaluated according to the standards of Greulich and Pyle, 1966.

In contrast to Leisti's case, we could observe quite a satisfactory response to this treatment, as shown by the eight auxologic measurements, three before therapy and five during therapy, carried out over a period of about 3 years (see Table I). The statural growth chart (Fig.5) clearly shows a sharp acceleration after the beginning of therapy, and a moderate catch-up growth, which seems to persist till now, even if the subject's stature is still much below the 3rd percentile.

The remarkable effectiveness of the replacement therapy is even better shown by the growth velocity chart (Fig.6): from a velocity much below the 3rd centile, the therapy has induced — still at present, after two years' treatment — a velocity on the level of the 90th centile, if related to chronological age, or of the 50th centile, if related (perhaps more properly) to statural age.

The quite favourable evolution of the bone age to chronological age ratio should also be noted. Before the beginning of treatment, this ratio (see Table I) was little below 0.35, while now, after a moderate increase observed during the first semester of therapy, it seems to be stabilized around values below 0.50, which makes us hope for a possible partial catch-up in our patient's final stature.

Without entering into a discussion of the specific nosologic classification, we should, however, note that our observation of an association of GH deficiency with 18p—syndrome, as already reported by Leisti et al. (1973) considerably reduces the possibility of a chance occurrence. Moreover, the phenotypic variability among affected subjects may be related to the different quantity of deleted chromosomal material and, in some cases, to an uncertain attribution of the damage between chromosomes 17 and 18, a difficulty which was overcome only in the seventies thanks to the chromosome banding technique (Caspersson et al.,

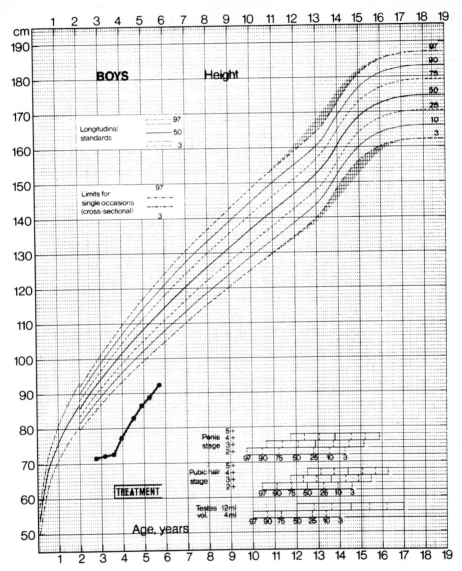

Fig. 5 Statural growth of the patient before and after beginning of treatment with GH.

1970).

Since the short arms of chromosome 18 have no bands (Paris Conference, 1971), it is unfortunately impossible, in our case, to determine the exact quality of deleted material, which might have allowed some analysis of the differences existing among the various cases of 18p−syndrome reported in the literature.

All this, anyway, could hardly explain the main problem roused by our observation and represented by the occurrence of a chromosomal aberration associated with stature-weight deficiency, which appears to be correlated, rather

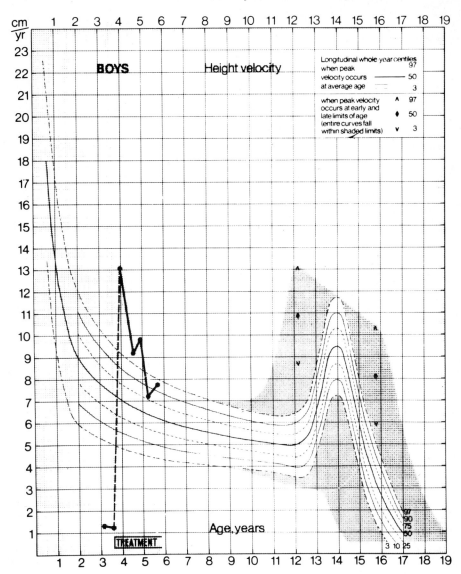

Fig.6 Statural growth velocity rate before and after beginning of treatment with GH.

than to an altered susceptibility of target organs, exclusively to an altered pituitary function concerning the growth hormone, as was clearly shown by the effectiveness of substitution therapy.

REFERENCES

Battin, J., Serville, F. and Marchand, J.C. (1973). *Arch. Fr. Pediatr.* **30**, 548.
Buffoni, L., Tarateta, A., Aicardi, G., Vianello, M.G. and Bonioli, E. (1976). *Minerva Pediatr.* **28**, 716-729.

Caspersson, T., Zech, L. and Johansson, C. (1970). *Exp. Cell. Res.* **60**, 315-319.
Feingold, M. (1973). *In* "Birth Defects. Atlas and Compendium." (D. Bergsma, ed), p.679. The International Foundation — Williams and Wilkins Co., Baltimore.
Greulich, W.W. and Pyle, S.I. (1966). "Radiographic Atlas of Skeletal Development of the Hand and Wrist". Stanford, California and London.
Leisti, J., Leisti, S., Perheentupa, J., Savilahti, E. and Aula, P. (1973). *Arch. Dis. Child.* **48**, 320-322.
Lurie, I.W. and Lazjuk, G.I. (1972). *Humangenetik* **15**, 203-222.
Paris Conference (1971). *Birth Defects Origi Artic. Ser.* 8.
Ruvalcaba, R.H.A., Thuline, H.C. and Kelley, V.C. (1972). *Arch. Dis. Child.* **47**, 307-309.
Tanner, J.M., Whitehouse, R.H. and Marshall, W.A. (1975). "Assessment of Skeletal Maturation and Prediction of Adult Height (TW2 Method)". Academic Press, London — New York — San Francisco.
Van Dyke, H.E., Valdamanis, A. and Mann, J.D. (1964). *Am. J. Hum. Genet.* **16**, 364-374.

5. AUXOLOGY, PSYCHOLOGY, AND PSYCHOSOCIAL PROBLEMS

FAMILY AND SCHOOL: SUBSYSTEMS IN INTERACTION

C. Pontalti and L. Ancona

Institute of General and Clinical Psychology, Catholic University
00168 Rome, Italy

SUMMARY

Family and school are to be considered as interdependent transitional systems. Research studies, carried out on samples of children experiencing shool-learning difficulties, as compared to controls, have shown the sociogram of the former to be characterized by lower expansiveness, higher number of rejections, lower number of positive choices, higher neglect, and lower explorative efficiency. Moreover, the school situation has been shown to be a continuation of the family situation, in which the relations between parents and children are perceived in a more negative way by the children in trouble, whose expansion and maturation has also been found to be smothered by a characteristic overstimulation. The family may actually fail to fulfil its task to compensate, through authentic affection and promotion of the Ego ideal, the necessary formulation of the school, tending to stress the Super-Ego.

This chapter discusses the results of a survey conducted at an elementary school in Rome. The purpose of the survey was to analyze the relationships within the family and within the class of both children having difficulties in learning and children who create no problems to their teachers in the teaching relationship. The problem of the mentally handicapped is a very complex one, indeed, as it has been dealt with by psychologists, pedagogists, sociologists and anthropologists.

Each of these approaches has resulted in a different pattern of intervention which places particular emphasis on one aspect of the problem. What seems to be lacking, however, is an attempt at understanding more in depth the interconnected variables and, above all, what happens in the relationship between the complex worlds of the family and of the school, which are both present in the common element, the child. In order to identify more accurately the cleavage point between the level of learning which the teachers consider as satisfactory and the unsatisfactory level, we have discarded children with clinically apparent neurological deficiencies and whose behaviour was clearly the competence of psychiatry.

The two samples we compared were chosen by teachers in each class on the basis of simple, but significant, instructions: "Choose one child who creates you learning difficulties and one child, not the most clever, who does not." The WISC test (Wechsler test for children) was applied to the two samples chosen on the basis of this criterion, and the IQs obtained differ significantly between both samples, even though the IQs of problem children are only slightly lower than average (oscillation range, 80-90). Both samples were homogeneous with regard to social class, according to the parameters of Hollingshead and Redlich (1958).

We have chosen to supply these data on behaviour in the beginning of this chapter to indicate that IQ differences between these two samples are minimal, even though they do differentiate the samples statistically. These minimal differences, however, are the expression of family and school structurings which are deeply different with regard to the interconnections in which the present aspect of both subsystems was examined. Finally, let us mention the size of these samples: 21 problem children, 21 children without problems in 21 school classes, 21 teachers, and 42 families.

After presenting the reference framework, we should now make a step backwards and outline the conceptual assumptions to which we were led by the most recent developments in the science of human relationships. We refer to the key concept of system, as outlined in the General Theory of Systems synthetized by von Berthalanffy and Miller and resumed by Watzlawick *et al.* (1971) in their fundamental work.

The General Theory of Systems can be best understood in its historical framework: it was preceded by the age of Descartes, during which science was dominated by physics and all scientists, whatever their scientific area of competence, tried to synthetize their knowledge by means of "reductive analysis".

This led to sacrificing research into the most vital types of human behaviour and obliged psychologists to reduce them to much less complex phenomena. Reductive analysis, however, ceased to be useful to physical sciences as well, as scientists became aware of the impossibility to conceptualize the world of very large phenomena and that of very small phenomena in a reductive manner. Quantum mechanics and the theory of relativity were born: the foundations of the conception of the world of physics and of biology in terms of systems and, indeed, in terms of relationships, were laid. More specifically, objects could no longer be studied monadically, but only in their functions and relationships with each other. The concept of linear causality was reversed into the cybernetic model of circular, and therefore not causal, retroaction. Scientists discovered that finalized behaviour, growth, creativity, could be evaluated scientifically and equalled to the dynamic interaction between the parts composing each living system.

The "organized complexities", or *Systems*, are the products of the dynamic interaction of their parts, rather than the sum of their absolute characters. In other words, neither the total result, nor the new characters can be explained by the nature of the parts themselves. The new characters can only be understood as a function of a continuous exchange of matter for energy, for information, between these parts. The world, the living organizations, are conceived as intercorrelated systems at different levels of logical abstraction. In this hierarchy, each system looks downwards to its subsystems and upwards towards its upper systems. Within this framework of epistemological reference, some concepts are fundamental; they are, indeed, laws which regulate the functioning of systems:

(1) The relationship is always a communication occurring at different level of abstraction and mediating contents and definitions on the relationship itself;

(2) The living system is highly homeostatic and the extension or the reduction of homeostasis defines the extent to which a system tends towards maturity, as well as its degree of openness or closure;

(3) The relationship takes place within a context which defines it.

In the light of these considerations, the family is defined as a "system undergoing constant change"; in other words, a system which adjusts itself to the different requirements of the different stages of development it experiences (requirements which change according to the changing of social demands on the family in time) in order to secure continuity and psychological growth for its members. Any kind of tension — be it caused by changes within the family (intrasystemic changes: the growth of children, the stages of the vital cycle, deaths, etc.), or without it (intersystemic changes: moving to another place, environmental changes, opening on other systems, etc.) — will be brought to bear on the functioning of the family and will require a process of adjustment, one the one hand, to maintain the continuity of the family and, on the other hand, to permit the growth of its members.

A crucial moment for the balance of the family system is the encounter with the reality of the school, a system clearly structured as distinct in relation to the family, not as an isomorphic entity with it. Over the years, from the child's birth to the first year of elementary school (when the first demand for efficiency is placed upon the child by society), the family has structured its own homeostasis, with redundance and automatism of relationship rules and of phantasmatic experiences. The creative interconnection of this structure with the school system — a system which, in other ways, is just as well defined and as preexistent — may succeed or fail, according to the stability of the processes of ripening and of differentiation of the child within the family and according to the possibility, on the part of the school, to change the family novel or to collude with the explicit or implicit definitions which circulate within the family.

Thus, a relationship is created which amplifies a preexisting deviation, which in turn conditions the survival of the family. This is possible because the school is a wide sytem, within which one can identify several subsystems with varying degrees of interconnections and communication with each other. Within the school system, the subsystem having the highest stability in time and the highest redundance is the "class".

The relationships between members of one school class, as any relationship within an interactive system, are circular; each response is a stimulus for the following event and it is absolutely arbitrary to refer to the sequence of behaviours in terms of cause and effect. The change from the linear model to the circular model has one fundamental consequence: there are no designated culprits or patients; the whole system suffers from malfunction, as well as the relationships between systems in close interaction. This is why the *pedagogic moralistic* assumption (according to which a student's performances are poor because he does not apply himself to his studies and lacks willpower) and the *clinical* assumption (stating that failure at school depends on lack of intelligence, on cerebral insufficiency) are obsolete and mystifying. The *social* assumption of cultural deprivation, to which much room was and is being devoted in pedagogic planning, is also being questioned because it lends itself to colonialistic-type manipulations in an attempt at "reclaiming" underdeveloped areas (Cordier, 1975).

The above considerations, however short, due to the limited time available, clearly indicate that the encounter of the child with elementary school and with the teacher is an extension of the family relationship which the child brings with himself. The real or phantomized relational conflicts which are present within the family are reexperienced in the class with the parent-teacher and with the siblings-classmates, and are complicated by the inner psychic conflicts of the teacher and by the latter's relational conflicts with regard to the system (the school). Within the family, and subsequently within the school which is its continuation, the growing child outlines and completes his identity by means of a constant confirmation of the role he feels he has to play and which is attributed to him by the family group or by its most authoritative member. In the normal pattern of things a child is helped to grow and to attain autonomy; he is gradually taught to discover within himself, regardless of the others' attitudes, original and creative potential to be implemented. There is a great number of families, however, within which this growth process is abnormally upset and even utterly deviated. This blocking process can already be noted at the level of biological growth: there are families, for example, in which physical growth is actively, however unconsciously, prevented, which results in a sort of pseudopituitary drawfing due to emotional causes.

It is evident that, if growing troubles are so strong that they can curb endocrine development by means of the neurohormonal play, the same may occur in the game of conditioning, learning, adjustments in a general sense. Within this framework, an unconscious collusion may occur within families, aimed at attributing the role of scapegoat to one of its members, in this case to one of the children. The subject thus chosen will not have many possibilities to escape his fate and will experience an impartial latently autistic depletion of his higher intellectual, cognitive, and linguistic capacities. The encounter with the school can either remodel this condition of marginal membership, or it can endorse it and aggravate it with the teacher's and the classmates' judgements which may define the child as inert, lacking willpower, and stupid. Thus, a vicious circle will be created, consisting of stimuli and responses at both the conscious and the unconscious levels, at the individual and social levels, which it will be practically impossible to break, unless one resorts to forces outside the system.

This research is based on these assumptions and is aimed at elucidating at the experimental level the way in which the above dynamics and relational systemic transactions occur.

Once the samples had been identified, the following methodology was chosen, at several converging levels.

(1) Level of relationships in the class: sociogram drafted according to the Theory of Signal Detection, as proposed by Maisoneuve and Carli.

(2) At the child's level

(a) Analysis of family phantomization, on the part of the child, in both samples, using the Family Relations Indicator (FRI) test by Howells and Lilzorish (a topical test), modified by the researchers of the Institute of Psychology of the Rome Catholic University, using the drawing of the family according to the Kinetic Family Drawings. Drawing permits accurate interpretation of the aspect of the child's phantasmatic world and its levels of differentiation; the identification of the child's self and the greater or lesser maturity of projective identifications and of the levels of autistic closure.

(b) Exploration of the cognitive style of the child in terms of field dependence and field independence, according to Witkin, by means of masked figures.

(3) At the teacher's level: application of the FRI to each of the 21 teachers, with the task of interpreting the tables imagining that they refer to scenes in the family life of the sample children, in order to acquire information as to the teacher's ideas about that family life.

(4) At the family level

(a) Study of the cognitive style of the members of the family by means of Masked Figures.

(b) Study of some thinking processes of all family members by means of Analogies and Wais cubes subtests.

(c) Analysis of the interactions of the entire family who had been given the task of agreeing a story on some FRI tables. This analysis was conducted with Bales' code, as modified by Mishler and Waxler (1968).

All the data collected, with the obvious exception of drawings, were treated statistically with models of multivariate analysis, to obtain a multidimensional gestalt of both samples. It is impossible to summarize here the bulk of these data, their analysis and the individual stages of interpretation. We shall confirm ourselves to indicating the most significant conclusions, always on the background of the slight IQ difference indicated above and therefore of the significance of the teachers' choice.

(1) Sociogram. The problem child feels perfectly well integrated into the class. He is, however, being systematically rejected by his classmates. A picture of "social marginality", as we have defined it, is thus created, which is denied and becomes a source of relational confusion and of mystification of the relationship which is highly pathogenic.

(2) Drawings. The problem child draws an incoherent, incomplete, and devitalized family with autistic nuclei and idealized projections, in a manner which differs sharply from that of the control child. The latter does not express *per se* a harmonious ripening of personality. Generally, he draws obsessive and distant relations with departmentalization of the family etological space, which, in themselves, are just as preoccupying, but which cannot be defined as symptoms in the family and in the school contexts, and are, in fact, enhanced and positively evaluated.

(3) Children's FRI. The problem child experiences his family as a whole as something very bad from the point of view of the general emotional climate. The parental subsystem is perceived as something positive in the mother-father relationship. The child perceives his relationship with his siblings as good; hence, we have a repetition of the same perception model noted by the sociogram with

regard to his classmates. It is the transgenerational relationship of his parents with him, and not with his siblings, which immediately takes on negative, violent, nonconformable connotations in his family intrapsychic phantomization. In short, he perceives himself as a source of intoxication of the family as a whole and he desperately tries, at the cost of seriously distorting reality, to recover his place in the siblings subsystem.

(4) Globally, the Masked Figures indicate that the cognitive style of the families of problem children is more field-dependent and less interconnected than that of the control families. This indicates the existence of a pathological fluctuation and overlapping of the relation boundaries within the family, something which Bowen described as "undifferentiated family ego mass". The thinking disorders identified by means of analogies and of the Wais cubes point to the same direction.

(5) The family interaction is characterized by a sharp fusional hyperprotectivity of mothers towards the problem children, without stimulation to promote autonomous expressiveness, in a general atmosphere of pseudoreciprocity in which no direct confrontation is possible, which is anxiety-generating but is personalized between the family members, with a total number of acts of communication significantly lower in problem families. It is interesting to note that, generally, in control families, the interactive communication attributes pragmatically to the child the role of parental child.

(6) *The teachers' FRI.* Teachers systematically denote the emotional climate of the problem family in the most negative manner, even though they phantomize healthy and partially creative areas of relations. There is, however, a clear identification of parents as culprits and of the child as a victim. This leads to reverse the scapegoat role and, paradoxically, to identify with the good parent, opening up the way to a paranoid condition of cross-over of responsibilities which, in its symmetry, leads in effect to a strict homeostasis in the interaction between the family system and the school system.

CONCLUSIONS

We would wish to emphasize two aspects. First, difficulties at school seem to be connected with intrapsychic and family problems, rather than with basic deficiencies. The school becomes, in actual fact, a collusion with the family system, by reproducing its relation rules and by amplifying its strict homeostasis in a redundant and automatic manner.

Second, the teachers' unawareness of systemic thinking leads them into the trap of helplessness, which is enacted by excluding the child and creating paranoid relationships with the families.

If we think that this is supported by the present cultural patterns which, by means of mass-media distortion, tend to culpabilize and to remove from any role the entire teaching class and to promote symmetrical paranoia on the part of parents with regard to teachers, we might perhaps understand the present feeling of confusion and of uneasiness, bordering on pathology which characterizes the relationships between school and family, apart from good sociological-moralistic intentions.

It is impossible to proceed any further. The intervention of outside operators trained in intervention strategies of systemic epistemology might perhaps start process of change. This is indicated very clearly in a book which we would strongly recommend to all those interested: "Il Mago Smagato" (The Helpless Sorcerer) by Palazzoli Selvini (1966).

REFERENCES

Cordier, J. (1975). "Une Anthropologie de l'Inadaptation". Editions de l'Université de Bruxelles.
Hollingshead, A.B. and Redlich, F.C. (1958). "Social Class and Mental Illness". Wiley, New York.
Minnchin, S. (1974). "Families and Family Therapy". Harvard Press.
Mishler and Waxler (1968). "Family Interactions". Wiley, New York.
Palazzoli Selvini, M. (1966). "Il Mago Smagato". Feltrinelli, Milan.
Rosenthal, R. and Jacobson, L. (1968). "Pygmalion in the Classroom". Holt, New York.
Watzlawick, P., Bevin, G.H. and Jackson, D.D. (1971). "Pragmatica della Communicazione Umana". Astrolabio, Milan.

CORRELATION BETWEEN THE AGE OF ONSET OF A CHRONIC ILLNESS AND THE CHILD'S PSYCHOLOGICAL ADAPTATION TO IT

J. Appelboom-Fondu

Medico-Psychological Department, Pediatric Service
St. Pierre University Hospital, Rue Haute, 1000 Brussels, Belgium

SUMMARY

The different factors influencing the psychological adaptation of children to chronic illness are considered. The characteristics of the illness itself seem to accentuate certain facets of personality; but the age of the child and his capacity to integrate stress and anguish are determining factors to reach a well-adapted behaviour. The family dynamics and the relationship between the medical staff and the child are essential in the control of the anxiety created by the illness.

I. INTRODUCTION

Chronic illness, which we shall arbitrarily define as a somatic disease lasting at least one year and requiring regular treatment, either curative or palliative, is an unexpected event, compelling, mysterious, coming on brutally and painfully into the life of the child and his family. Chronic illness implies the intervention in the family structure of new factors: the doctor and the treatment team in a hospital.

The purpose of this study is to define the role of the child's relationship with his family, his doctor, and the hospital structure, in his adaptation to his disease.

The child's adaptation depends on different factors:
(1) the characteristics of the illness itself;
(2) the inherent characteristics of the child;
(3) the characteristics of the family environment;
(4) the characteristics tied to the medical organization.

The interaction of these factors will determine the child's adaptation to his illness.

II. THE CHARACTERISTICS OF THE ILLNESS ITSELF

A. Heredity

The hereditary trait plays an important role in the parents-ill child relationship. We were impressed, in a study concerning the personality of 17 preadolescent hemophiliacs, to have discovered how much these subjects' relationship with their mothers is unique: they feel their mothers to be someone dangerous, invading, inaccessible, authoritarian, susceptible to transmit to them a multilating and deforming defect; their fathers often are their refuge against the anguish they feel toward their mothers. Their difficulty to identify, as adolescents, with this father, who himself is exempt of any illness, is even much stronger when he is an image objectively inaccessible to them.

B. The Moment of Onset of Symptoms

The moment when the diagnosis is made plays an essential role in the adaptation of the child and his family to the disease.

We have been impressed, during an earlier study on diabetic adolescents, to see that the number of patients presenting psychopathological problems was greater than among hemophiliacs. In our study, out of 30 diabetics, aged 9 to 17, 18 have a fragile personality (60%); 26 have depressive symptoms, anxiety, or aggressiveness, with the use of their symptomatology or treatment in a dynamic context. Furthermore, one adolescent, who was already schizophrenic, does not treat himself, nor does he examine his urines. The social adaptation after the onset of the diabetic symptomatology is relatively poor in 23/30 cases.

Our criteria of appreciation have been defined, on the one hand, by such clinical evidence as the psychiatric examination and interviews with parents and teachers concerning the adaptation in family, group, and society; on the other hand, by such objective data as the results of tests such as the TAT, Rorschach, and SIT (Appelboom-Fondu *et al., in press*).

This methodology was applied to a population of 17 adolescent hemophiliacs aged 9 to 14. Ten of them have a fragile personality; 14 adolescents present minor manifestations of anxiety related to the disease; 3 of these have poor social adaptation.

These data suggest the following hypothesis: when the child is less than six years old, the onset of a chronic symptomatology has the value of a new experience, to be integrated with other new experiences in the development of his personality. It is well known that the conjunction of the congenital effective potential and the early experiences during the first six years of life create the basic personality of the child. One can imagine that the personality of the child who is sick early in life would be formed by integrating the illness as a constant and inherent element of his individuality.

On the other hand, when the illness appears later, the brutal stress that constitutes the onset of a chronic symptomatology weakens the personality and can provoke a psychopathological decompensation, especially during a critical period, as is the case with adolescence.

We also can imagine that, no matter what the age of onset of the illness, the constraints inherent to its nature put the child and his family in a difficult situation, since he must incessantly adapt his tendencies to the reality of his disease.

C. The Curability

The imminence of death creates, as we have been able to verify among leukemic children at an advanced state of illness, a global fissuring of the defense mechanisms with the manifestation of an anguish of abandonment and behavioural regression (Alby and Alby, 1971; Lanski, 1974).

This anguish is generally more intense, the more the family and the treatment team are incapable of verbalizing the feelings of the child, being themselves inhibited by their own problems in regard to death. This anguish is felt by the child very early, even before his logical intellectual processes allow him to understand what is happening. The child's behaviour is very variable, ranging from aggressiveness towards his doctors and his family to self-destruction, psychiatric decompensation, and suicide.

D. The Type of Treatment

During our study concerning 30 diabetic children, either treating themselves or not, we were struck by the number of those (23) who "cheated" in their urine analyses or in the adaptation to their diet (Appelboom-Fondu *et al.*, 1974). This therapeutic sabotage had for most of the children a different worth, depending upon the personality and/or the family context. For some of them, it was a symptom, a call for help; for others, it was a demonstration of independence or of total indifference.

Likewise, among young patients with a corticotherapy for renal disease, or among young cardiac patients treated for several years by antibiotics, a symptomatology of rejection of treatment is frequently found.

On the other hand, hemophiliacs treat themselves adequately; patients who have to undergo open-heart surgery follow their medical treatment, no matter how constraining it may be. These findings seem to show that, no matter what the age or the implication of the patient in his treatment, the acceptance or non-acceptance depends on the hope of a cure and the existence of a pain that the patient wishes to be eased. Thus, the tangible hope of a curative intervention encourages the preliminary treatment.

On the other hand, the daily injection of insulin, even though no physical sensation indicates an illness, is, in the mind of the young patient, a tedious obligation, at best useless.

The anxiety of a child who does not feel ill yet, and who is forced by an outside agent to treat himself daily, is greater than that of a patient in an acute situation. The former has but little control of his state and thus of his treatment, which increases his insecurity.

E. The Value of the Disease

We think that a disease that changes the external physical aspect of the patient is more likely to perturb an adolescent, in the phase of restructuring his body image, than a younger child. Thus, the risk that psychological problems be focused on self-affirmation is, in our opinion, more important in a patient with adrenogenital syndrome, or a dwarf, than in other patients. For Leonard (1972) all changes in body image create a problem of personality that often justifies psychological help.

Table I.

	Diabetics (N = 14)	Hemophiliacs (N = 17)	
Oral impulses			
Food concern	7	1	$p \leqslant 0.001$
Dependency tendencies	9	5	$p = 0.01$
Aggressive impulses			
Self-aggressive tendencies	6	7	ns
Hetero-aggressive tendencies	7	16	$p \leqslant 0.01$
Retention problems	1	1	ns
Oedipal impulses (accepted/feared)			
Desire to seduce the mother	0/8	3/5	ns
Rivalry with the father	0/6	2/7	ns
Identification as a man	0/6	3/8	ns

During a comparative study of 14 preadolescent diabetics and 17 preadolescent hemophiliacs matched for age, socioeconomic level, and sex, we were able to show, with the help of projective tests (TAT, Rorschach), that the diabetics especially care for food and are dependent upon their mothers, toward whom they have an infantile behaviour pattern. The hemophiliacs, instead, are for the most part aggressive toward the society, their mother, father, siblings, equals, without any discrimination. For them, the outside world is considered to be dangerous because they are susceptible to the slightest trauma; preventive attack is their defense mechanism (Table I).

These findings allow us to think that the intimate nature of the disease constitutes a focus of fragilization in personality development, but it does not necessarily determine the behaviour of the child. Obviously, not all children with the same illness necessarily have the same behaviour, and the nature of the illness only accentuates certain (often unconscious) facets of personality.

III. THE CHILD'S OWN CHARACTERISTICS

The diverse aspects, previously developed concerning the disease itself, tend to specify the important, but not necessarily essential, role of the disease in the development of personality. Each child reacts to chronic illness via his own effective potential. The conscious or unconscious anguish created by the appearance of the illness expresses itself in different fashions, such as phobias, nightmares, physical signs of anguish, concern about the future, concern about a constantly better understanding of the illness, overinvestment in external values.

We can ask ourselves how the sick child behaves. Some are cooperative, intellectualizing, normally functioning in school and at home, and impose upon themselves only the limits caused by their disease. Optimist during remissions, they are sad or angry at the time of a relapse. Others adapt less well to their illness: some are depressive dependent, indifferent; others bring on the known danger in order to control it better by doing forbidden acts or sabotaging their treatment. Still others are aggressive and hostile to any outside help. These different types of reactional behaviour are a function of the strength of each patient's personality, his personal problems, and his capacity to adapt to the illness and the stress.

The age at which the symptomatology appears is an important factor in the child's adaptation to his illness: thus, if the illness appears early, the personality will be found, in spite of this permanent element, to have some special characteristics linked to the nature of the illness. The social adaptation of the child is better in this group, because the integration of the illness takes place progressively, while at the same time the child's defense mechanisms are developing. On the other hand, if the illness appears later, the personality being already developed, the inherent stress weakens the personality and will eventually provoke the emergence of earlier psychological problems that had been compensated. Certain problems will be recruited more intensively depending upon the nature of the disease. Psychosocial nonadaptation is more frequently observed in this group.

IV. THE CHARACTERISTICS INHERENT TO THE FAMILY ENVIRONMENT

The parents' role in the child's adaptation to his illness seems primordial to us. During a study concerning the family dynamics as influenced by hemophilia, we recognized an obvious concordance between the psychosocial inadaptation of the child and the inadaptation of the parents, independently of the severity of the disease (Weinstock-Paquay and Appelboom-Fondu, *in press*).

The functioning of 25 families could be studied and analyzed before and after the onset of the illness. Apparently, 10 of them, that had functioned by open communication, i.e., where conflicts were openly approached before the illness, were restructured, even consolidated, after having lived through a more or less long period of mourning and of important depression at the time of the diagnosis. In these families, the children are the best effectively adapted and the most well balanced. In 8 other families, that previously functioned in a non-decisive manner, the appearance of the disease provoked a withdrawal from the outside world, with a restructuring of the family into clans, and a limitation of intellectual or effective interests. No matter what the age of the patients, the children are socially adapted at the cost of an important investment of psychic energy. For the last 7 families, the familial dysfunctioning was manifested before the diagnosis of the disease; the conflicts were major, with frequent opposition. We found, in a short time after the diagnosis, a dislocation of the family unit, or the appearance of important psychiatric troubles in one of the parents. It is in these families that we most frequently find a psychosocial inadaptation of the child and the more serious personality troubles.

V. THE CHARACTERISTICS RELATED TO THE MEDICAL STRUCTURE

A. The Doctor

The role of the doctor is ambiguous because, for the patient, his intervention risks to lead to a feared verdict (diagnostic or therapeutic), but at the same time his intervention is reassuring. For the doctor, to treat a child implies empathy for the patient and his family: conciliation, confidence, but also painful intervention. Also the attitude of the doctor towards the child will reflect his own problems with respect to the illness and to death (Balint, 1960; de Ajuriaguerra, 1970).

Some will be paternalistic, which will favor adaptation in young children, but which will handicap the adolescent striking for independence. Others will have a relation based on reciprocal seduction, which risks increasing the patient's feeling of solitude if his relations change because of an aggravation of his disease. Still others will act as treatment technicians, rejecting the patient and his family as soon as a problem separated from the illness itself is introduced. Some, by their evasive attitude or their excessive meticulosity, increase the anguish and the inadaptation of their patient. The relationship of the child to his doctor depends on his age, his personality, and his earlier experiences with the doctor.

Thus, certain children will be very attached to their mother, appreciating the intervention of the doctor as particularly aggressive and even destructive. Other children actively refuse medical intervention. Older children become the accomplice of their doctor, and identify with him in order to control the anguish that weighs on their already compromised future (35 out of 45 diabetic children want to become doctors or nurses).

The attitude of the doctor toward his patient should depend upon the age, personality, and the evolutive stage of the illness, and not upon the personal problem of the doctor with respect to his patient.

B Hospitalization

Hospitalization is a painful experience for a child already affected by his illness, separating him from his family and establishing new relations in another social structure. This rupture of the family-child tie has different repercussions depending on the age of the patient (Belmont, 1970).

Thus mother-child separation will have negative psychological repercussions in the neonatal period only if the nurse, the preferred maternal substitute, does not respond to the immediate physical needs of the child, Therefore, in every possible way, it is best to respect the child's own feeding and sleeping cycle in order to avoid obliging the child to learn new habits. If the neonate has little chance to be perturbed, the mother, on the other hand, may feel this separation painfully and the link that she has established with her infant may be broken by the anxiety and guilt that she is experiencing.

Later on, when the child can distinguish his mother from others, the separation increases the anguish of abandonment. This privation of maternal affection may take on dramatic proportions and cause a serious depression, called the anaclitic depression of Spitz.

During adolescence, the physical problems of integrity and sexual capacity are exacerbated by the illness and the hospitalization. The patient is forced into a situation of dependence and segregation. He is an integral part of the "weak" group with respect to the "strong" group made up of the doctors and nurses.

New relationships are formed in this structure for the child and his family, with the treatment team as well as with the other patients. The adaptation of the child to his hospitalization depends upon the quality of these relations.

VI. CONCLUSION

The age and personality of the child are the determining factors in the adaptation to chronic illness. However, the characteristics of the disease, the quality of the child/family, child/doctor, and the family/treatment-team relationships, favor this integration. Also, psychological assistance for the parents in sensitivity groups, the acknowledgement of the relational problems of the treatment team by Balint groups, and the acknowledgement of the common problems in summer camps for ill children, appear to us to be efficient methods of assistance allowing a better adaptation of each to the chronic illness, and therefore a better life.

REFERENCES

Alby, N. and Alby, J.M. (1971). *Psychiatr. Enf.* 14, 465-501.
Appelboom-Fondu, J., Verstraeten, F. and Dopchie, N. (1974). *Rev. Neuropsychiatr. Inf.* 12, 725-735.
Appelboom-Fondu, J., Verstraeten, F. and van Loo-Reynaers *(in press)*.
Balint, M. (1960). "Le Médecin, Son Malade et la Maladie". Presses Universitaires de France, Paris.
Belmont, H.S. (1970). *Clin. Pediatr.* 9, 472-483.
de Ajuriaguerra, J. (1970). "Manuel de Psychiatrie", pp.888-911. Masson et Cie, Paris.
Lansky, S.B. (1974). *J. Child. Psychiatr.* 13, 499-508.
Leonard, B.J. (1972). *Nurs. Clin. North Amer.* 7, 687-695.
Weinstock-Paquay, M. and Appelboom-Fondu, J. *(in press)*.

PSYCHOLOGICAL AND REHABILITATION ASPECTS OF SHORT STATURE AND DELAYED PUBERTY

A. Galatzer, E. Rosenblith and Z. Laron*

*Institute of Pediatric and Adolescent Endocrinology
Beilinson Medical Center, Petah Tikva
and
Sackler School of Medicine, Tel Aviv University, Israel*

SUMMARY

A study has been carried out on the influence of growth and sexual retardation on the self-concept, social interaction, and emotional stability. Two groups of patients — 59 with retarded sexual development with or without growth retardation, and 64 with marked shortness of stature resulting from pituitary deficiencies or bone diseases — were studied and compared to normal controls. Interviews with subjects and parents, teacher's report on school performance, and the results of the Tennessee Self-Concept Scale test, were used for assessment. The results clearly confirm that adolescents with growth and sexual retardation suffer from social, emotional, and scholastic problems. A closer collaboration between physician and psychologist is proposed in the treatment of these cases.

I. INTRODUCTION

Adolescence was long ago recognized as a critical period in the developmental processes of the mature personality. There is a large amount of theoretical and empirical literature dealing with the consequences of the rapid physiological, biological, and psychological changes that occur during this short period of life (Mussen, 1969).

The body's characteristics and body image are essential to the adolescent. MacCandles (1970) claims that, to a great extent, the adolescent "is his body, and his body is he". The body and its physical changes as a major aspect in personality

---

* Established Investigator of the Chief Scientist's Bureau, Ministry of Health.

development are emphasized also by Erikson in his theory on the development of self-identity (Erikson, 1968). The search for self-identity is a crucial aspect in the individual's development. Each adolescent must prove himself as to his past, present, and future states, i.e., his biographical past, his present status (the "Who Am I?" question), and what the future holds for him. The search for identity is common to society as a whole, and every individual has to use all his mental and physical assets in order to build up his own identity. This search cannot be accomplished only by introspective soul-searching. Societal acceptance or nonacceptance is also needed. The answer to the question "Who Am I?" depends to a great extent on the individual interaction with society, i.e., on the social feedbacks received from others and the way in which these feedbacks are incorporated into the self image. For the adolescent, this is provided by his peer group, as well as adults with close familial or social ties. Positive feedback is important in the establishment of a well-defined structure of identity which will lead to the creation of a healthy and integrated personality. Negative feedback from society, in addition to problems in pubertal development, on the other hand, will cause a role confusion which endangers ego development.

When there is a significant lag in one or more of the crucial biological parameters, whether growth or maturity of sexual organs, we can expect to find difficulties in the formation of an integrated and healthy self-identity or self-concept. Several studies (Mussen and Jones, 1957; Jones and Bayly, 1950) pointed out the possibility that late maturers are apt to be restless, tense, and less popular than early maturers. The study presented here was designed to evaluate the influence of delayed sexual development and short stature in boys and girls on the self-concept, social interaction, and emotional stability.

II. SUBJECTS AND METHODS

The patient material of this study comprised 59 patients (37 males and 22 females) with retarded sexual development with or without growth retardation, and 64 patients (39 males and 25 females) with marked shortness of stature resulting from pituitary deficiencies or bone diseases. For the former group, a group of 59 normal adolescents matched for age, socioeconomic status and intelligence served as controls. For the latter group, a group of 20 selected normals served as controls. Patients with gonadal dysgenesis were excluded since the additional physical problems further complicate the child's psychological development.

Extensive interviews were conducted with both the patient and his parents, so as to determine the social and emotional adjustment of the individual patient. Each patient was subjected to the Tennessee Self-Concept Scale Test (Fittes, 1965). An evaluation of school performance was obtained by the patient's teacher.

Definitions

Growth retardation: height between 2 to 4 SD below the mean for chronological age. *Marked growth retardation:* height shorter than 4 SD below the mean for age. *Delayed puberty:* no signs of initial puberty (such as pubic hair and enlargement of the testes in the boys, or breast buds in the girls) by the age of 14 in the boys and 13 in the girls. These limits were set in accordance with the standards in our population.

III. RESULTS

The group of 59 adolescents with both sexual and growth retardation will be discussed first. Table I presents the social problems encountered in these boys and girls. The most common problem was found to be the difficulty these patients had in building stable relationships with the opposite sex. These problems were seen to increase as they mature from early to late adolescence and are common to both sexes. The boys evidenced shyness in starting conversations with the opposite sex despite their neat appearance and, often, physical attractiveness. Those who dared ask a girl for a date were rejected, each rejection reinforcing the fear of social contacts with females. The female patients described themselves as "wall flowers", who would not usually be asked to dance when they attended parties. Even when they were asked, the contact was brief and embarrassing. These negative feedbacks caused many of them to avoid participating in the social activities of their peers. Some developed rationalization defence mechanisms against these "dull" social activities. They tended to form friendships with a

Table I. Type of social problems reported by adolescents with growth and sexual retardation (N = 37 boys and 22 girls).

1. Difficulties in interrelationship with the other sex
   Girls: Avoiding approach of boys
   Boys: Afraid to approach girls
2. A tendency to have much younger friends
3. Avoidance of social activities — staying at home alone most of the time
4. Being afraid to go to the beach or on outings with friends so as not to reveal parts of the body in public
5. Still being called "boy" or "girl" by peers, family, teachers in late adolescence
6. Boys — sometimes being approached as girls by strangers

younger age group, sometimes with a difference of more than two years in age, and with these new friends, who were more similar in their physical appearance, were able to form better relationships. Their higher intellectual and academic levels enabled them to become leaders of these younger groups, which gave them the positive feedbacks which were almost impossible to receive from their own age group. These adolescents with sexual retardation tended to refrain from joining school trips or gymnastics because of the frightening possibility of embarrassment in the showers and locker rooms. This was especially true of those 18-year-olds who had reached the age for recruitment into the army and were afraid that their military service would be postponed because of their insufficient development, thus further singling them out and separating them from their peers.

One of the problems with which these adolescents had to cope was their childish physical appearance. Often people took them to be much younger than they really were, addressing them accordingly. In extreme cases, boys were distaken for young girls and girls for boys. This negative feedback from society enhanced the confusion these adolescents already felt concerning their self-identity.

Table II. Presence or absence of social problems in adolescents with growth and sexual retardation.

	Problems	No problems
Boys (N = 37)	30	7
Girls (N = 22)	18	4

Out of the 59 adolescents in this group, 48 reported having been disturbed by at least some of the above-noted social problems (Table II). Most of them (37 of the 48) had suffered from more than one aspect of this problem. This constant negative feedback from society left its mark on the emotional status of these adolescents.

Table III. Presence or absence of emotional problems in adolescents with growth and sexual retardation.

	Problems	No problems
Boys (N = 37)	30	7
Girls (N = 22)	18	4

Table IV. Emotional problems presented by adolescents with growth and sexual retardation (N = 37 boys and 22 girls).

1. Problems of sexual identity (expressing fears of being of, or changing into, the other sex)
2. Problems of body image (fixation on sexual organs)
3. Emotional liability (depression, withdrawal, euphoria)
4. Overoccupation with sexual fantasies (day dreaming, embarrasing parents with intimate and sexual questions)

Some kind of emotional problem was detected in 48 of the 59 patients (Table III). The most common problems are presented in Table IV.

The formation of a positive body image and a congruent gender identity, so important at this age, was problematic for these youngsters. When asked to give a "yes" or "no" answer to the statement "I would like to change some parts of my body", 24 of the 37 males gave a positive answer, as compared to 9 of the 37 in the control group; while 12 of the 22 females gave a positive response, as compared to 5 of the 22 controls. This problem was also manifested indirectly by these patients in the question asked of their physician concerning the reasons for their sexual retardation, and by the skepticism which they exhibited towards the explanations given and their chances for future development. In some severe cases (males with gynecomastia), the fear of changing into the other sex was openly expressed.

Changes in the genital organs are very important to every normal adolescent. In some of our patients, however, the anxious expectation for these changes appeared to overwhelm their behaviour and thinking. In some cases, it was manifested in a constant search for physical changes: looking for enlargement of the sexual organs, or, in cases of growth retardation, the taking of daily measurements of height. Other patients became obsessively occupied with day-dreaming about sexual activity. Two of the boys used to embarrass their parents by asking questions about the parents' sexual activity, as a means of dealing with the confusion they felt about their own potential for engaging in such activity. There were also signs of depression and withdrawal in a few cases.

Table V. School achievements of adolescents with growth and sexual retardation in relation to the intellectual level (N = 37 boys and 22 girls).

School achievement		Intellectual level		
		Above average (IQ > 110)	Average (IQ = 90-110)	Below average (IQ < 90)
No problems	Boys (N = 13)	5	7	1
	Girls (N = 9)	5	3	1
Difficulties — underachievement	Boys (N = 24)	2	17	5
	Girls (N = 13)	0	9	4

This group of patients was subdivided into three groups in accordance with their IQ levels: above average, average, and below average. The patients of the above-average group had almost no problems at school, but often used their intellectual capacity to compensate for their physical inferiority and thus earned the admiration and friendship of their peers. The children with average or lower-than-average IQ levels were more apt to manifest problems in scholastic performance and, in many cases, their actual achievements were far behind their intellectual underachievement (Table V). Out of the 59 patients in this group, 37 were described by their teachers as potential. In each of the groups, no difference was found between males and females in their tendency towards underachievement. The explanation which these patients gave was that, as a result of the tension which they felt, their attention span and ability to concentrate were affected.

The above observations were further strengthened by the results obtained with a structured personality questionnaire. The Tennessee Self Concept Scale (TSCS) consists of 100 self-descriptive statements to which the subject gives one of five responses, ranging from "completely true" to "completely false". The questions are organized in the form of a rectangular matrix divided into rows and columns. The three horizontal rows contain items descriptive of the individual's identity, self-satisfaction, and behaviour. There are five vertical columns which describe the physical, moral-ethical, personal, family, and social selves. A three-way analysis of variance between sex, sexual development, and growth development was carried out (Table VI). Of the 9 scales in the self area, 6 showed significant differences between the group of patients and the control group. The group of patients, on the whole, had lower scores for self-identity and self-satisfaction. They described themselves less positively than did the controls in the physical, personal, and social selves. They also provided evidence for a higher number of conflicts in their self-description, their score being inferior on the general-maladjustment, psychotic, personality deviant, and neurotic scales. The only variable which accounted for this difference was the sexual development. It

Table VI. Tennessee Self Concept scores in adolescent with growth and sexual retardation as compared to normal controls.

	Retarded (N = 59)		Controls (N = 59)		F
	X̄	SD	X̄	SD	
Self concept					
Total self	337.1	33.0	356.4	28.9	0.001
How I am	118.7	12.2	125.6	8.7	0.001
Accept	105.9	12.2	115.6	14.1	0.001
Act	112.5	12.2	115.2	11.1	ns
Physical self	66.2	8.2	74.0	7.2	0.001
M/E self	68.5	7.9	70.3	8.1	ns
Personal self	68.5	9.0	72.0	7.3	0.022
Family self	68.1	8.9	70.1	8.3	ns
Social self	65.8	8.5	70.1	7.7	0.005
Empirical scores					
Total conflicts	40.0	12.3	34.7	9.7	0.010
General maladjustment	93.3	11.1	98.0	7.5	0.009
Psychotic	57.5	7.4	52.4	6.6	0.001
Personality development	71.2	9.8	75.5	10.5	0.023
Neurotic	80.5	11.7	88.4	8.6	0.001

Table VII. Wechsler intelligence test in five groups of children and adolescents with short stature as compared to normal.

Group No.	Diagnosis	N	Verbal IQ		Performance IQ		Total IQ	
			X̄	SD	X̄	SD	X̄	SD
1.	Laron-type dwarfism	14	80.7	14.9	83.5	16.0	80.6	14.8
2.	Isolated growth hormone deficiency	14	107.9	18.9	109.4	16.5	109.4	17.4
3.	Multiple pituitary hormone deficiency	17	93.1	16.0	91.3	11.6	92.1	13.5
4.	Multiple craniopharyngioma	7	104.3	12.6	94.8	11.3	100.7	10.9
5.	Bone disease	12	105.2	14.6	97.0	21.8	102.0	18.0
6.	Controls	20	102.3	15.4	102.7	18.7	102.4	17.0

would appear that, for this group, the retardation of sexual development constituted a more serious problem than that of the growth retardation.

The findings of the group of 64 patients with shortness of stature illustrates the importance of growth as a dominant factor in adjustment. These patients were subdivided into five diagnostic groups which were compared to the control group of 20 normals. Table VII presents the IQ scores of these children. It is evident that, with the exception of the group of pituitary dwarfism with the high plasma immunoreactive growth hormone (Laron-type dwarfism), which has a relatively low IQ level, all of the groups are within the average range.

Table VIII. Parents' expectations for their child's future in five groups of children and adolescents with short stature.

Group No.	Diagnosis	I want my child to marry		I believe he can marry			The future of my child		
		Yes	No	Yes	No	Hope	Independent	Institution	Don't know
1.	Laron-type dwarfism	13	1	1	9	4	1	1	12
2.	Isolated growth hormone deficiency	13	—	8	2	4	10	—	4
3.	Multiple pituitary hormone deficiency	17	—	4	5	8	7	—	10
4.	Multiple craniopharyngioma	7	—	3	2	2	3	—	4
5.	Bone disease	12	—	6	2	4	7	—	55

Many of these adolescents had difficulty in adapting themselves to the scholastic and social demands made upon them at school. This difficulty was due in part to the fact that their teachers and fellow pupils did not know how to react to their shortness. At times they were rejected and at times overprotected but never were they accepted for what they were. These uncertainties were further reinforced by their parents' attitudes (Table VIII). The parents were highly concerned about the future of these children and unconsciously tended to transfer this anxiety to the children. When parents were asked if they wanted their child to get married, all except one answered positively, but only one-third believed this to be actually possible. Only 43% of the parents expected their children to achieve independence later in life.

Table IX. Self-concept scale in five groups of children and adolescents with short stature.*

	Positive	Medium	Negative
Conflicts	6	2, 4, 5	1, 3
Total Self Concept	6	2, 5	1, 3, 4
Physical Self Concept	6	2, 5	1, 3, 4
Social Self Concept	6	2, 5	1, 3, 4
Emotional Stability	6	2, 5	1, 3, 4
Personal Integrity	6	2, 5	1, 3, 4

* Based on the Tennessee Self-Concept Scale.

These uncertainties left their marks on the children's sense of identity and self-image. A comparison of the TSCS profiles of these six groups revealed three distinctive groups (one-way analysis of variance) (Table IX). The most positive profile was that of the normal control group. Groups 2 (isolated growth hormone deficiency) and 5 (bone disease) had an intermediate profile, and Groups 1 (Laron-type dwarfism), 3 (multiple pituitary hormone deficiency) and 4 (pituitary-tumor), the lowest profile. This was particularly evident in the total self, physical self, social self, and other empirical scales measuring conflicts, emotional stability, and personality integration. Patients of group 2 received substitutional growth-hormone therapy with good results, and those of group 5, unlike the others, had no problem of sexual maturity. This may explain the better self concept in these two groups, each of which showed normal development in either growth or sexual development, or at least had a good chance for future development.

It is of interest that within group 5, three females have attained university degrees; one of these, who is married for the second time, both husbands having been of normal height, is the mother of a normal baby.

IV. DISCUSSION

The results of this study further confirm that adolescents with growth and sexual retardation suffer not only from a physiological handicap, but also from social, emotional, and scholastic problems. This is probably true for both sexes, although the problem of delayed puberty is more commonly encountered in males than in females. When females do appear for treatment, the influence of the delay in sexual development upon their social and emotional lives proves to be similar to that in boys.

The findings of this study indicate that, from the emotional point of view, it is probably more difficult to cope with sexual retardation than with growth retardation. To clarify this point, however, additional research on a larger sample, including patients with other syndromes as well (such a study is presently in progress in our Institute), is needed.

When dealing with severe growth retardation, the growth factor obviously becomes the more dominant feature in general adjustment. The objective problems of daily life for the short-statured adolescent often prove overwhelming, particularly since society contributes little or nothing towards their solution.

There is as yet insufficient evidence as to the influence of these problems during adolescence on the future behaviour of these patients as adults in social, emotional, and sexual areas. A number of follow-up studies carried out on late maturers do throw some light on this aspect. Peskin (1967), Ames (1957) and Corbox (1967) found that late maturers tend to show a significant lag in their social relationships in later life. They marry later, have fewer children, and hold less important jobs.

The fact that these youngsters need considerably more attention than normal adolescents is well recognized and has generally been accepted by most medical authorities (Bailey, 1974). A thorough explanation needs to be given, not only to the adolescent, but to his parents, concerning his physical status, his chances for future development, and the reasons for the timing of treatment administered. It is most important, however, to keep in mind the fact that a single explanation is not sufficient. When the adolescent and his parents appear for an initial meeting with the doctor, they are usually very tense and it is almost impossible for them to understand everything being said to them. They tend to distort the information being given to them in such a way that it will corroborate the unconscious fears or hopes which they have.

It is thus of utmost importance to provide a framework within which the patient and his parents may receive additional information as needed, and be given the opportunity freely to express their fears and hopes, both necessary in the process of adjusting to, and accepting the reality of, the situation. This process, however, is time-consuming and the treating physician is often unable to provide such comprehensive treatment. It is therefore proposed that a multidisciplinary approach, based on the collaboration between physician and psychologist, similar to that used with juvenile diabetics (Laron, 1974), be instituted for the treatment of patients with short stature and delayed puberty.

REFERENCES

Ames, R. (1957). *J. Educ. Res.* **8**, 69-75.
Bailey, J.D. (1974). *Pediatr. Clin. North. Am.* **21**, 1029.
Corboz, R.J. (1967). "Spatreife und bleibende Unreife". Springer Verlag, Heidelberg, New York.
Erikson, E.H. (1968). "Identity, Youth and Crisis". Norton, New York.
Fittes, W.H. (1965). "Tennessee Self Concept Scale, Manual". Dept. of Mental Health, Nashville, Tennessee.
Jones, M.D. and Bayly, N. (1950). *J. Educ. Psychol.* **41**, 129-148.
Laron, Z. (1974). *Ped. Annals* **3**, 63-77.
MacCandles, B.R. (1970). "Adolescents". Bryden, Hindsdale, Illinois.
Mussen, P.H. and Jones, M.C. (1957). *Child.Dev.* **28**, 243-256.
Mussen, P.H. and Conger, J.J. (1969). *In* "Child Development and Personality", Int. Ed. (3rd Ed.), pp.605-763. Harper, New York.
Peskin, H. (1967). *J. Abnorm. Psychol.* **72**, 1.

SUBJECT INDEX

A

Aarskog facial-digital-genital syndrome, 232
Achondroplasia, 253-258
Adipose tissue, genetic component in, 45
Adolescence, psychological problems of, 303-304
Adolescent growth spurt, 111, 112, 123-132, 135
Adoption studies, 217-223
Africa, height and weight in, 141, 142
Alanine storage, 215
American Negroes
  arm length in, 147
  height and weight in, 140-144
  leg length in, 147
  sitting height in, 147
  skeletal maturation in, 150
Amniocentesis, 63-64
Amniography, 63
Amnioscopy, 64
Anabolic steroids, 208
  growth response to, 209
Andro-FSH, 99
Androgens, 208
Anencephalus, 67
Anthropometry, 51, 57, 199-200, *see also* measurement techniques
Arctic conditions, growth in, 167-172
Argentina, height and weight in, 140-144
Arm length, population variability of, 147-149
Assortative mating
  and heritability 41
  and spouse similarity, 37-38
Australia, height and weight in, 140-144
Autoaggressive disease, 8-10, 16-20
  age of onset of, 10

B

Biorhythms and maturation, 39
Birth rank and growth, 184
Birth weight, secular trends in, 168
Blood glucose regulation, 214-216
  and age, 214-216
Blood pressure
  father-offspring *v.* mother-offspring correlation for, 36, 37
  genetic component in, 45
Body image and personality changes, 298
Body shape, 146-149
Bone age and socioeconomic factors, 57, 58
Bone disease
  parents' expectation in, 309
  Wechsler Intelligence Test in, 308
Brain maturation, 72
Brasil, height and weight in, 140-144
Breast feeding, 85, 155
  v. formula feeding, 85

C

California, height and weight in, 142
Castration, hormonal response to, 90, 91, 98
Caucasians, height and weight in, 161-164
Cell turnover, 7
Central system of growth control, 9, 19
Chinese
  arm length in, 148-149
  height and weight in, 141-143, 161-164
  sitting height in, 147
  skeletal maturation in, 150
Chondroectodermal dysplasia, 228, 229
Chronic illness, 295
  and doctor's attitude, 299-300

and family dyanmics, 299
and hospitalization, 300
and personality, 295-301
Chronogenetics, 8-10
of retinoblastoma, 10, 16
Circadian rhythms, 95-96
Clinical standards, see growth standards
Coffin-Siris syndrome, 229
Community as growth accelerating factor, 176
Congenital anorchia, and testosterone, 210
Congenital malformations, 67
Costa Rica, urban-rural differences in body size, 145
Craniopharyngioma, 105-107, 240, 241, 242, 243, 250, 251, 308-309
Cross-sectional standards, see growth standards
Curve smoothing, 135
Czechoslovakia, secular growth trend in, 175-185
age at menarche in, 184-185

D

De Lange syndrome, 231
Denmark, height and weight in, 168
Developing countries, social class discrepancies in growth, 145
Diabetes
and achondroplasia, 258
and Turner's syndrome, 267
personality in, 296, 298
Diabetes insipidus, 240

E

Ear shape, father-offspring v. mother-offspring correlation for, 36, 37
Ear size, father-offspring v. mother-offspring correlation for, 36, 37
Ecologic factors, 33-46
Ecosensitivity
of homozygotic v. heterozygotic subjects, 42-44
of males v. females, 42-45
Embryogenesis, 8
Emotionally retarded growth, 290
Endurance fitness, father-offspring v. mother-offspring correlation for, 36, 37
Energy intake
and age, 50, 51
and sex, 51
and socioeconomic status, 50-51, 52
Enzyme levels, father-offspring v. mother-offspring correlation for, 36, 37
Erythroblastosis fetalis, 67

Eskimos, 167-168
Greenland Eskimos, height and weight in, 167-168
Canadian Eskimos, height and weight in, 168
Estrogen/creatinine ratio, 73-77
Europe
height and weight in, 141, 142
urban-rural differences in body size, 145

F

Face breadth, environmental susceptibility of, 42
Face height, father-offspring v. mother-offspring correlation for, 36, 37
Factor analysis, 23-30
Family
and school interactions, 287-293
as growth accelerating factor, 176
dynamics, 289-290
size, 155
and growth, 155, 183-184
Family Relations Indicator Test, 291
at child's level, 291-292
at family level, 291
at teacher's level, 291, 292
Far East, height and weight in the, 141, 142
Fast, response to, 213-216
Fetal biopsy, 64, 67
Fetal growth, 72-73
Fetal surgery, 64, 67
Fetal transfusion, 67
Fetography, 63
Fetoscopy, 63-69
risks of, 67, 68, 69
Filipino, height and weight in, 161-164
Finland
height and weight in, 140-144
secular growth trends in, 170
growth in northern v. southern populations, 171
obesity in, 172
urban-rural differences in body size, 145
Follow-up studies, 187-198
Foot breadth, father-offspring v. mother-offspring correlation for, 36, 37
Forbidden clone, 9
Forehead breadth, father-offspring v. mother-offspring correlation for, 36, 37
Free fatty acids, and blood glucose, 214-216
Full-term infants, perinatal growth of, 79-85

G

Gene mutation in stem cells, 15-21
General Theory of Systems, 288
Genetic factors in body measures, 45

Subject Index

Genetic v. environmental similarity, 45, 46
Germans
  leg length in, 147
  sitting height in, 147
Gestational age, 75, 79-85
Girth factor of growth, 27
Glucagon, response to, 215
Goldenhar syndrome, 279-285
  association with chromosomal aberration, 279-285
  growth hormone treatment of, 279-285
Gonadotropin deficiency, 240, 241, 242, 243
  and testosterone, 209-210
Gonadotropin releasing hormone, 102-104
Gonadotropin secretion
  and Turner's syndrome, 270-277
  negative feedback control of, 89
Gonadotropins, synthesis and release of, 93, 95
Greece, urban-rural differences in body size in, 145
Grip strength, 45
Growth
  acceleration factors, 175-176
  and maturity, 23-32
  and response to fast, 214-216
  charts, 133-138
    US population, 133-138
  chronogenetic factors in, 1-6
  circulation factor of, 27-29
  control, 16-21
  definition of, 7
  ecologic factors in, 33-48
  factors, 23-30
  genetic factors in, 33-48
  girth factor of, 27
  inhibiting factors, 154, 155
  initiating impulses of, 33-34
  length factor of, 29
  masculinity factor of, 29-30
  normal v. neoplastic, 7-22
  nutritional factors in, 49-62
  population variability of, 132-206
  potential, 54
  psychosocial factors in, 287-294
  secular trend in, 153-156, 175-198
  velocity, 81-85, 140
    and testosterone, 210
    in full-term infants, 82-85
    in preterm infants, 82-85
    in small-for-dates infants, 82-85
Growth deficiencies
  primary, 225-237
    automatic identification of, 234-237
    classification of, 230, 232-234
    diagnostic problems of, 229-230, 232-233
    etiology of, 226-227
    family history in, 227, 229

intelligence in, 307, 308
parents' expectation in, 309
physical examination in, 229
psychological problems in, 303-311
school achievement in, 307
self-concept in, 309
secondary, 225-226
Growth hormone
  and blood glucose, 214-216
  and estrogens, 208
  and prenatal growth, 242
  and testosterone, 208-211
  deficiency, 101-108, 239-252
    and puberty, 101-108
    and testosterone, 210
    familial, 243
  in achondroplasia, 253-258
  in Turner's syndrome, 261-267
  sex differences in secretion, 208
Growth standards
  clinical, 110-120
  cross-sectional, 110-120, 124, 157-158, 187-198
  height for age, 201
  height velocity, 110-115
  height, 110-115
    parent-allowed-for, 120
  limb length, 118-119
  longitudinal, 110-120, 124-131, 157-158
    factor analysis in, 124-125
    nonlinear models in, 126-131
    polynomial models in, 125-126
  pubic hair, 113, 185, 186
  trunk length, 118-119
  weight for age, 115
  weight-for-height/height for age, 201
  weight for height, 115, 201
  weight by length, 135
Guatemala, height and weight in, 140-144
Gyno-FSH, 99

H

Hallermann-Streiff syndrome, 231
Haptoglobin levels
  father-offspring v. mother-offspring correlation for, 36, 37
  genetic component in, 45
Harelip, 67
Hawaiian, height and weight in, 161-164
Head breadth, environmental susceptibility of, 42
Head shape
  environmental susceptibility of, 42
  genetic component in, 45
Head size, genetic component in, 45
Health care as growth accelerating factor, 176
Height
  and nutrition, 153, 155

and socioeconomic status, 51, 53, 54, 55
father-offspring v. mother-offspring correlation for, 36, 37
genetic threshold, 154
increase and morbidity, 153-154
increase and mortality, 153
population variability of, 140, 141, 142, 144, 146, 147, 157-164
standards, *see* growth standards
Hemophilia
 mother-child psychological relation in, 296
 personality in, 296, 298
Hip width, population variability of, 149
Homozygotic v. heterozygotic subjects, 42, 43, 44
HOP index and socioeconomic factors, 57, 58
Hormone concentration, circadian rhythms of, 95-96
Hungary
 biacromial weight in, 189-191
 bicristal width in, 189-191
 height in, 189-191
 chest circumference in, 189-191
 lower extremity length, 189-191
 secular growth trend in, 187-198
 upper extremity length, 189-191
 weight in, 189-191
Hybrid vigour and growth, 54, 57
Hydrocephalus, 67
Hygienic conditions and growth, 57
Hypochondroplasia, 233
Hypoglycemia
 and age, 214-216
 and alanine, 215
Hypopituitary dwarfism, and fast, 216
Hypothalamic-pituitary-gonadal axis, 87-88
 differential sensitivity of, 89, 91, 96

I

IQ and school performance, 288
Immunoreactive insulin
 in achondroplasia, 253-258
 in Turner's syndrome, 261-267
Inborn errors of metabolism, 225-226
India
 height and weight in, 140-144
 urban-rural differences in body size, 146
Industrialization and growth, 154, 175, 177, 183, 188-198
Industrialized countries, social class discrepancies in growth in, 145
Iran, height and weight in, 140-144
Isolated growth hormone deficiency
 growth in, 239-252
 LH-RH administration in, 101-104
 parents' expectation in, 309

puberty in, 101-104
Wechsler Intelligence Test in, 308
Italian v. British children, 51, 53, 201
Italy
 growth in, 49-60, 199-204
 nutrition in, 49-60, 199-204
 socioeconomic development of, 50

J

Jamaica
 height and weight in, 142
 urban-rural differences in body size in, 143 146
Japanese
 height and weight in, 141-143, 161-164
 leg length in, 147
 sitting height in, 147

K

Korean, height and weight in, 161-164

L

Lapps, 169-170
 Inari Lapps, secular growth trend in, 169-170
 Skolt Lapps
  genetic isolation in, 169
  growth rate in, 169
  secular growth trend in, 169
  socioeconomic conditions in, 169
Laron-type dwarfism, 240, 241, 242, 243, 250
 parents' expectation in, 309
 Wechsler Intelligence Test in, 308
Lebanon, height and weight in, 140-144
Leg length, 117
 population variability of, 147, 148
Length factor of growth, 29
Leukemia, personality in, 297
Limb length, 118-119
Linear v. circular relations, 288, 289-290
Longitudinal standards, *see* growth standards
Longitudinal studies, 84, 110-120, 124-131, 157-158

M

Malate dehydrogenase, 45
Masculinity factor of growth, 29-30
Maternal drug intake, 227
Maternal environment, 36-37
Maternal infection, 227
Maternal urinary estrogens, 71-77
Maturity factor of growth, 30

Measurement techniques, 116-117, 134-135, 139-140
  head circumference, 134
  leg length, 117
  parallax error, 135
  recumbent length, 134
  sitting height, 116
  stature, 135
  weight, 134
Menarche
  age at, 35, 38, 39, 184-185
    in mother and daughter, 39
  and nutrition, 39
  and season of year, 36, 38, 39
  in urban v. rural populations, 38, 39-41
Meningomyelocele, 67
Menstrual cycle, in mother and daughter, 39
Midparent height and growth, 184
Migration and growth, 54, 57
Multiple pituitary hormone deficiency, 105-108, 240, 241, 242, 243, 250-251
  administration in, 105-108
  and craniopharyngioma, 105-107, 240, 241, 242, 243, 308-309
  parents' expectation in, 309
  puberty in, 106-108
  Wechsler Intelligence Test in, 308

N

Nasal breadth, environmental susceptibility of, 42
Neoplastic growth, 8, 19
Netherlands
  height and weight in, 140-144
  height secular trend in, 153-156
Neuter-FSH, 99
New Guinea, height and weight in, 140-144
Nigeria
  height and weight in, 140-144
  urban-rural differences in body size, 146
Norway, height and weight in, 168
Nutrition
  and environment, 199-204
  and parental education, 204
  and socioeconomic status, 49-60
  in a Roman population, 49-60
  status and self-perception, 60
  status assessment, 200
  surveillance, 199-200, 204

O

Obesity
  and alanine, 216
  and fast, 216
  and growth hormone, 216
  and mother's occupation, 204
  in Britain, 145
  in lower socioeconomic groups, 145
  in the US, 145
  social inheritance of, 217-223
Occupation of parents and diet, 204
Omphalocele, 67
Ontogenesis, 33-34

P

Paragenetic factors, 36-37
Parent-child correlation, 36-37, 39-41, 43
  and socioeconomic factors, 39-42
Parent-child similarity, 217-223
Parental height, 54
  and hypopituitarism, 241-242
Perinatal growth, 71-77, 79-85
Personality development, 303-304
Peru, height and weight in, 140-144
Philippines, height and weight in, 140-144
Pituitary function, 101
Pituitary hormone
  deficiency of, 239-251
    heredity of, 241, 250
  feedback control of, 98-99
  immunological v. biological activity of, 239-240
Placenta, 80
Placental growth, 72
Plasma cortisol and blood glucose, 214-216
Poland
  height and weight in, 140-144
  urban-rural differences in body size in, 145
Polynomial models, 125-126, 135, 158-159
Premature infants, 79-80
Prenatal diagnosis, 65
Preterm infants, perinatal growth of, 79-85
Prolactin secretion in Turner's syndrome, 270-277
Proportions and growth, 192-198
Psychomotor traits
  father-offspring v. mother-offspring correlation for, 36, 37
  genetic component in, 45
Puberty, 33-34, 111-115
  and weight, 276
  hormonal control of, 87-99
  in isolated growth hormone deficiency, 101-104
  in multiple growth hormone deficiency, 106-108
Pubic hair, 113, 185, 186
  and height, 185, 186
Puerto Rican, height and weight in, 161-164

R

Reaction time, genetic component in, 45
Reference population, 110
Respiratory instability, 71-77
Respiratory traits, genetic component in, 45
Retinoblastoma, 10-16
  age of onset in, 12-16
  chronogenetics of, 8-10
  familial cases of, 15
  heredity of, 10-16
  in France, 14
  in the UK, 11
  in US Whites, 12
  in US Negroes, 13
Rubinstein-Taybi syndrome, 228, 229
Rumania, urban-rural differences in body size in, 145
Rural environment
  growth in, 199-204
  nutrition in, 199-204

S

School
  as growth accelerating factor, 176
  dynamics, 289-290
Secular growth trends, 134, 153-156, 175-185
  in the US, 137-138
Self identity, 303-304
Sex ratio, 34-36
  and maternal menarche, 35-36
Sex steroids, 88-89
Sexual dimorphism, 34-36
Sexual maturation, 96-98
  in the male, 96-97
  in the female, 97
Sexual retardation
  and school achievement, 307
  and intelligence, 307, 308
  parents' expectation in, 309
  psychological problems of, 303-311
  self-concept in, 309
Shoulder muscles, father-offspring v. mother-offspring correlation, 36, 37
Shoulder width, population variability of, 149
Sickle-cell anemia, 67
Silver-Russell syndrome, 230
Sitting height, 116
  population variability of, 147, 148
Skeletal defects, 67
Skeletal maturation, 149-151
  and altitude, 151
  population variability of, 149-151
Skinfold thickness, 51, 217-223
Small-for-dates infants, perinatal growth of, 79-85

Socioeconomic factors
  and energy intake, 50, 51
  and growth, 49-60
    in African children, 200
    in Indian children, 200
    in Italian children, 200
  and mental development, 60
  and micropathology, 59, 60
  and skinfold thickness, 51, 56
  in body size, 144-145
Sociogram, 290, 291
Somatomedin activity
  in achondroplasia, 253-258
  in Turner's syndrome, 261-267
Spina bifida, 67
South Africa, urban-rural differences in body size, 146
Sweden
  height and weight in, 140-144
  secular growth trend in, 170

T

Taiwan, height and weight in, 140-144
Tanzania, height and weight in, 140-144
Tennessee Self-Concept Scale in growth and sexual retardation, 308, 309
Testosterone
  as growth-stimulating agent, 208, 209
  and hypophysectomy, 208
Thalassemia, 67
Thyroid hormone, 242, 251
Treatment, rejection of, 297
Trunk length, 118-119
  father-offspring v. mother-offspring correlation for, 36, 37
Tumours, growth curve of, 8
Turner's syndrome, 261-267
  and diabetes, 267
  and hypothalamus-pituitary axis, 270-277
  gonadotropin secretion in, 270-277
  immunoreactive insulin in, 261-267
  prolactin secretion in, 270-277
  somatomedin activity in, 261-267
  thyrotropin secretion in, 270-277
Twin placenta, 80
Twins
  developmental stages in, 10
  environment in, 39
  perinatal growth of, 79-80

U

Ultrasonography, 63
Unisex phantom, 188, 191, 192-198
United Kingdom, height and weight in, 140-144
Urbanization and growth, 50

Urban-rural differences in body size, 145-146, 147
USA
 height and weight in, 140-144
 urban-rural differences in body size, 145

W

Wechsler Intelligence Scale, 287-288
 in growth and sexual retardation, 308
Weight
 and socioeconomic status, 51, 53, 54, 55
 population variability of, 140, 141, 142, 157-164
 standards, *see* growth standards